Artificial Intelligence and Hardware Accelerators

Ashutosh Mishra • Jaekwang Cha • Hyunbin Park •
Shiho Kim

Editors

Artificial Intelligence and Hardware Accelerators

 Springer

Editors
Ashutosh Mishra
School of Integrated Technology
Yonsei University
Incheon, South Korea

Jaekwang Cha
Yonsei University
Incheon, South Korea

Hyunbin Park
School of Integrated Technology
Yonsei University
Incheon, South Korea

Shiho Kim
School of Integrated Technology
Yonsei University
Incheon, South Korea

ISBN 978-3-031-22172-9   ISBN 978-3-031-22170-5   (eBook)
https://doi.org/10.1007/978-3-031-22170-5

This Springer imprint is published by the registered company Springer Nature Switzerland AG
The registered company address is: Gewerbestrasse 11, 6330 Cham, Switzerland

Preface

Artificial intelligence (AI) is designing new genesis around the globe and garnering great attention from industries and academia. AI algorithms are indigenously intensely computational and data ambitious, which entails appropriate AI accelerators to comply with the computational efficiency. In general, AI professionals and hardware designers are different persons having no expertise in either area. This gap between the algorithmic and hardware experts brings an intense demand for a bridge to subdue this technological slit. This book aims to bridge this gap by assimilating state-of-the-art technologies and architecture. The chapters are arranged by considering *computing system platforms*, *technological development*, *emerging applications*, and the *existing issues* in AI hardware accelerators.

Chapter 1 comprehends the basics of AI, algorithms, and processor designs. It specifies the demands and requirements of AI and the hardware. It summarizes the entire book in a glance and induces the desire for an in-depth understanding of the technologies involved in recent accelerator designs. The computing system platforms are considered in Chaps. 2, 3, 4, 5, and 6. Chapter 2 deals with "single-tenant," i.e., single owner of the compute. The standalone personal computer is an example of single-tenant computing. This chapter describes CPU and GPU-based accelerators, including training and inference accelerators. Chapter 3 incorporates "multiple tenants," i.e., the owners associated with the servers or cloud computing independently. It also includes training and inference accelerations. Chapters 4 and 5 are dedicated to the AI accelerators in smartphones. Further, Chap. 6 provides the AI acceleration for embedded computing systems.

The trending technological developments in the AI accelerators have been considered in Chaps. 7 and 8. Chapters 7 consists of the reconfigurable and programmable AI accelerators (FPGA and ASIC dedicated hardware accelerators). The neuromorphic approach of the hardware design for AI accelerations is considered in Chap. 8. It discusses the state-of-the-art devices, circuits, and architectures used in neuromorphic computing. An emerging application of AI accelerators is elicited in Chap. 9. It contains accelerators for AVs. It describes AV computing platforms such as NVIDIA Jetson and Tesla's full self-driving (FSD) computer.

Energy is a crucial parameter for designing any efficient hardware. The AI models are highly hungry for energy due to extensive calculations and voluminous data. Chapter 10 includes the demand for energy in an AI accelerator.

Every chapter presents the requirements of hardware corresponding to their theme and provides comprehensive explanations of the recently available solutions and developments. In this way, the book will be helpful to design engineers, researchers, hardware experts, and industry employees to obtain insight into the AI models and algorithms. Therefore, they can develop an enhanced and efficient computing platform to accelerate AI applications and implement intelligence in devices. This volume can also serve as a textbook or reference book for undergraduate and graduate students in either domain (AI or hardware). The literature and survey covered in this book are very recent and relevant to the theme of the corresponding chapter. Therefore, a refined and inclusive survey of the AI hardware accelerator designs is encapsulated in this book, and it will be an add-on advantage to its readers.

As the editorial board members, we are grateful to all the contributors to this book. We are also indebted to organizations for their funding and support, due to which this assimilation of work became possible. This work was supported by Korea Institute for Advancement of Technology (KIAT) grant funded by the Korean Government (MOTIE) (P0020535, The Competency Development Program for Industry Specialist).

Incheon, South Korea   Ashutosh Mishra
Incheon, South Korea   Jaekwang Cha
Incheon, South Korea   Hyunbin Park
Incheon, South Korea   Shiho Kim

Contents

Artificial Intelligence Accelerators

Ashutosh Mishra, Pamul Yadav, and Shiho Kim

1 Introduction

Smart technologies and an intelligent society are the demand of this era. Artificial intelligence (AI) and deep learning (DL) algorithms are playing a vital role to cater these demands to meet the expectations of the smart world and intelligent systems [1, 2]. Increased computing power, sensor data, and improved AI algorithms are driving the trend toward machine learning (ML) in cloud-based and edge-based intelligence [1, 3]. Both are applicable through intelligent devices, wearable electronics, smartphones, automobiles, robots, drones, etc. However, efficient hardware can solely accomplish the required performance of implementing these algorithms. Therefore, AI accelerators are the state-of-the-art research area for circuits and systems designers and academicians. AI accelerators are needed to cater to the insatiable demands of compute-intensive AI applications [4].

Generally, four fields of research initiate the requirement of AI accelerators (as depicted in Fig. 1). Neuroscience comes with an idea of how the human brain attains intelligence. The AI researchers try to mimic those ideas to develop AI algorithms to implement intelligence into machines. Researchers in cyber-physical systems (CPS) or intelligent systems try to incorporate these innovations to create intelligent solutions for a smart society [5]. These solutions may be software-based or on some wearable devices. Anyhow efficient hardware is essentially required to accomplish these innovations in real applications. Therefore, AI acceleration is a demanding area of research for the AI research society.

The on-device AI is another emerging system on chip (SoC) technology making connected devices, including automobiles, high-definition (HD) cameras, smart-

A. Mishra (✉) · P. Yadav · S. Kim
School of Integrated Technology, Yonsei University, Incheon, South Korea
e-mail: ashutoshmishra@yonsei.ac.kr; pamul@yonsei.ac.kr; shiho@yonsei.ac.kr

© The Author(s), under exclusive license to Springer Nature Switzerland AG 2023
A. Mishra et al. (eds.), *Artificial Intelligence and Hardware Accelerators*,
https://doi.org/10.1007/978-3-031-22170-5_1

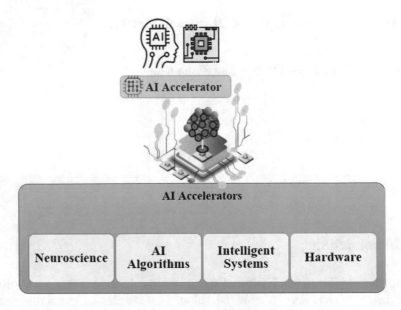

Fig. 1 Areas of research that prevail for developing efficient AI accelerators

phones, wearables, and other Internet of things (IoT) devices, smarter plus faster. The popular on-device applications are face detection, object detection and tracking, pose detection, language identification, etc. Moreover, the on-device inference is popular due to its lower latency and increased privacy compared to the cloud-based paradigm. However, executing such compute-intensive tasks on small devices can be more challenging due to a lack of computation power and energy consumption [6, 7]. Further, the global pandemic has imposed the situation to set a "new normal" that requires accelerating the digital transformation in many areas [8]. Such as digital economy, digital finance, digital government, digital health, digital education, etc. Indeed, multiple initiations have already existed on digital platforms and digital solutions. AI and SoCs have shown these possibilities into reality.

In addition, on-device AI offers immediate response, enhanced reliability, increased privacy, and utilizing efficient network bandwidth, and supports everything from superhuman computer vision and natural language processing to wireless connectivity, power management, photography, and more [6, 7]. However, the algorithmic superiority of intelligent systems is achieved by highly high computing capacity and memory, or in a nutshell, with high-performing hardware. It imposes a significant amount of challenge in designing suitable hardware platforms that can seamlessly execute the SOTA AI and ML algorithms over them. Moreover, such devices should have lower latency and higher reliability and can preserve the users' privacy. Therefore, on-device AI accelerators are still required to satisfy these insatiable demands.

Further, automation in transportation is also crucial for satisfying the increasing urbanization and traffic. AI has elevated the intelligent transportation system (ITS) to its next level, i.e., high driving automation (level 4) and full driving automation (level 5). However, various hurdles exist in implementing level 4 and beyond AVs into the real world. In addition, recognition and the corresponding fallback system are required to provide a countermeasure under irregular situations in such AVs. Such as human-non-human discrimination, traffic sign and traffic gesture recognition [9], in-cabin monitoring [10], personal-privacy protection [11], suspect recognition [12], adverse weather conditioning [13] road damage, pass-over object detection, etc.

Therefore, AI accelerators are in demand to cater to developing efficient and intelligent systems. This chapter mainly focuses on the introduction and overview of the AI accelerators. Here, we have explained the demands and requirements of the AI accelerators with their multifaceted applications. Here, we will focus broadly on the hardware aspects of training, inferences, mobile devices, and autonomous vehicles (AVs)–based AI accelerators. We start with a general overview of AI approaches. After that, the hardware accelerators are described in brief. Further, we have explained the requirements of training and inference. Finally, we have provided insight into the SOTA technological developments in AI accelerators.

1.1 Introduction to Artificial Intelligence (AI)

Artificial intelligence (AI) is a technique to introduce intelligence in devices, machines, software, hardware, etc. It is an enormous canopy covering machine learning (ML) and brain-inspired approaches such as spiking neurons, neural networks (NN), and deep learning (DL) techniques (depicted in Fig. 2). AI is witnessing continuous growth with SOTA research and development in ML and DL algorithms. Deep neural networks (DNNs) surpass humans in various cognitive tasks [1, 14, 15]. Few well-known examples where AI proved itself better than humans are art and style imitation, image and object recognition, predictions, video games, voice generation and recognition, website design modifications, etc.

It has enabled intelligence in machines. Neuroscience is behind this development. The ideas and findings of neuroscientists have been incorporated into the AI approaches to develop brain-inspired algorithms. Artificial neural networks (ANNs), spiking neural networks (SNNs), and DL are some such typical examples.

ML is a way for machines to learn problem-solving skills. An ML model must undergo two phases: training and inference. In the training phase, the ML model tries to learn the skills while it performs actual predictions in inference. The ML algorithms have been further subdivided into four broader categories: supervised, semi-supervised, unsupervised, and reinforcement learning (Fig. 3 represents the same).

- *Supervised learning:* A labeled dataset is used to train the algorithm. During training, the model tries to identify features or characteristics of elements with the same label used for inference to classify a given input into an adequate class.

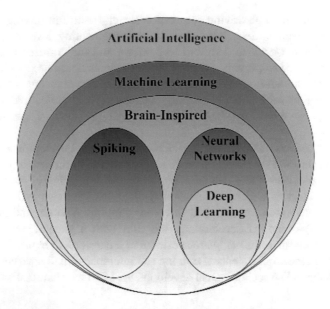

Fig. 2 The canopy of Artificial Intelligence covers ML and brain-inspired learning such as spiking, NN, and DL. (Figure adapted from Ref. [1])

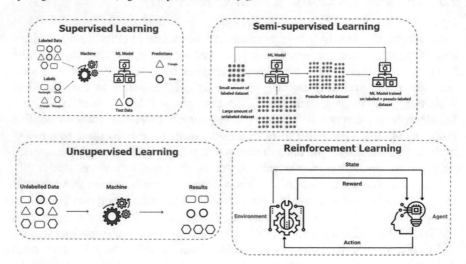

Fig. 3 ML: Supervised, semi-supervised, unsupervised, and reinforcement learning. (Figures adapted from Ref. [1])

- *Semi-supervised learning:* In a scenario where there are datasets containing only a few labeled samples and other items with no labels. With a few labeled datasets, the semi-supervised learning method tries to pseudo-label a larger number of unlabeled datasets. Further, training of the ML model is performed using the labeled and pseudo-labeled dataset. The algorithm should be able to predict new samples through the learned features.

- *Unsupervised learning:* It is used to cluster the items when there are no defined classes or given labels. This approach attempts to learn the existing similarities between samples in the dataset and cluster the dataset according to their similarity features.
- *Reinforcement learning:* No dataset is present in this category of ML algorithms. An agent tries to find the best policy in a simulated environment to achieve a goal. The agent attains a reward while interacting with the said environment. The reward can be positive for a correct decision and negative in terms of penalties for incorrect behaviors or actions. The reward mechanism helps an agent attain policy to make the best actions.

1.1.1 AI Applications

Industries and academia incorporate AI in various forms. Many applications involve AI techniques in one way or another [1, 16]. Figure 4 depicts the areas where AI has rigorously been applied. In aerospace, it performs commercial flight autopilot, weather detection, etc. In sports, wearable techs, smart ticketing, automated video highlights, and various computer vision–based applications are involved. Mobile (smartphone) phones incorporate AI to increase smartness in their applications. Likewise, workplaces, entertainment, hospitality, media, gaming, education, shopping centers including retail and online, transport, banking and finance, government and politics, event, insurance, cybersecurity, smart homes, defense, social networks, real estate, agriculture, healthcare, etc., use AI techniques in various forms.

A few of the popular domains of AI are:

- *Computer Vision (CV):* It has further subdivisions in machine vision, video/image recognition, etc.
- *Machine Learning (ML):* Performed in supervised, semi-supervised, unsupervised, and reinforcement learning.
- *Natural Language Processing (NLP):* Chatbot, classification, content generation, and content/semantic recognition are typical applications involving NLP.
- *Expert Systems:* It is a kind of knowledge-based system. They are designed to solve complex problems with the professional matter by knowledge reasoning to emulate a decision-making ability of a human expert.
- *Recommendation Engines*: It utilizes a data filtering tool to recommend the most relevant items to an individual. Netflix, YouTube, Amazon, etc., are typical recommender systems examples.
- *Robotics:* Robots are supposed to replicate human actions, and AI facilitates them to acquire such seamless replications.
- *Speech:* AI-enabled speech recognition provides the capability of speech to text or text to speech.

The AI domains help achieve the urbanization goals depicted in Fig. 4.

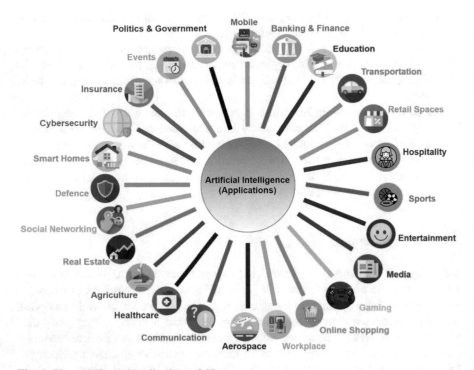

Fig. 4 The multifaceted applications of AI

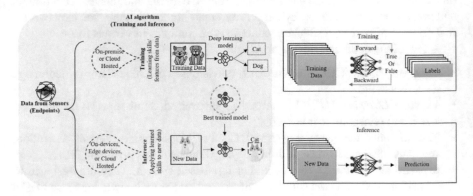

Fig. 5 An illustration of AI algorithm training (i.e., learning) and inference

1.1.2 AI Algorithms

There is an increasing demand for AI algorithms to help implement industrial applications. DL-based AI algorithms generally follow the trend of training and inference, as depicted in Fig. 5.

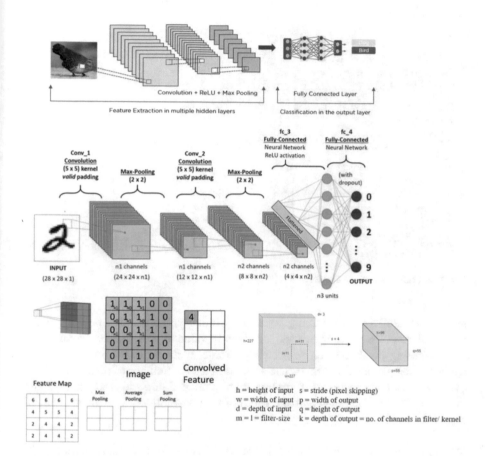

Fig. 6 An illustration of CNN

Sensors are the endpoints of AI systems, providing the required database for training and inference. Training and inference can be hosted on the cloud or implemented on board, on-chip, or edge devices. A DL model is supposed to learn the skills or features by extracting information from the database available during training. The training dataset is generally labeled. Therefore, the AI algorithm tries to identify a feature/pattern during a "forward" pass by adjusting its learning parameters in terms of weights and biases associated with corresponding neurons and connections. In the "backward" pass, these weights and biases are updated by counting the errors. Convolutional neural networks (CNN) are used in DL algorithms. Figure 6 illustrates the concepts of a typical CNN [1, 15]. Features of the input image are extracted by using convolving the kernel (filter) layers, pooling layers, and applying the activation functions (also called the transfer functions, e.g., rectilinear unit (ReLU), etc.).

The fully connected layers are used for classification at the output layer. There may be max-pooling in which the maximum value among the kernel values is taken

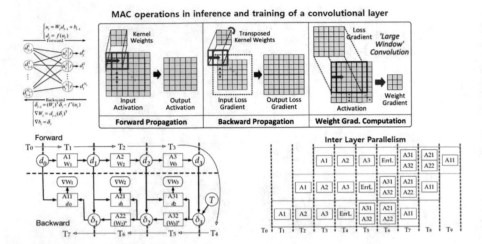

Fig. 7 A brief illustration of CNN

as an output of the max-pooling layer [1]. Similarly, the average pooling or the sum pooling corresponding average, or the sum value, represents the output of the corresponding pooling layer, respectively. Some of the standard notions used in CNN are also depicted in Fig. 6. The stride defines the number of pixels that shifts over the input matrix while performing convolution; padding adds zeros to maintain the input size, etc. Output height and width of the convolution layer can be obtained from (1) and (2) as:

$$\text{Output (width)} = \lfloor (\text{Input (width)} + 2 \times (\text{padding}) - \text{filter (width)})/\text{stride} + 1 \rfloor. \tag{1}$$

$$\text{Output (height)} = \lfloor (\text{Input (height)} + 2 \times (\text{padding}) - \text{Filter (height)})/\text{stride} + 1 \rfloor. \tag{2}$$

The forward and backward propagation is described with the multiply and accumulate (MAC) operations in the training and inference of a convolutional layer in Fig. 7 [1, 17, 18].

In forward propagation, the kernel weights (W) are convolved over input with the addition of respective bias (b) to produce the intermediary output (u) before activation for the lth layer. An activation function is denoted as (f) in the forward and backward propagation. The forward and backward propagations are defined by (3) and (4), respectively, as:

Forward propagation : −

$$\text{Intermediary output} : u_l = (W_l) \times (d_{l-1}) + (b_{l-1}). \tag{3}$$
$$\text{Final output} : d_l = f(u_l).$$

Backward propagation : −

Error for the previous layer $\left((l-1)^{\text{th}} \text{ layer}\right) : \delta_{l-1} = (W_l)^{\text{T}} \times (\delta_l) \times f'(u_l)$.
Weight gradient of the lth layer : $\nabla W_l = f(u_l)$.
Bias gradient of the lth layer : $\nabla b_l = \delta_l$.

$$(4)$$

The forward and backward propagations are implemented at the hardware using registers, multipliers, and adders as depicted in Fig. 7 in seven timing cycles (T_1 to T_7). Interlayer parallelism illustrates the acceleration in computation by incorporating parallelism in the same time cycle.

The concept of intelligent machines was initiated way before, in the 1940s. Fig. 8 represents a timeline of a few breakthrough algorithms along with the two winters in AI (before the deep neural networks (i.e., DL)). An electronic brain was introduced by the famous duo McCulloch and Pitts [19]. It learned the basic logic functions such as AND, OR, and NOT operations. Rosenblatt has introduced "Perceptron," which can mimic these logical operations [20].

Similarly, ADALINE was introduced in 1960 with the capability of the learnable weights and threshold [21]. In 1969, the XOR problem started the FIRST winter in the research of AI [22]. Although it has caused sluggish research in AI, it has resulted in the demand for efficient algorithms. Therefore, it can be regarded as a SOFTWARE (algorithmic) need in AI research. Researchers were in search to find a way out of XOR and nonlinear problems. Rumelhart et al. have resolved the classical XOR problem using multi-perceptron learning (MLP) [23]. However, it had initiated another winter in AI because of the heavy computations involved in MLP models for complex problems. This SECOND winter in AI had again slowed down the findings of AI research, however, it has resulted in the demand for an efficient computing system. Therefore, it can be regarded as a HARDWARE (accelerator) need in AI research.

Due to the two winters in AI, the research in AI reached pseudoscience status. Fortunately, a few researchers have continued their research on AI and DL, which significantly maintained the advances in AI. In 1995, Cortes and Vapnik developed the support vector machine (SVM) [24]. It was a system for mapping and recognizing similar data. In 1997, Hochreiter et al. developed the long short-term memory (LSTM) for recurrent neural networks [25]. A jerk in AI research came after the commencement of faster processing through graphics processing units (GPUs) in 1999 [26]. It has surged the computational speeds of pictures and graphics by multiple times. Therefore, AI research was re-energized in the early 2000s with the emergence of GPUs and larger datasets. In 2006, Hinton et al. significantly elevated their research in DL to ensure its potential and outcomes for future AI research [27, 28].

Further, developing open-source, flexible software platforms such as Theano, Torch, Caffe, TensorFlow, PyTorch, etc., has given the appropriate thrust to the current AI research. Therefore, more advanced AI algorithms were introduced later.

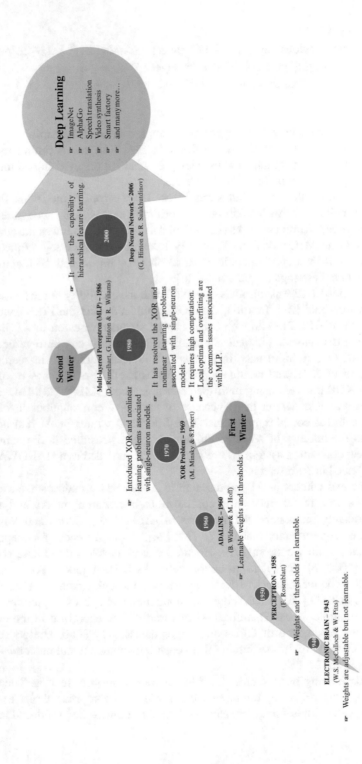

Fig. 8 A timeline of AI approaches with the two winters and the commencement of DL

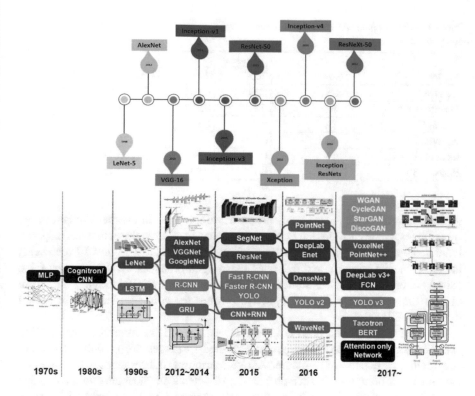

Fig. 9 A timeline of popular DL algorithms

A timeline with some of the popular algorithms of AI approaches used in DL is summarized in Fig. 9.

There are typically two ways of measuring the accuracy of an AI algorithm. One is Top-1 accuracy, and the other is Top-5 accuracy. In Table 1, Top-1 and Top-5 accuracy indicates the model performance on the ImageNet validation dataset. Depth indicates the topological depth of the network. It includes the Convolutional layers, pooling layers, activation layers, batch normalization layers, etc. Top-1 accuracy is the conventional version of accuracy. It considers a single class with the highest probability. Top-5 accuracy uses top-5 classes instead of a single class. For example: for an image of blueberry the predictions by an AI algorithm with prediction probabilities are as follows: cherry: 0.35%; raspberry: 0.25%; blueberry: 0.2%; strawberry: 0.1%; apple: 0.06%; and orange: 0.04%. According to the Top-1 accuracy measurement, the prediction (cherry: 0.35%) is wrong. However, according to the Top-5 accuracy measurement, the prediction is correct as blueberry is still in the five higher prediction probabilities.

After developing efficient AI approaches, several applications exist to shift machines' intelligence into augmented intelligence. In other words, AI is currently augmenting intelligence on devices. Therefore, AI on edge devices and on-device

Table 1 A few popular AI approaches include size, accuracy, number of parameters, and depth comparison

Model	Size (MB)	Accuracy		Parameter	Depth
		Top-1	Top-5		
VGG16	528	0.713	0.901	138,357,544	23
Inception-V3	92	0.779	0.937	23,851,784	159
ResNet50	98	0.749	0.921	25,636,712	
Xception	88	0.790	0.945	22,910,480	126
InceptionResNetV2	215	0.803	0.953	55,873,736	572
ResNeXt50	96	0.777	0.938	25,097,128	23

AI have continuously shown tremendous developments. In another domain of AI research, especially for unseen environments and scenarios, the active learning and federated learning approaches have contributed significantly. Figure 10 represents the basic blocks of active and federated learning approaches. Active learning (also called "query learning," or sometimes "optimal experimental design") is a form of semi-supervised learning. Here, active implies the continual learning of the AI model [29–32].

In multiple sophisticated tasks such as speech recognition, information extraction, classification, filtering, etc., labeled instances are complicated, time-consuming, or expensive to obtain. Active learning (AL) provides a self-labeling method and is often used in such problems. AL selects the most uncertain unlabeled example (query) for labeling by a human (experts) and iteratively labels the remaining similar data. The key feature of AL is curiosity, which is why it achieves greater accuracy with fewer labeled instances. In the AL loop, there are four aspects: Train, Query, Annotate, and Append. In the AL loop, "Train" is to train the model on a labeled dataset. The "Query" is to select the unlabeled examples from the dataset using some acquisition function. The "Annotate" is to label the examples chosen by the oracle (subject matter experts). Finally, "Append" is to add the newly labeled examples to the training dataset.

The AI-based AL architecture generally compares the prediction confidence of an unlabeled dataset and accordingly decides whether to ask the expert (such as a human). Depending upon the prediction confidence, the unlabeled dataset is added to the labeled dataset for utilizing it in further training. Therefore, "Query" is an essential task of the AI-based AL architecture. Another important task is "Annotate," which is accomplished by an expert. Now, who should be an expert? The expert should be the subject matter expert (SME). Therefore, it is better to use diverse knowledge from different sources depending on the query.

Federated learning (FL) is a decentralized approach in ML termed by Google in Refs. [33–36]. It enables AI models to obtain multiple experiences from different datasets located at various sites. These sites may be a local data center or a central server. There is no need to share training data with the main central server. It is based on iterative model averaging [33, 36]. Therefore, it has the inherent potential for privacy protection of the individual user. Yang et al. have further categorized FL as

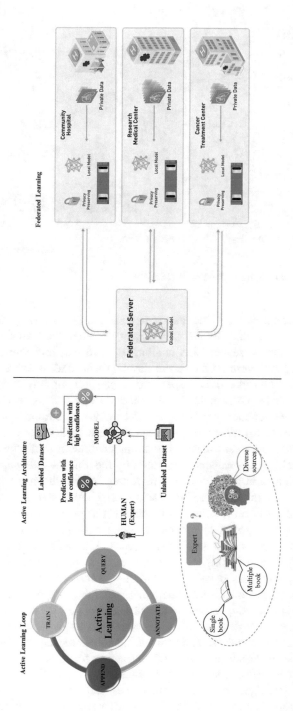

Fig. 10 Basic blocks of active and federated learning approaches

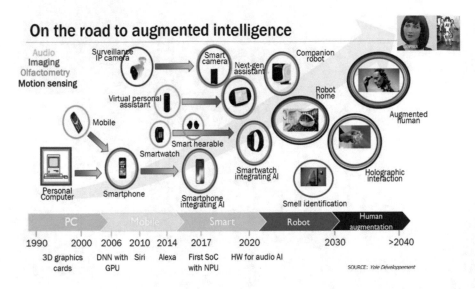

Fig. 11 A roadmap illustrating augmented intelligence

horizontal FL, vertical FL, and federated transfer learning [34]. They classified FL according to data partitioning among various party types in the feature and sample spaces. Li et al. have presented FL's challenges, methods, and future directions [35]. Zhang et al. have surveyed the recent developments and research on FL [36]. In their survey, they considered five aspects: data partitioning, privacy mechanism, machine learning model, communication architecture, and systems heterogeneity to summarize the characteristics of existing FL and analyze its current applications.

Various market survey agencies and experts have forecast the trend in this augmentation to move forward toward human augmentation in the future. A road to augmented intelligence has been illustrated by the *Yole Développement* in Fig. 11 [37]. Like humans, in machines, the main areas of intelligence augmentation are audio, vision, olfactory, and motion sensing. Intelligence in machines has been embedded since the 1990s after the commencement of personal computers (PCs). The graphic cards and graphical processing units have become more advanced and converted cellular phones into smartphones. DL algorithms and efficient hardware helped develop intelligent assistants such as Siri, Alexa, Cortana, etc. In the meantime, there has been an evolution of surveillance Internet protocol (IP) cameras, virtual personal assistance, smart wearables such as smartwatches, etc. With the system on chip–based neural processing unit in 2017, AI has been integrated into electronic devices to make them even more intelligent. In 2020, we witnessed smart cameras, smarter virtual personal assistance with enhanced features, wearable electronic devices with attractive features, smarter consumer electronic items and gadgets, smart homes, and many more. Soon, we are about to witness more intelligent products such as companion robots, robotic homes, electronic noses for smell identification, etc. It is expected that by 2040, there will

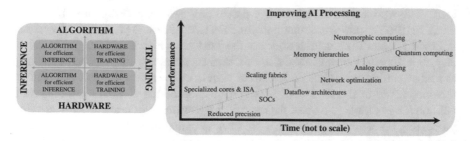

Fig. 12 Requirements of AI algorithms and AI accelerators for training and inference, and the roadmap of improvements in AI processing

be the augmentation of intelligence for humans, i.e., augmented humans instead of currently demonstrated simple robots.

Such a massive development is manifest by the amalgamation of efficient AI algorithms and compatible AI hardware. Still, there is a demand for continual expansion in future research areas (AI algorithms and hardware). Figure 12 represents this demand for AI algorithms and hardware in training and inference.

It also represents the improvement in AI processing with the emerging AI techniques. The scaling fabrics, memory hierarchies, dataflow architectures, and network optimizations are in recent research areas of AI. Neurocomputing, analog computing, and quantum computing are emerging as future research on AI processing.

1.2 Hardware Accelerators

Hardware accelerators are computer hardware specially designed to perform some functions more efficiently than is possible in software running on a general-purpose processing unit alone [1, 38–51]. It combines the flexibility of general-purpose processors, such as central processing units (CPUs), with fully customized hardware, such as graphical processing units (GPUs) and application-specific integrated circuits (ASICs), to increase the efficiency by orders of magnitude. For example, visualization processes may be offloaded onto a graphics card to enable faster, higher-quality playback of videos and games. In this way, an AI accelerator is a class of specialized computer systems designed to accelerate AI applications, such as ANNs, computer or machine vision, ML, etc. [4, 38, 50, 51]. It is the current requirement that AI and ML applications lead to meet the expectations of the smart world and intelligent systems. The algorithmic superiority of intelligent systems is achieved by extremely high computing capacity and memory, or in a nutshell, with high-performing hardware [39, 40].

However, it imposes significant challenges in designing suitable hardware platforms that can seamlessly execute the SOTA AI and ML algorithms over them. The well-known aspects of any processing hardware design are SIZE (chip area), ENERGY (power consumption), and DELAY (latency) [1, 38–48]. Therefore, destined AI hardware accelerators are the best choice to excel the performance and meet the requirements. GPUs, specialized accelerators, etc., are the commonly available hardware designs as hardware accelerators. However, enhancement in the complexity of the SOTA AI, ML, and DL algorithms and the increasing data volume always leads to a quest for the next improvement in designs for improving processing efficiency. The AI hardware designers and engineers regularly explore new ideas for efficient architecture designs to improve performance. Popular design approaches are based on the approximate computing design, in-memory computing design, etc.

Further, there are different approaches, such as quantum computing-based machine intelligence designs. These are the commonly explored computing approaches for AI workloads. Renowned companies in hardware designs are involving their research and developments (R&Ds) in innovating new hardware architectures to achieve the required processing power with unprecedented speed for satisfying the needs of efficient AI workloads [1, 41–48].

2   Requirements of AI Accelerators

To elicit the requirements and problems of AI accelerators, we need to understand Fig. 13. It reflects the accuracy of AI algorithms with their model complexity [49]. NVIDIA Titan X Pascal GPU has been used to obtain this comparative result. It shows the computational complexity associated with the AI models (the higher the ball size, the higher the computational complexity of the AI model). The most complex model does not need to have the highest accuracy. As the NASNet-A-Large has the highest accuracy despite its moderate computational complexity. In contrast, versions of the VGG models have moderate accuracy at very high computational complexity.

2.1   Hardware Accelerator Designs

This comparative chart illustrates that efficient computing platforms are necessary to accomplish the tremendous computational requirements of the AI models. Fig. 14 summarizes the classical computing architectures [52–56].

The hardware acceleration journey embarks through the simpler processor design consisting of random-access memory (RAM), a set of registers, a processor, and the buses (single bus and bidirectional bus) for data transfer. It was limited to a single integer operation per cycle. An out-of-order (OoO) scheduler has been introduced to overcome this limitation. It has alleviated the number of operations in a single cycle. Through the OoO, there was a possibility of processing the integer and floating-

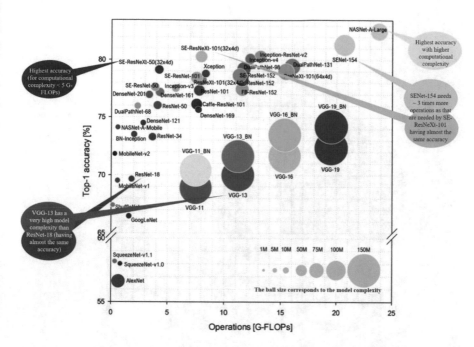

Fig. 13 Accuracy versus model complexity of popular AI algorithms. (Figure adapted from Ref. [48])

point data simultaneously at arithmetic and logic unit (ALU) and floating-point unit (FPU), respectively. In the Pentium simplified architecture, the OoO has been used to process the instructions via two ALUs and one FPU. It has extended the number of simultaneous operations. However, there was a drawback of non-optimal utilization of processing elements in case of ALU saturation. Because there was a single instruction stream in this architecture, the next instruction can never be fetched unless the previous has been processed. Through hyperthreading, a dual-core concept has been embedded in the Pentium processors. It was implemented by replicating two cores using two separate instruction streams. This concept was termed simultaneous multi-threading (SMT) [55]. This dual-core architecture had an issue with latency during blockage of either instruction stream. It reduces the processing efficiency to 50% in blockage conditions. Later, the concept of vector processing was introduced. It has the facility of executing multiple data in a single instruction. Therefore, it was called a single instruction multiple data (SIMD) architecture [49, 56, 57].

Here, multiple data can be processed in a single instruction. It was an implementation of multi-threading. There can be six instruction streams in this architecture. This architecture has boosted the processing speed multiple times with the set of ALUs, FPUs, and registers. However, this architecture's fundamental issue remains during a blockage in the instruction stream. With a newer concept of

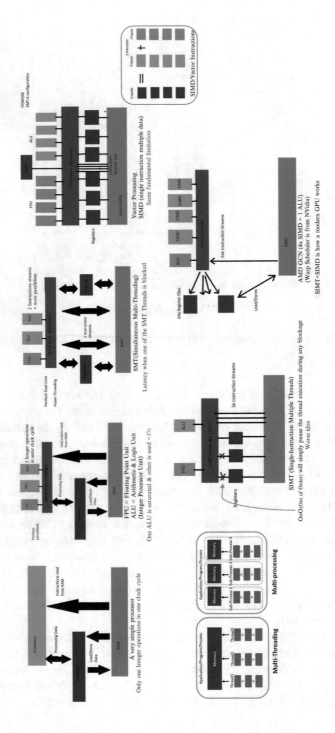

Fig. 14 A few examples of processor designs used for hardware acceleration

single instruction multiple threads (SIMT), a SIMT scheduler has been added to the OoO scheduler to overcome this limitation. SIMT is to implement multiprocessing. It utilizes the concept of sub-divisions of the task and parallelism. Therefore, an optimized performance is possible through the SIMT architectures [58]. However, the idea has even worsened the performance as the OoO pauses the thread execution during any blockage.

SIMD and SIMT processing individually was a failure. Therefore, the processor designers have combined these processing ideas to develop a powerful processor. In this way, the idea of a modern GPU evolved. Based on this idea, NVIDIA has introduced the warp schedular to replace the OoO schedular, which is how modern powerful computer systems cater to the demands of complex computing. A comparison of multicore and multithread processing is discerned in Fig. 15. In a multicore processor, the CPU has multiple co-processors to accelerate the processing task by subdividing it and processing it in parallel [59]. Each co-processor, called a core, has a control unit, ALU, and the cache memory containing the L1 and L2. L3 cache memory is a shared cache with each core. Every co-processor accesses the data from the main memory using a common system bus through the memory controller and the system interface.

In contrast, the GPUs have streaming multiprocessors (SMs). An SM has multiple streaming processors (SPs), also called cores. Each core is connected through a shared memory in an SM. Simultaneously the SPs and SMs are connected to the GPU device memory to access the data from the device memory. The execution of the instructions is performed in a multithreading fashion. Each thread contains its per-thread local memory to execute the subtasks assigned to the corresponding thread [60, 61].

A multicore processor is a processing unit with two or more separate processing cores. Each core reads or executes instructions (such as add, move data, and branch) in parallel. However, the multicore processor can run instructions on separate cores simultaneously, increasing the overall speed. In multithreading computing techniques, the individual thread has an associated local memory, and the block of such thread is called a thread block. A group of thread blocks forms the grid. A typical GPU architecture contains SMs with SPs, i.e., cores connected through the shared memory and the device memories, as exhibited in Fig. 15. The thread distribution is responsible for distributing the tasks to different thread blocks to process them parallelly. There are various device architectures in processors, namely, temporal and spatial architectures. Figure 16 represents the generic block diagram of the temporal, spatial, and DNN accelerator architectures [1, 62]. The temporal architectures, in general, are followed in CPUs and GPUs. It employs various techniques to improve parallelisms, such as vectors, i.e., SIMD, or parallel threads, i.e., SIMT. Such temporal architectures use centralized control for many arithmetic logic units (ALUs). These ALUs can only fetch data from the memory hierarchy and cannot communicate directly with each other.

In contrast, spatial architectures use dataflow processing, i.e., the ALUs form a processing chain so that they can pass data from one to another directly. ALUs have their control logic and local memory called scratchpad or register file (Reg File).

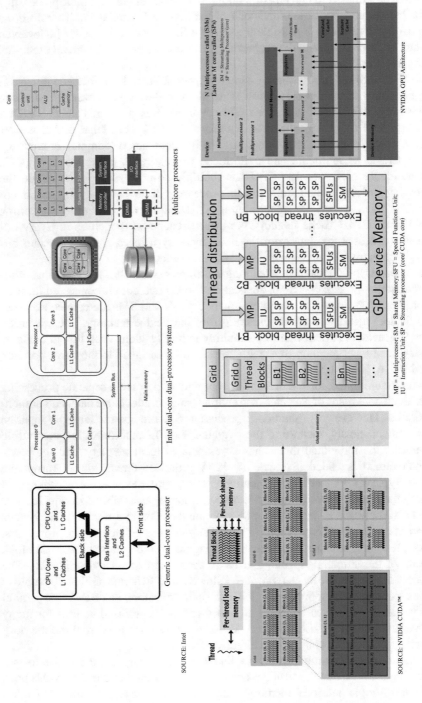

Fig. 15 Multicore and multithread processor designs

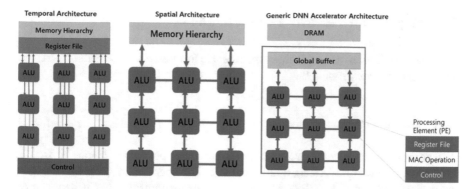

Fig. 16 Temporal, spatial, and DNN architectures. (Figures adapted from Refs. [1, 62])

The ALU with its local memory as a processing engine (PE). Spatial architectures are commonly used for deep neural networks in field-programmable gate arrays (FPGAs) and application-specific integrated circuit (ASICs)–based designs.

2.2 Domain-Specific Accelerators

Accelerators have been designed for various tasks such as Graphics rendering, DL training, inference accelerators, simulation accelerators, bioinformatics, image processing, etc. A domain-specific accelerator is specialized for a particular domain of applications. Therefore, a domain-specific hardware accelerator results much better than a general hardware accelerator. A domain-specific hardware accelerator commonly exploits the following techniques for its performance and efficiency gains:

- *Data specialization*: It is the core technique to accelerate the processing capability. The task can be performed in one cycle using specialized operations on domain-specific data types compared to several cycles taken by a conventional computer.
- *Parallelism*: A high level of parallelism (often exploited at different levels) is required to obtain higher performance.
- *Local and optimized memory*: A very high memory bandwidth can be achieved with low cost and energy by optimizing the memory allocation. In other words, storing the required data structures in local memories enhances the memory bandwidth.
- *Reduced overhead*: A specializing hardware eliminates the overhead of program interpretation.

The metrics used to measure the performance of AI accelerators are described below [1].

2.3   Performance Metrics in Accelerators

There are various methods to measure the performance of accelerators. A few popular metrics are given below.

2.3.1   Instructions Per Second (IPS)

It measures the speed of a processor in instructions per cycle. It can be in kilo instructions per second (KIPS), a million instructions per second (MIPS), or billion instructions per second (GIPS). However, this measurement scheme depends on the instruction sequence, the data, and external factors. Therefore, an accurate measurement is difficult using IPS.

2.3.2   Floating Point Operations Per Second (FLOPS, flops, or flop/s)

It involves floating-point calculations to measure the performance of a computing system. Therefore, a more accurate measurement is possible using FLOPS compared to the IPS. Still, the AI processors require something more relevant as the computation in AI processors is in MAC.

2.3.3   Trillion/Tera of Operations Per Second (TOPS)

It measures the maximum achievable throughput instead of calculating the actual throughput. Most of the AI operations are MAC. The TOPS measurements are performed as (1):

$$\text{TOPS} = (\text{number of MAC units}) \times 2 \ (\text{frequency of MAC operations}). \qquad (5)$$

2.3.4   Throughput Per Cost (Throughput/$)

Even this measurement is not enough information for the performance measurements of an AI accelerator. Other factors also decide the performance of a processor, such as chip area and the cost. A portion of silicon area (or circuitry) on an IC cannot be powered at the nominal operating voltage. This silicon area is called dark silicon. Therefore, performance measurement should be based on the throughput and the cost, i.e., the Throughput/$. The throughput and the cost of an AI accelerator depend on four key components: MACs; static random-access memory (static RAM or SRAM); dynamic random-access memory (DRAM); and the interconnect architecture [1].

Power requirements

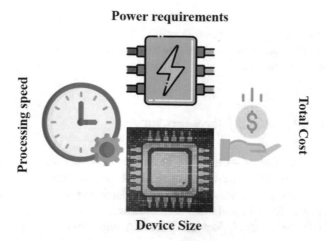

Device Size

Fig. 17 Key metrics and design objectives of the AI accelerators

2.4 Key Metrics and Design Objectives

SRAM is mainly used as a cache memory in a processor as it is faster than DRAM. Depending on size and speed, the cache memory has three levels, L1, L2, and L3. DRAM accelerates performance in various processors. The interconnection is responsible for connecting the processing elements and memory units. Therefore, increasing any of these components (MACs, SRAM, DRAM, and interconnect) improves both throughput and cost. Therefore, optimizing the throughput with the least numbers of MAC, SRAM, DRAM, and interconnect units is recommended. It will eventually maximize the Throughput/$. Interestingly, the power dissipation increases with more numbers of any of these (MACs, SRAMs, DRAMs, and interconnects) components as well. Therefore, the cost and power are roughly correlated. Following are the key metrics and design objectives of any electronic device (also in AI accelerators) [1]. These are also depicted in Fig. 17.

- *Processing Speed*: An AI accelerator requires to enable faster training and inference. It facilitates an ML expert to:

  - Faster training: An ML expert may attempt multiple DL approaches.
  - Hyperparameter optimization: Optimize the structure of their ML algorithms.
  - Faster inference: It is crucially required in applications like autonomous driving.

  Therefore, the processing speed of an AI device should be very high.

- *Power Requirements*: Lesser power consumption means higher device on-time, and a user can avail of more applications through a device. Therefore, there is always a requirement for an energy-efficient device in AI applications.

- *Device Size*: Increasing the requirements of wearable devices, IoT applications, smartphones, etc., puts size limitations on AI devices.
- *Total Cost*: The cost of the device should be lesser. Because it is highly crucial to any procurement decision.

3 Classifications of AI Accelerators

A general classification of AI accelerator architectures has been exhibited in Fig. 18. The AI accelerator architectures can be classified into three categories, namely, general-purpose AI accelerator architectures, special-purpose AI accelerator architectures, and emerging AI accelerator architectures.

Below are the brief descriptions of each accelerator type:

- *Central Processing Units (CPU)*: It is the general-purpose processor primarily used in standalone computes (Intel-Core, Advanced Micro Devices (AMD), Ryzen, etc.) [59].
- *Graphical Processing Units (GPUs)*: It was designed for graphical rendering using parallel computing. Parallel computing allows processing multiple data pieces simultaneously, making the GPUs useful for ML applications, video editing, gaming applications, etc. [59–61]. It is very effective to train complex DL algorithms. GPUs are considered the current workhorses for DNNs for both training and inference. However, some limitations exist with GPUs, such as bandwidth, latency, and branch prediction. The GPU becomes even slower for a chaotic code flow than the CPU and may need a CPU to be controlled. However, the latest GPUs have been more independent than earlier. It can execute a new tasks after the first task given by the CPU.

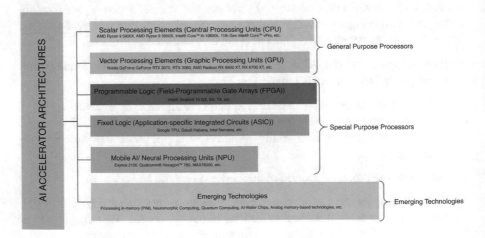

Fig. 18 General classification of AI accelerator architectures

- *FPGA and ASIC*: They are spatial architectures. The primary purpose of FPGAs is programmability to implement any possible design. They are relatively cost-effective with a short time to market, and the design flow is simple. However, FPGAs cannot be optimized for the various requirements of different applications, are less energy-efficient, and have lower performances than ASICs. On the contrary, ASICs need to be designed and produced for a specific application that cannot be changed over time. The design flow is consequently more complex, and the production cost is higher, but the resulting chip is highly optimized and energy efficient. Challenges in ASIC & FPGA-based AI accelerators are a significant amount of storage, external memory bandwidth, and computational resources on the order of billions of operations per second [63].
- *Tensor Processing Unit (TPU)*: Google has developed this ASIC-based AI accelerator for Google cloud applications [64, 65]. It has enabled businesses with cloud-based AI applications. Google utilizes the cloud TPU in Google Translate, Google Photos, Google Search, Google Assistance, Gmail, etc., to incorporate AI on these applications. TPU uses TensorFlow (an open source) platform for implementing ML. It has small digital signal processors with local memory on a network. It uses on-chip high bandwidth memory (HBM), with every core having scalar, vector, and matrix units (MXUs).
- *Neural Processing Unit (NPU)*: Smartphones or mobile applications become ubiquitous by applying DL approaches. These are resource-constrained devices and require immediate responses simultaneously maintaining user privacy. Employing the on-device ML technologies over these devices requires considerable computation and communication. NPU resolves these issues and serves as a neural-network specialized hardware accelerator for the mobile application processor (AP). The NPU architecture uses data-driven parallel computing to increase overall performance and is dedicated to energy-efficient DNN acceleration [65, 66]. NPUs can be reconfigured to switch between models in real-time. Therefore, it can be used as the optimized hardware per the demand.
- *AI-Wafer Chips*: With tremendous package density on silicon wafers, the AI-chip die contains trillions of transistors (e.g., Cerebral). AI-Wafer chip consists of three parts, namely, computing (processor), storage (memory), and networking (interconnects). Recently, Cerebras has developed their gigantic AI-Wafer chips to accelerate AI and ML workloads [67, 68]. It has introduced two wafer-scale engines (WSEs), WSE-1 and WSE-2. WSE-2 is the largest AI-Wafer chip ever built (containing 2.6 trillion transistors). It has 123 times more compute cores than the largest GPU, with a thousand times more on-chip memory (high-performance). AI-Wafer chips are higher performing than general-purpose chips. The AI-Wafer chips bring faster computation and high bandwidth memory, making them perform better than the general-purpose hardware.
- *Neuromorphic Architectures*: Research in neuroscience and materials brought the capabilities to mimic brain cells. These chips can have an advantage in terms of speed and efficiency training neural networks. The neuromorphic architectures render these advantages by mimicking human brains [69, 70]. Therefore, AI accelerator designers are fascinatedly researching the hardware scopes in this

area. A few example projects for neuromorphic architectures are TrueNorth, SpiNNaker, SyNAPSE, BrainScaleS, etc. [71, 72]. These are a few of the state-of-the-art developments in neuromorphic architectures.

- *Analog Memory-Based Accelerators*: Digital systems rely on 0's and 1's centric computing. However, analog techniques contain constantly variable signals without specific ranges. Analog memories can store a continuous range of values in terms of voltage. The nonvolatile memories (NVMs) in crossbar arrays have pioneered the in-memory computing system. These crossbar arrays of analog synaptic devices are a faster and energy-efficient alternative to traditional (i.e., the von-Neumann) computing systems. The crossbar array can be used for training and inference purposes without a separate memory access routine in the von-Neuman architecture. These NVM crossbar arrays have high storage density and can efficiently perform massively parallel MAC operations. Therefore, it can overcome the memory bottleneck faced by the von Neumann–based architectures in the ML workloads. Recently, the IBM research team demonstrated that large arrays of analog memory devices could achieve similar levels of accuracy as that of GPUs for DL applications [73].

4 Organization of this Book

Considering the above categories of computing hardware, we have divided the further chapters of this book into different themes such as computing system platform–based chapters, technological development–based chapters, emerging application–based chapters, and the existing issue–based chapter.

The computing system platforms–based discussions have been compiled in subsequent Chapters: Chaps. 2, 3, 4, 5, and 6. Chapter 2 focuses on "single-tenant," i.e., single owner of the compute. The standalone personal computers are an example of single-tenant computing. It is a self-sufficient and independent compute platform for performing every application without essential network connectivity. Therefore, it has not much vulnerability to privacy and security. There is usually no requirement for cloud service in a standalone computing system. Though it is possible to utilize cloud services on standalone systems, it is a bit expensive. This chapter describes CPU- and GPU-based accelerators, including training and inference accelerators. PC has been a primary stand-alone compute platform for simple neural network development. Chapter 3 includes the cloud- and server-based accelerator architectures, i.e., "multiple-tenants." Multiple tenets refer to owners associated with the cloud servers independently. Cloud or server services allows sharing of resources and networks, with the primary focus on the privacy and security of the users. Cloud computing and data centers are considered two different applications in a multi-tenant-based computing system. There is a difference in the system requirements as well in computing and data center applications. In computing applications, the cloud or servers require efficient processing capabilities along with robust networking and connectivity. It also includes the training and inference accelerations.

In contrast, the data center requirements mostly focus on effective data storage capabilities with robust networking and connectivity. Therefore, this chapter targets the relevant issues and developments on servers and cloud-based computer systems, including their privacy and security concerns. Chapters 4 and 5 have descriptions of the AI accelerators in smartphones. Smartphones are considered resource-constrained devices. They rely on cloud and edge computing techniques through an efficient networking environment. The on-board mobile is typically consisting of a weak computing resource of CPU, GPU, or NPU. Therefore, the specific task must be cleverly allocated to a cloud or on-board device, such as the training being performed using a cloud server. At the same time, the inference is implemented using a server or on-device resource, depending upon the application. These chapters encompass the issues of the smartphone systems, including their hardware, software, and benchmarking requirements. Chapter 6 is consisting of the embedded computing systems that are inference dominant. The embedded computing systems generally combine the CPU, NPU, and GPU. Here, training can be performed in a standalone or cloud server environment. The trained neural network is transferred to the target system's memory using the standards and formats of interchanging neural networks. ONNX (Open Neural Network Exchange standards) and NNEF (Neural Network Exchange Format) are the de-facto standards formats of interchanging neural networks. Embedded systems are used in electronic devices for specific applications such as consumer electronics (i.e., electronic equipment used in household applications, e.g., television, set-top box, air-condition, washing machines, etc.), robotic platforms, embedded computing boards (e.g., Arduino boards, Raspberry-pi board, Texas instruments boards, Google Coral Dev boards, Rockchip boards, etc.), and the autonomous vehicle computing platforms.

Chapters 7 and 8 are comprised of descriptions of trending technological developments in the AI accelerators. The reconfigurable and programmable AI accelerators have been discussed in Chap. 7. It includes the reconfigurable FPGA accelerators as well as the application-specific dedicated hardware accelerators. The problems and issues of the system bus of the von-Neumann architectures have been addressed. Chapter 8 discusses the neuromorphic approach of the hardware design for AI accelerations. It contains the state-of-the-art devices, circuits, and architectures used in neuromorphic computing.

Accelerators for autonomous vehicles (AVs) have been described in Chap. 9. It includes the AV computing platforms (such as NVIDIA Jetson, full self-driving (FSD) computer from Tesla), etc. This chapter targets insight into the embedded computing platforms. Although AVs and smartphones are specialized classes of embedded computing platforms, we have separated these into separate chapters because of their extended computational and connectivity requirements than the other embedded hardware accelerators. It will help in the better understanding of the concepts for future readers. Energy is an essential requirement of the hardware. Especially, the DL-based AI models are highly hungry for energy as they rely on more extensive calculations and voluminous data. Memory access operations for data loading and storage dissipate most of the power of hardware acceleration. Performing complex computations using voluminous data further elevates energy consumption. Therefore, there must be efficient utilization of energy in AI processing. Chapter 10 discusses the demand for energy in computing AI models and the efficient ways to design an accelerator to optimize energy consumption.

5   Popular Design Approaches in AI Acceleration

Most of the computing of AI algorithms is related to the MAC operation. A comparison of MAC architecture is discerned in Fig. 19, considering conventional and systolic-array-based neural computing units [74]. In conventional hardware, the multiply operation is carried out between the operands stored in two different registers. Then the value of the multiplication logic is added via the accumulate adder to the third operand. In this way, it is a cumbersome process for voluminous AI model calculations consuming plenty of time. Therefore, a systolic-array-based neural computing unit is used for MAC operation in AI computations.

It saves enormous time and provides ease of MAC computations. A typical AI accelerator architecture of FPGA and ASIC is shown in Fig. 20 [63, 75, 76]. The AI accelerators on ASICs or FPGAs have an array of processing elements (PEs) connected to the network-on-chip (NOC). The off-chip memory and the array of PEs are interconnected via a global buffer (GB) and the NOC. Each PE has a registered file (RF). In FPGA accelerator architectures, configurable logic blocks (CLBs) and programmable interconnects (PI) are interconnected to provide reconfigurability to the entire system. However, the ASIC accelerators are destined for specific applications and are non-reconfigurable architectures. There are different blocks

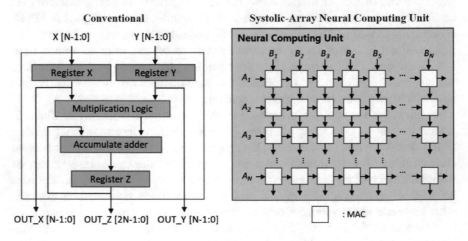

Fig. 19 A comparison of MAC operations performed on conventional and neural computing units.

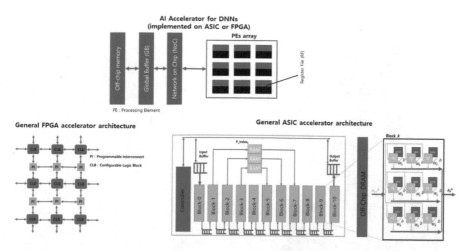

Fig. 20 General AI accelerator architectures of FPGA and ASIC. (Figures adapted from Refs. [63, 75, 76])

between the controller and the off-chip dynamic memory, along with different buffers to manage the data flow. Each block has a set of PEs with the associated memory. ASIC accelerators are inherently faster than that FPGA accelerator designs at the cost of their fixed applications.

Figure 21 consists of various ASIC accelerator architectures of the tensor processing unit (TPU). Google has introduced the TPU as an ASIC architecture for accelerating the AI application on the Google cloud services [64]. TPU has multiple versions over the period (TPU v2, v3, v4, and edge TPU) to assimilate further advancements to satisfy the demands of complex training and high-speed inference through the Google cloud. Cloud TPU is a custom ASIC design from Google to power their cloud products with AI. Some of the typical examples offered using TPU are Google Translate, Google Photos, Google Search, Google Assistant, and Google mail services. The TPUs were developed for inference acceleration purposes.

The TPU has a systolic data flow engine. It is a set of horizontal and vertical arrays of 256 sizes forming a two-dimensional pipeline to implement the MAC operations between the weights. Every weight in the array shifts by a single step to be multiplied by the weights in the cell further, the product gets a partial sum by moving down at each cycle in a systolic function. Moreover, an ML model in TPU can also be reconfigured, like the CPU or GPU. It executes complex instruction set computer (CISC) instructions on multiple networks such as CNN models, LSTM models, etc. In this fashion, it has boosted the data-centric applications on Google using TPU Pods. In TPU Pod, multiple TPUs are connected through dedicated high-speed network interfaces. It can have up to 2048 TPU cores to distribute the processing load across multiple TPUs.

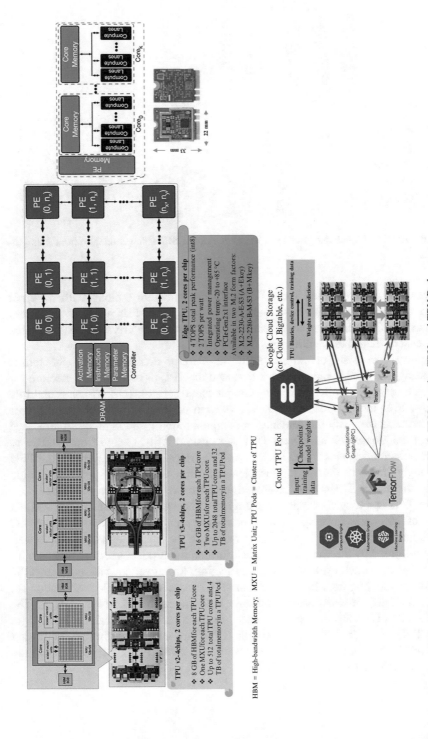

Fig. 21 The ASIC accelerator architectures of TPU: TPU v2, TPU v3, edge TPU, and TPU Pod

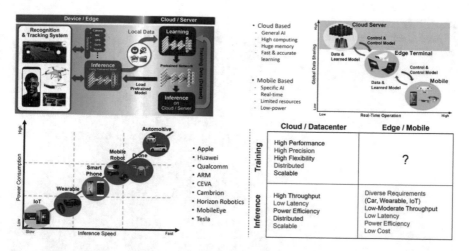

Fig. 22 A comparison of cloud-based AI and AI on the edge devices

Further advancements in algorithms and devices have shown us the shift of AI applications from the cloud or servers towards the edge or devices. A comparison of the same is briefly presented in Fig. 22.

Figure 23 represents the AI accelerator architectures for edge applications. It consists of two modules: the processing system and the programmable logic. The OpenStack worker module supports the standard cloud computing platform deployed as infrastructure-as-a-service clouds in the processing system. Virtualized network virtualization combines the hardware and software network resources and functionality into a single virtual network as a software-based administrative entity. Further, a full-featured integrated development environment and open libraries enable the embedded Linux system to perform the computation. Advanced extensible interface (AXI) interconnect helps connect a memory-mapped master device to the slave devices. The programmable logic is subdivided into static and dynamic areas. The static area enables the reconfiguration of the dynamic area through the interconnection of the processor system, AXI interconnects, and the packet capture (PCAP) [77].

Trending research and development of the on-device AI has brought intelligence to the devices [6, 78, 79]. It has multiple advantages over cloud AI. Figure 24 demonstrates the on-device AI and its advantages. The cloud-based inference mainly suffers the privacy, network bandwidth, reliability, and latency issues. In the example shown in Fig. 25, a pedestrian has a lesser distance to save themselves due to lower latency in the case of cloud-based inference. However, the margin of collision increases with the on-device inference. Again, the on-device AI requires almost 105 times lesser bandwidth for the network bandwidth than the cloud-based AI inference.

Moreover, the on-device AI-based inference makes the privacy concerns almost insignificant to the user. It is because there is no user data transfer between the device and cloud during on-device inference. In contrast, cloud-based AI inferences

Fig. 23 An illustration of the
AI accelerator architecture for
edge devices

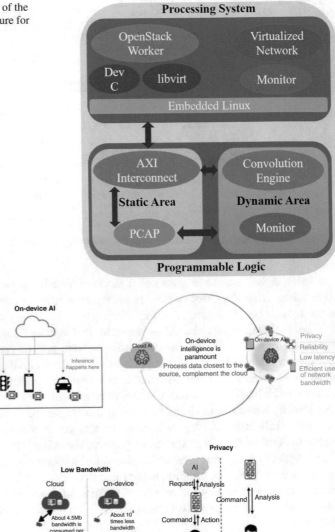

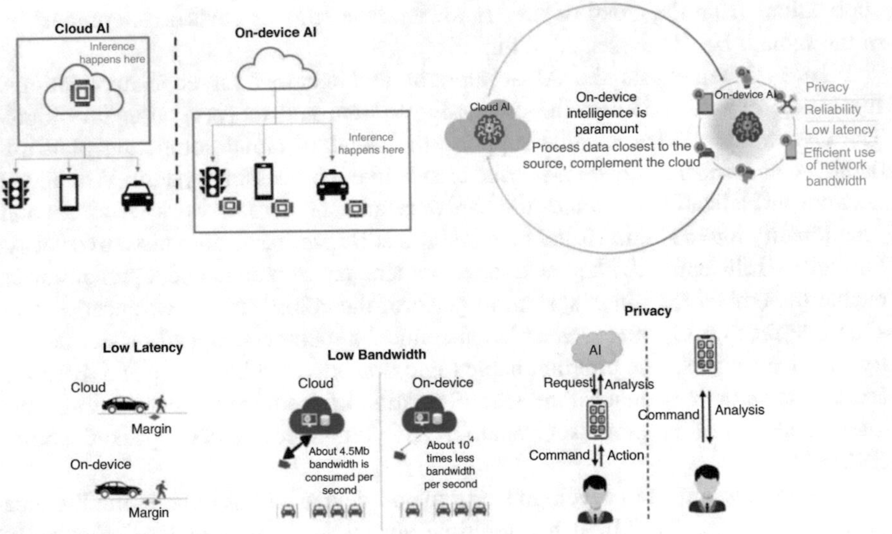

Fig. 24 Illustration of the on-device AI with its advantages over cloud-based AI

are subjected to severe privacy concerns. Despite these worthful advantages of the
on-device AI, there is an issue with training in such hardware caused by their limited
hardware resources. Therefore, efficient resource-constrained training platforms are
in demand to accelerate the on-device AI.

Neural processing units (NPUs) are another emerging category that falls under
the on-device AI [65]. They have led the pathway for AI acceleration on smart-

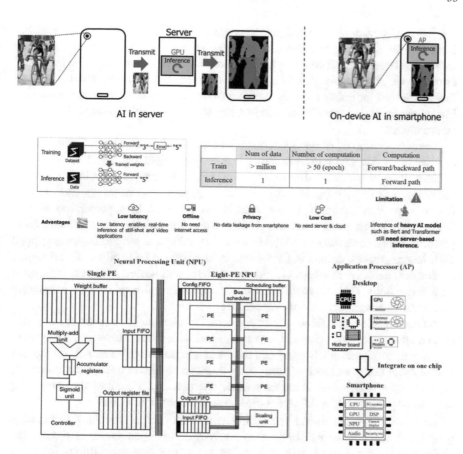

Fig. 25 An illustration of AI acceleration on the smartphone. (Figures adapted from Refs. [1, 51])

phones. Figure 25 represents the on-device AI in smartphones. In this illustration, an image through the smartphone's camera is captured for object detection by applying a semantic segmentation approach. The server-based inference intends to transmit the captured image to the server for further processing and inference. Server or cloud-based AI is assisted with heavy computing systems (i.e., powerful GPUs). The segmentation output is further transmitted to the smartphone as an inference output. This operation requires sufficient network bandwidth to transfer the image from device to server and retrieve the result back to the device. In addition, it consumes a higher time for inference.

In contrast, if the smartphone (i.e., the device) is equipped with the on-device AI via the application processor (AP), then inference can be performed at the device without any involvement of the server. The on-device AI can save inference time and preserve the end-user's privacy. Moreover, it does not require any network connection for inference. Tabulated information in this illustration renders the size of the data involved in training and inference through smartphones. However, smartphones are inherently resource constrained. Therefore, training on such devices still has issues because training requires almost twice as much computation as needed in inference. Training must compute the forward feature extractions and the backward error for weight and bias optimization. In addition, the size and depth of the AI model selected for training are also subject to the training computation complexity. For example, transformer-based models are still untrainable through the APs of smartphones.

The acceleration of AI applications in smartphones is generally performed using the NPU and AP. The NPU has multiple PEs (e.g., eight-PE NPU). A single PE of the NPU consists of a buffer to store weights, MAC unit, accumulator registers, activation unit (e.g., sigmoid unit), and the registers to store inputs and outputs. It computes very similarly to the calculation involved in a perceptron model. Series of such PEs connected through the system bus and the bus schedular, NPU performs faster computations [65]. However, the APs of smartphones are equipped with heterogeneous computing systems. It has CPU, GPU, NPU, digital signal processors, network accelerators, display devices, and secure hardware integrated on a single chip. Table 2 provides a comparative description of various computing architectures commonly used for processing.

As discernible from Table 2, every processing element (processor or core) has certain advantages and disadvantages associated with them. Therefore, a heterogeneous platform is more advantageous than a single processing element. It employs two or more different types of computing cores to obtain an optimal (maximized) AI performance. However, it causes another open issue of allocating the computing core for a specific task as different cores have different capabilities.

Intel has suggested thread scheduling with intel® hyper-threading technology [80] for a software-based solution to overcome the open issue. It directs the cores suitable for a given task, enhancing the computing capabilities. Intel has developed a hybrid computing architecture consisting of vector, scalar, spatial, and matrix processors. They introduced their performance core (P-core or Cove) and the efficient core (E-core or Mont) to accomplish optimal performance by utilizing the hybrid computing architecture. The P-cores are destined to elevate the performance in terms of lower latencies, and the E-cores are developed to excel throughput via enhanced processing. Sapphire rapids series are an example of such hybrid computing architectures. Intel has involved an OS scheduler directing the corresponding thread to improve their computation performance (Fig. 26).

6   Bottleneck of AI Accelerator and In-Memory Processing

The AI computations are highly data hungry. They require voluminous data in training the models. The Von-Neumann architecture fetches data serially from the storage (shown in Fig. 27).

Table 2 Commonly used processors in AI accelerations and their characteristics

Processor	Power Consumption	Strengths	Limitations
CPU	High	→ Flexible → General-purpose processing → Complex instructions and tasks → System management	→ Possible memory access bottlenecks → Few cores (4–16)
GPU	High	→ Parallel cores (~1000 s of cores) → High-performance AI processing	→ Power consumption → Large footprint
FPGA	Medium	→ Configurable logic gates → Flexible → In-field re-programmability	→ Programming complexity
ASIC	Low	→ Custom logic designed with libraries → Faster processing → Small footprint	→ Fixed function → Expensive custom design
Vision Processing Unit (VPU)	Ultra-low	→ Dedicated image and vision co-processor → Small footprint	→ Limited dataset and batch size → Limited network support
Tensor Processing Unit (TPU)	Low to medium	→ Specialized tool support → Optimized for TensorFlow	→ Proprietary design → Limited framework support

Therefore, a typical Von-Neumann architecture is the main bottleneck of AI computation and power efficiency because of memory access for obtaining and storing data. The non-Von Neuman is a way out of this bottleneck [1]. It converges logic and high bandwidth memory. Therefore, there requires considerably lesser data movement between the processing element (i.e., logic) and the memory. It supports data-intensive workloads. Due to the lesser data movement, this approach also serves the demand for energy-efficient architectures as each data movement accounts for high energy consumption. Processing in memory has the potential to tackle the issues raised via the Von-Neumann architecture. Figure 28 discerns some of such approaches. It represents the issues of the Von-Neumann architectures and the capacities of memory for processing enhancement [81–83].

In early computers, the non-volatile memory (NVM) and dynamic random-access memory (DRAM) were directly connected to the processing core, leading to system blockage and mismatches in memory and core speeds. Later, cache memories were developed to match the speed of core and memory. Further, the high bandwidth memory shows the capabilities of memory elements in accelerating the processing performances. Thereafter, the concept of near-memory computing

A. Mishra et al.

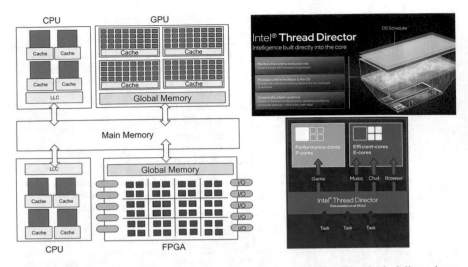

Fig. 26 An illustration of heterogenous computing architecture with a thread scheduling scheme for AI accelerations

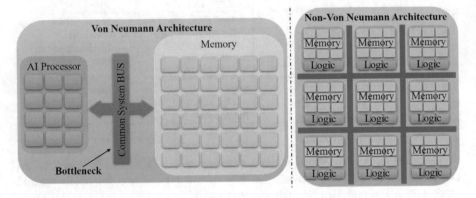

Fig. 27 Von Neumann versus non-Von Neumann architecture

has been introduced to accelerate processing performances. In such approaches, processing cores are placed closer to the memories. Recently, in-memory computing has been emerging to excel processing performances further. Processing-in-memory (PiM) combines both the processing (logic operations) and a memory element. PiM accelerators offer many potential benefits, including:

- Reduced data movement: weights
- Higher memory bandwidth: reading multiple weights in parallel
- Higher throughput: performing multiple computations in parallel
- Lower input activation: delivery cost due to increased density of compute

However, several key design challenges and decisions need to be considered in practice. Analog processing is typically required to bring the computation into the

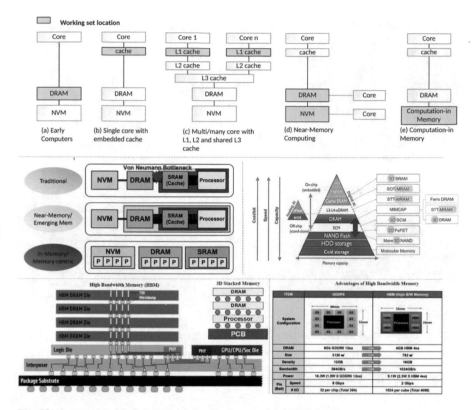

Fig. 28 Approaches involving processing in memory for AI accelerations

array of storage elements or its peripheral circuits. Therefore, significant challenges for processing in memory (PiM) are its sensitivity to circuit and device non-idealities (i.e., nonlinearity and process, voltage, and temperature variations). Figure 29 exhibits the capabilities of in-memory processing over conventional processing [1].

Conventional processors activation and weight memory are separately interconnected to the MAC unit. Therefore, a latency in processing occurs due to a mismatch in the memory bandwidths and has lower memory bandwidth. The activation memory was interconnected to the weight memory via a digital to analog converter (DAC) to eliminate the mismatch. Further, the MAC unit (i.e., compute) was integrated into the weight memory for faster processing. This memory (weight memory) and compute (MAC unit) integration constituted the concept of in-memory processing. Recent developments in devices and circuit designs have rendered various alternatives to accelerate AI processing.

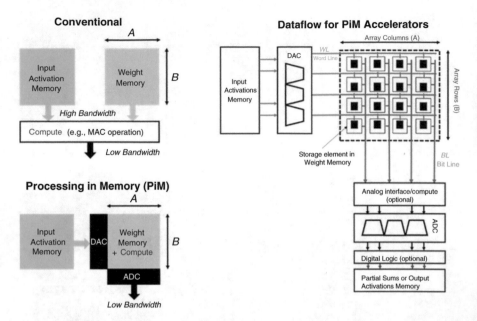

Fig. 29 Conventional versus processing-in-memory (PiM) dataflow. (Figure adapted from [1])

7   A Few State-of-the-Art AI Accelerators

Following are a few recent AI accelerator designs (described in brief).

- Lightspeeur® 5801S AI Accelerator is Gyrfalcon's fourth generation best-of-breed AI accelerator (shown in Fig. 30) [84]. It offers a 12.6 TOPs/Watt and 2.8 TOPs peak performance at 200 MHz/224 mW. It is fabricated using a 28 nm process and offers an inference time of less than 4 ms. It is developed to provide high-performance capabilities to edge AI devices. Due to its low-cost design, it is a suitable product for mass production and application in embedded, IoT, and mobile devices. Lightspeeur uses CNN-based architecture for performing AI tasks such as image classification, object detection, speech recognition, natural language processing, robotics, AR/VR, etc. Therefore, it offers support for industry-level architectures such as VGGNet, ResNet, MobileNet-v1, and similar custom networks. It uses a patented APiM™ (AI Processing in Memory) neural network core to support on-chip parallelism to help tackle crucial bottlenecks such as memory to provide efficient performance for industry-level applications. On the software side, it is known to support various deep learning libraries such as Caffe and TensorFlow. It also has a home-built SDK (Software Development Kit) for faster system development.
- Amazon has solutions for both training (AWS Trainium) and inference (AWS Inferentia). It is represented in Fig. 31. AWS Trainium is the second custom machine learning chip designed by AWS [85]. Trainium uses a systolic array

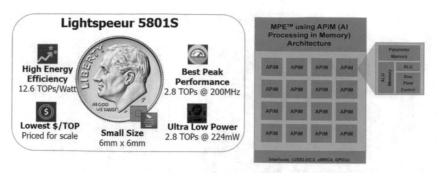

Features:

➤ 12.6 TOPs/ Watt
➤ 2.8 TOPs peak at 200MHz & 224mW
➤ 28nm process
➤ APiM™ neural network core
➤ Inference time: < 4ms

Applications:

☞ Face Detection/Recognition
☞ Speech/Voice Recognition
☞ Natural Language Processing
☞ Deep Learning Enabled Devices
☞ Edge Inference Systems
☞ Mobile Edge Computing
☞ Vision Systems
☞ Smart Toys/Robotics/Home
☞ Augmented Reality/Virtual Reality

Fig. 30 Lightspeeur® 5801S AI Accelerator. (Figures imported from Ref. [84])

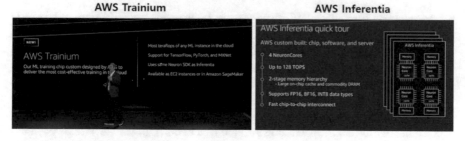

Fig. 31 AWS Trainium and Inferentia. (Figures imported from Refs. [85, 86])

matrix multiply engine. Systolic array matrix means that many closely related computational units exist that process the data that is received from their neighbors and transmit the output to the next neighbor. This approach differs from GPUs, which use several registers and memory accesses. Trainium has targeted training AI models in the cloud. AWS Trainium preserves flexibility by utilizing 16 fully programmable handlers. It performs 3.4Pflops low-precision calculations and up to 840 Teraflops in FP32 calculations. AWS Trainium can be accessed using AWS Neuron SDK for model development. It incorporates popular deep learning libraries such as MXNet, TensorFlow, and Pytorch.

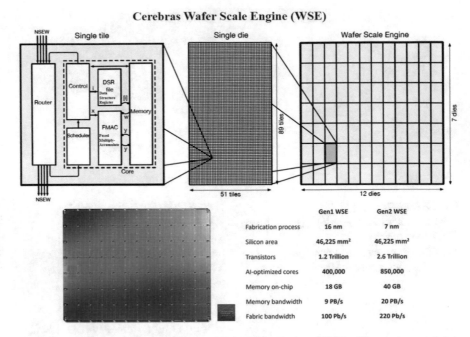

Fig. 32 A gigantic AI chip by Cerebras (Wafer Scale Engine (WSE)). (Figures imported from Ref. [68])

- On the inference side, AWS Inferentia is built using large amounts of on-chip memory capable of catching large AI models [86]. In contrast to storing the models off-chip, this feature significantly improves inference latency. Neuron Cores form the "core" of Inferentia, have high-speed access to the models stored on-chip, and are not limited by the off-chip memory bandwidth. Each AWS Inferentia chip can perform 128 TOPs calculations.
- Figure 32 discerns the gigantic AI wafer chip. Cerebras Systems (CS-2) is a system solution for deep learning acceleration at the hardware level. It is powered by the Cerebras second-generation Wafer Scale Engine (WSE-2) [68]. WSE-2 is the largest industry-grade chip ever built, consisting of approximately 1.2 trillion transistors. The Cerebras Systems (CS)-2 wafer is a multiple instruction, multiple data (MIMD), distributed-memory machine with a 2D-mesh interconnection fabric. The fundamental element of the wafer is called a tile. Each tile contains one processor core, its memory, and the router for the connection. Its router links to the routers of the four neighboring tiles. The wafer contains a 7×12 array of 84 identical dies, whereas a die holds thousands of tiles. It has 40 Gigabytes of on-chip SRAM, all accessible within a single clock cycle, and provides 20 petabytes/sec of memory bandwidth and 220 petabits/sec of interconnect bandwidth. WSE-2 comprises over 100 times more compute cores, 1000 times more high-speed on-chip memory, and over 12,000 times more fabric bandwidth than any graphical processing accelerators developed. The size of

WSE-2 spans around 46,000 mm2. Sparse linear algebra compute (SLAC) cores are responsible for computation inside of WSE-2. There are over 850,000 SLAC cores in WSE-2, primarily designed for sparse linear algebra calculations. Such an enormous design is capable of almost any changes in the fundamentals of linear algebra used for machine learning modeling, achieving industry-grade applications. SLAC cores allow WSE-2 to skip zero-to-zero multiplications in the big data to prevent wastage of compute resources and improve the system's overall efficiency. Cerebras Software platform allows researchers to train their machine learning models without disrupting their workflow. It supports popular deep learning libraries such as TensorFlow and Pytorch. It consists of a graph compiler that translates the ML models into optimized executables for CS-2.

- Nervana Systems was acquired by Intel and produced their AI chip as the Intel-Nervana neural network processor (NNP-T) [87]. It was a standalone accelerator for DL and AI applications and was available for PCIe and OAM cards. It was fabricated on TSMC's 16 nm process; the chip utilizes a single large 680 square-mm of die area with over 27 billion transistors and typical workload power ranging from 150 W to 250 W. The chip integrates 24 tensor processor cores (TPCs) with 2.5 MiB of local scratchpad memory. Figure 33 presents the architectures of the NNP-T accelerators. However, the experts at Intel realized that the Intel-Nervana chips were not at par with the other prominent players in the AI hardware market, such as Nvidia, Samsung, AMD, etc. Therefore, Intel has decided to discontinue the NNP-T processors and proceed to the Habana labs.
- Habana provides an AI accelerator capable of both training (Gaudi) and inference (Goya), given in Fig. 34. A completely programmable TPC cluster and a matrix multiplication engine (MME) comprise the heterogeneous compute architecture of Gaudi. It uses a cluster of eight TPC 2.0 cores and is the first AI Processor that integrates on-chip remote direct memory access (RDMA) over Converged Ethernet (RoCE v2) engines. These engines play a critical role in the inter-processor communication needed during the training process. With the native integration of 24×100 Gbps RoCE V2 RDMA NICs on-chip, the Gaudi2 Processor provides 2.4 Terabits of networking bandwidth and allows inter-Gaudi communication via direct routing or through conventional Ethernet switching. The Gaudi2 Memory subsystem consists of 96 GB of HBM2E memories with 2.45 TB/sec bandwidth and 48 MB of local SRAM with enough bandwidth to support the simultaneous operation of MME, TPC, DMAs, and RDMA NICs. All popular data formats needed for deep learning are supported by Gaudi2, such as FP32, TF32, BF16, FP16, and FP8 (both E4M3 and E5M2). The software package consists of the graph compiler and runtime from Habana, the TPC kernel library, firmware, and drivers, as well as developer tools, including the SynapseAI Profiler and TPC programming tool kit for creating custom kernels. SynapseAI is performance-optimized for Habana's Gaudi2 AI processors and integrates with the well-known TensorFlow and PyTorch frameworks [88].
- The Goya Inference Processor has a cluster of eight programmable cores and is based on the scalable architecture of the Habana Tensor-Processing Core (TPC) [89]. TPC is a special core created by Habana to assist deep learning workloads.

Intel Nervana Neural Network Processor-T (NNP-T)

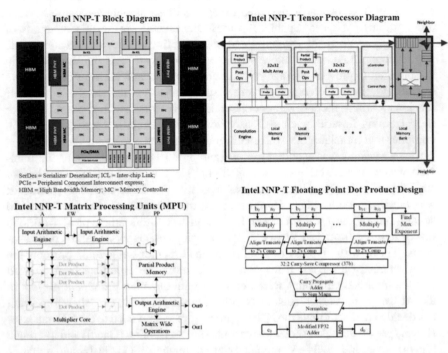

Fig. 33 Intel-Nervana NNP-T AI accelerator. (Figures imported from Ref. [88])

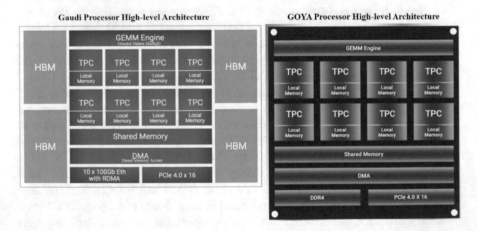

Fig. 34 Gaudi and GOYA processors from Habana. (Figures imported from Refs. [88, 89])

It is a VLIW SIMD vector processor with hardware and an Instruction-Set Architecture designed to handle deep learning tasks effectively. Goya engines are capable of concurrent operation and shared memory communication. The

Graphcore Intelligence Processing Unit (IPU)
(Colossus MK2 GC200 IPU)

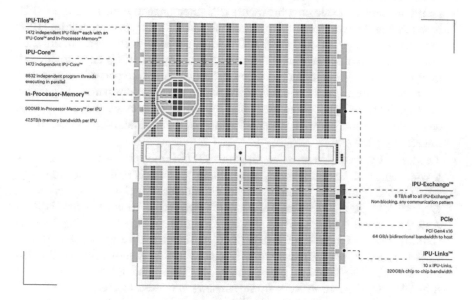

Fig. 35 An intelligent processing unit (IPU) from Graphcore. (Figures imported from Ref. [90])

processor employs PCIe Gen4 ×16 for the external interface, enabling connectivity with any host of choice. The processor has two 64-bit DDR4 memory interface channels with a 16 GB maximum memory capacity. Under quantization controls that the user can choose, the Goya architecture offers mixed precision both for integer and floating points, enabling it to serve a variety of workloads and applications flexibly.

- Graphcore has also produced their AI accelerator GC200 (in Fig. 35). It contains 59.4 billion transistors and is built using the latest TSMC 7 nm process [90]. Each MK2 IPU has 1472 IPU cores, running 8832 independent parallel program threads. Each unit holds 900 MB of in-processor memory with 47.5 TB/s bandwidth and can deliver up to 250 TFLOPS of AI compute at FP16.16 and FP16.SR (stochastic rounding). IPU memories have a substantially larger aggregate bandwidth (45 TB/s) and lower latency (6 clock cycles) than DRAMs because they are built as SRAM. When the addresses of their memory processes or their control flows diverge, IPU cores are unaffected. No cost is associated with performing unrelated instruction flows that display uncorrelated memory accesses. When the addresses of their memory processes or their control flows diverge, IPU cores are unaffected. No cost is associated with performing unrelated instruction flows that display uncorrelated memory accesses. IPUs increase efficiency by oversubscribing threads to cores like CPUs and GPUs do. Each IPU tile provides hardware support for six threads in a way that is functionally equivalent to the SMT (simultaneous multithreading) method used in CPUs. The tight cooperation and effective data interchange between tiles on an IPU system are made possible by the IPU interconnect. It also causes an IPU-based system to function as a single, cohesive unit. A system with multiple IPUs exposes both single IPU and multi-IPU devices separately. The term "multi-IPU" refers to a virtual device that combines several physical IPUs and provides all their memory and computing resources as though they were part of a single unit. For the developers, the multi-IPU programming model is transparent. In general, the scalability of applications onto big IPU systems does not need additional development effort because of the efficiency of the abstraction provided by the underlying hardware.

- The new NVIDIA® Ampere architecture GPU family is meant to speed up a wide range of computationally demanding workloads and applications [91, 92]. GA102, the most potent Ampere architecture GPU is employed in all GeForce RTX 3090, GeForce RTX 3080, NVIDIA RTX A6000, and the NVIDIA A40 data center GPUs. GA102 is composed of Graphics Processing Clusters (GPCs), Texture Processing Clusters (TPCs), Streaming Multiprocessors (SMs), Raster Operators (ROPS), and memory controllers. It is shown in Fig. 36. The full GA102 GPU contains seven GPCs, 42 TPCs, and 84 SMs. The 8 nm 8 N NVIDIA Custom Process from Samsung was used to create the NVIDIA Ampere architecture-based GA102 GPU, which has a die size of 628.4 mm2 and 28.3 billion transistors. 32-bit floating-point (FP32) operations make up most graphics demands. The Ampere GA10× GPU architecture's streaming multiprocessor (SM) has been developed to allow double-speed processing for FP32 operations. The new GA10× SM enables the concurrent execution of RT Core and graphics workloads or RT Core and computes workloads, greatly speeding various ray tracing procedures. Additionally, they accelerate the rendering of raytraced motion blur for quicker outcomes with improved visual correctness. The third-generation Tensor Cores from NVIDIA are incorporated into the GA10× SM for increased performance, efficacy, and programming flexibility.
- Sparsity in deep learning shows the importance of individual weights evolved during the learning process and by the end of network training. Only a subset of weights acquires a meaningful purpose in determining the learned output. The remaining weights are often useless in the context of the deep learning model's problem-solving ability. In contrast to the previous generation Turing Tensor Cores, a new sparsity feature is added that can quadruple the throughput of Tensor Core operations by utilizing the fine-grained structural sparsity in deep learning networks. Coarse-grained sparsity explores zeroing out specific weights distributed across the neural network, and fine-grained sparsity explores zeroing out the entire sub-networks of a neural network. Both coarse and fine-grained sparsity are represented in Fig. 37.

NVIDIA GA102 GPU with 84 SMs

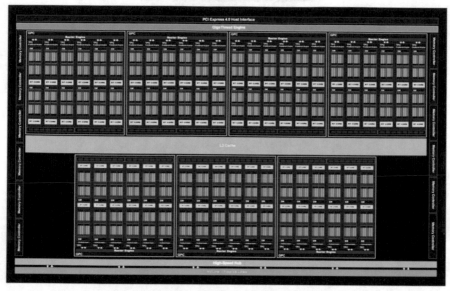

Fig. 36 An illustration of the NVIDIA GA 102 AI accelerator. (Figures imported from Refs. [91, 92])

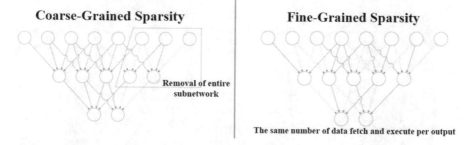

Fig. 37 An illustration of coarse-grained and fine-grained sparsity for AI accelerations

- Figure 38 presents the tensor core GPU (NVIDIA A100) from NVIDIA [91, 92]. Figure 39 exhibits the newer models of NVIDIA datacenters [93]. The A30 is designed for data analytics and AI server applications; however, the A10 is developed for mixed computing and graphics workloads. NVIDIA's first data center Arm-based CPU is designed to address the computing requirements for the world's most advanced applications. It combines energy-efficient Arm CPU cores with low power, high bandwidth memory subsystem to deliver high performance with great efficiency, $10\times$ faster performance versus today's state-of-the-art NVIDIA's deep learning GPU accelerator (DGX) system, running on $\times86$ CPUs Sampling in 2022; Shipping 2023. ALPS is 20 Exaflops of

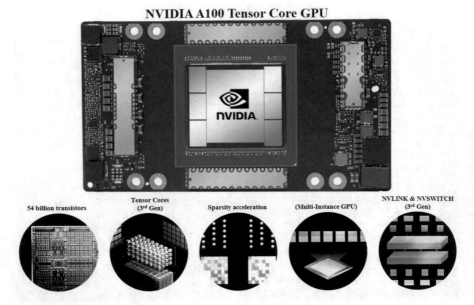

Fig. 38 An illustration of NVIDIA A100 tensor core GPU for AI accelerations. (Figures imported from Refs. [91, 92])

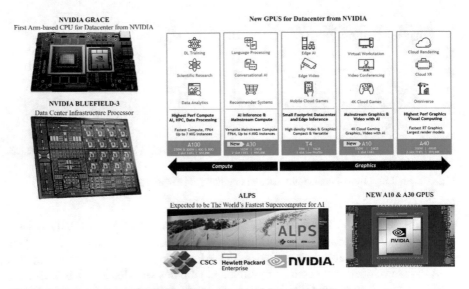

Fig. 39 AI acceleration in NVIDIA. (Figures imported from Ref. [93])

AI-powered by NVIDIA Grace CPU, and next-generation NVIDIA GPU can train generative pre-trained transformer (GPT) 37 times faster than NVIDIA Selene, currently recognized as the world's leading supercomputer for AI high performance computing (HPC) and AI for Scientific and Commercial Apps in 2023. The A30 graphics card represents the true little sister of the A100 and is based on the same compute-oriented Ampere architecture. It supports the same features, a wide range of formats for AI and HPC workloads (FP64, FP64TF, FP32, TF32, bfloat16, FP16, INT8, INT4), and even multi-instance GPU operations (MIG) with 6 GB instances. From a performance standpoint, the A30 offers just over 50% of the performance of the A100.

- Regarding memory, the drive has 24 GB of RAM with a bandwidth of 933 GB/s. The subsystem appears to lack ECC support, which could be a limitation for those who need to work with large datasets, who should upgrade to the more expensive and powerful A100. NVIDIA A30 will be available in a dual-slot FHFL (full-height, full length) form factor, with a PCIe 4.0 ×16 interface and a thermal design power (TDP) of 165 W, higher than the A100's 250 W. Additionally, the A30 supports an NVLink at 200 GB/s (a decrease in consistency compared to the A100's 600 GB/s). Instead, A10 is a completely different product that can be used for graphics, AI inference, and video encoding/decoding. The unit supports FP32, TF32, blfoat16, FP16, INT8, and INT4 formats for graphics and AI, but not the FP64 required for HPC. This is a single slot FHFL graphics card with a PCIe 4.0 ×16 interface that will be installed in servers running NVIDIA RTX Virtual Workstation (vWS) software and equip workstations that require both AI and graphics capabilities remotely. To a large extent, A10 should be a viable proposition for artists, designers, engineers, and scientists (who do not need FP64). NVIDIA A10 appears to be based on the GA102 GPU (or a derivative thereof), but since it supports INT8 and INT4, we cannot be 100% sure this is the same processor found on the GeForce RTX 3080/3090 and RTX A6000 cards. Finally, the performance of A10 (31.2 FP32 TFLOPS, 125 FP16 TFLOPS) is like that of the GeForce RTX 3080, and on board, there are 24GB of GDDR6 memory capable of offering a bandwidth of 600 GB/s.
- Huawei has recently introduced its training and inference processors. Figure 40 depicts the Ascend (for training) and Atlas (for inference) series of AI accel-

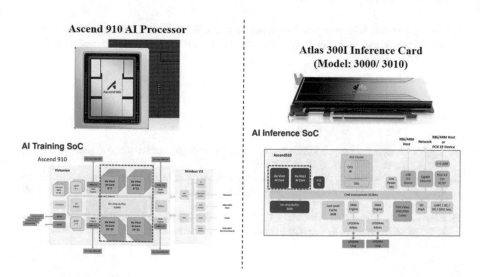

Fig. 40 AI training and inference processors from Huawei. (Figures imported from Ref. [94])

erators from Huawei. Huawei has its solutions for both training (Ascend 910) and inference (Ascend 310) [94]. Ascend 910 is a high-integration SoC-based AI processor. It is suitable for heavier AI model training. It delivers a speed of 320TFLOPS for FP16 (i.e., half-precision 16-bit floating-point format) and 640TOPS for INT8 (i.e., 8-bit integer data type) of computing performance. Ascend 910 requires only 310 watts of max power consumption. Atlas 300I is an inference card. It has Ascend 310 AI processor to unlock superior AI inference performances. It delivers 22TOPS for INT8 and 11TFLOPS for FP16 operations. It has just 8 watts of power consumption.

8 Conclusions

Advancements in AI algorithms have manifested the requirement for efficient AI accelerators. These AI accelerators have facilitated the shift of AI from cloud-based services to edge devices and on-device AI applications. The active learning and federated learning frameworks bridge the cloud-AI, and on-device AI applications with enhanced performances and augmented application domains. Evolutional high bandwidth memory (HBM) and revolutionary innovations in processing in memory (PiM), neuromorphic, etc., have meliorated AI hardware acceleration. Energy efficiency is a major concern of AI accelerators. There is a requirement for energy-efficient hardware, and AI algorithms should also consider the energy consumption issues on the computing hardware. Hybrid and heterogeneous computing architectures have been used recently for obtaining the optimized performances of the accelerators in terms of throughput and energy consumption. Selective scheduling using the thread director approach has been used to cater to such demands. Considering the market trends in AI accelerators, the developments can be assumed as the solutions for such demands.

Acknowledgments This work was supported by the Brain Pool Program through the National Research Foundation of Korea (NRF), funded by the Ministry of Science and ICT (NRF-2019H1D3A1A01071115).

References

1. Sze, V., Chen, Y.H., Yang, T.J., Emer, J.S.: Efficient processing of deep neural networks. Synth. Lect. Comput. Archit. **15**(2), 1–341 (2020)
2. Injadat, M., Moubayed, A., Nassif, A.B., Shami, A.: Machine learning towards intelligent systems: Applications, challenges, and opportunities. Artif. Intell. Rev. **54**(5), 3299–3348 (2021)
3. Rosendo, D., Costan, A., Valduriez, P., Antoniu, G.: Distributed intelligence on the edge-to-cloud continuum: A systematic literature review. J. Parallel Distrib. Comput. **166**, 71–94 (2022)

4. Akhoon, M.S., Suandi, S.A., Alshahrani, A., Saad, A.M.H., Albogamy, F.R., Abdullah, M.Z.B., Loan, S.A.: High performance accelerators for deep neural networks: A review. Expert. Syst. **39**(1), e12831 (2022)
5. Silva, G.A.: A new frontier: The convergence of nanotechnology, brain machine interfaces, and artificial intelligence. Front. Neurosci. **12**, 843 (2018)
6. Janapa Reddi, V., Kanter, D., Mattson, P., Duke, J., Nguyen, T., Chukka, R., Shiring, K., Tan, K.S., Charlebois, M., Chou, W., El-Khamy, M.: MLPerf mobile inference benchmark: An industry-standard open-source machine learning benchmark for on-device AI. Proc. Mach. Learn. Syst. **4**, 352–369 (2022)
7. Su, W., Li, L., Liu, F., He, M., Liang, X.: AI on the edge: a comprehensive review. In: Artificial Intelligence Review **55**, 6125–6183. Springer (2022). https://doi.org/10.1007/s10462-022-10141-4
8. Vyas, L.: "New normal" at work in a post-COVID world: Work–life balance and labor markets. Policy Soc. **41**, 155–167 (2022)
9. Mishra, A., Kim, J., Cha, J., Kim, D., Kim, S.: Authorized traffic controller hand gesture recognition for situation-aware autonomous driving. Sensors. **21**(23), 7914 (2021)
10. Mishra, A., Lee, S., Kim, D., Kim, S.: In-cabin monitoring system for autonomous vehicles. Sensors. **22**(12), 4360 (2022)
11. Mishra, A., Cha, J., Kim, S.: HCI based in-cabin monitoring system for irregular situations with occupants facial anonymization. In: International Conference on Intelligent Human Computer Interaction, pp. 380–390. Springer, Cham (2020)
12. Mishra, A., Cha, J., Kim, S.: Privacy-preserved in-cabin monitoring system for autonomous vehicles. Comput. Intell. Neurosci. **2022**, 1 (2022)
13. Jhung, J., Kim, S.: Behind-the-scenes (Bts): Wiper-occlusion canceling for advanced driver assistance systems in adverse rain environments. Sensors. **21**(23), 8081 (2021)
14. Goodfellow, I., Bengio, Y., Courville, A.: Deep Learning. MIT Press, Cambridge, United States (2016)
15. What is the convolutional neural network architecture?: https://www.analyticsvidhya.com/blog/2020/10/what-is-the-convolutional-neural-network-architecture
16. Zhang, C., Lu, Y.: Study on artificial intelligence: The state of the art and future prospects. J. Ind. Inf. Integr. **23**, 100224 (2021)
17. Choi, S., Sim, J., Kang, M., Choi, Y., Kim, H., Kim, L.S.: An energy-efficient deep convolutional neural network training accelerator for in situ personalization on smart devices. IEEE J. Solid State Circuits. **55**(10), 2691–2702 (2020)
18. Song, L., Qian, X., Li, H., Chen, Y.: Pipelayer: A pipelined reram-based accelerator for deep learning. In: 2017 IEEE International Symposium on High Performance Computer Architecture (HPCA), pp. 541–552. IEEE (2017)
19. McCulloch, W.S., Pitts, W.: A logical calculus of the ideas immanent in nervous activity. Bull. Math. Biophys. **5**(4), 115–133 (1943)
20. Rosenblatt, F.: The perceptron: A probabilistic model for information storage and organization in the brain. Psychol. Rev. **65**(6), 386 (1958)
21. Widrow, B., Hoff, M.E.: Adaptive Switching Circuits. Stanford Univ Ca Stanford Electronics Labs, United States (1960)
22. Minsky, M., Papert, S.: Perceptrons: An Introduction to Computational Geometry. The MIT Press, Cambridge, MA (1969)
23. Rumelhart, D.E., Hinton, G.E., Williams, R.J.: Learning representations by back-propagating errors. Nature. **323**(6088), 533–536 (1986)
24. Cortes, C., Vapnik, V.: Support-vector networks. Mach. Learn. **20**(3), 273–297 (1995)
25. Hochreiter, S., Schmidhuber, J.: Long short-term memory. Neural Comput. **9**(8), 1735–1780 (1997)
26. Famous graphics chips: Nvidia's GeForce 256, https://www.computer.org/publications/tech-news/chasing-pixels/nvidias-geforce-256
27. Hinton, G.E., Salakhutdinov, R.R.: Reducing the dimensionality of data with neural networks. Science. **313**(5786), 504–507 (2006)

28. Hinton, G.E., Osindero, S., Teh, Y.W.: A fast learning algorithm for deep belief nets. Neural Comput. **18**(7), 1527–1554 (2006)
29. Settles, B.: Active Learning Literature Survey. University of Wisconsin-Madison, United States (2009)
30. Konyushkova, K., Sznitman, R., Fua, P.: Learning active learning from data. Adv. Neural Inf. Proces. Syst. **30** (2017). https://proceedings.neurips.cc/paper/2017/file/8ca8da41fe1ebc8d3ca31dc14f5fc56c-Paper.pdf
31. Ghai, B., Liao, Q.V., Zhang, Y., Bellamy, R., Mueller, K.: Explainable active learning (xal) toward ai explanations as interfaces for machine teachers. Proc. ACM Hum.-Comput. Interact. **4**(CSCW3), 1–28 (2021)
32. Anahideh, H., Asudeh, A., Thirumuruganathan, S.: Fair active learning. Expert Syst. Appl. **199**, 116981 (2022)
33. McMahan, B., Moore, E., Ramage, D., Hampson, S., y Arcas, B.A.: Communication-efficient learning of deep networks from decentralized data. In: Artificial Intelligence and Statistics, pp. 1273–1282. PMLR (2017)
34. Yang, Q., Liu, Y., Cheng, Y., Kang, Y., Chen, T., Yu, H.: Federated learning. Synth. Lect. Artif. Intell. Mach. Learn. **13**(3), 1–207 (2019)
35. Li, T., Sahu, A.K., Talwalkar, A., Smith, V.: Federated learning: Challenges, methods, and future directions. IEEE Signal Process. Mag. **37**(3), 50–60 (2020)
36. Zhang, C., Xie, Y., Bai, H., Yu, B., Li, W., Gao, Y.: A survey on federated learning. Knowl.-Based Syst. **216**, 106775 (2021)
37. Artificial intelligence computing for consumer: Market and Technology Report. https://s3.i-micronews.com/uploads/2019/10/Yole_YD19045_Artificial-Intelligence-Computing-for-Consumer_October_2019_Sample.pdf
38. Bavikadi, S., Dhavlle, A., Ganguly, A., Haridass, A., Hendy, H., Merkel, C., Reddi, V.J., Sutradhar, P.R., Joseph, A., Dinakarrao, S.M.P.: A survey on machine learning accelerators and evolutionary hardware platforms. IEEE Des. Test. **39**(3), 91–116 (2022)
39. Machupalli, R., Hossain, M., Mandal, M.: Review of ASIC accelerators for deep neural network. Microprocess. Microsyst. **89**, 104441 (2022)
40. Tao, Y.: Algorithm-architecture co-design for domain-specific accelerators in communication and artificial intelligence. Doctoral Dissertation (2022)
41. Du, L., Du, Y.: Hardware accelerator design for machine learning. Mach. Learn.-Adv. Tech. Emerg. Appl., 1–14 (2017)
42. Batra, G., Jacobson, Z., Madhav, S., Queirolo, A., Santhanam, N.: Artificial-Intelligence Hardware: New Opportunities for Semiconductor Companies. McKinsey Co, United States (2018). https://www.mckinsey.com/~/media/McKinsey/Industries/Semiconductors/Our%20Insights/Artificial%20intelligence%20hardware%20New%20opportunities%20for%20semiconductor%20companies/Artificial-intelligence-hardware.pdf
43. Dally, W.J., Turakhia, Y., Han, S.: Domain-specific hardware accelerators. Commun. ACM. **63**(7), 48–57 (2020)
44. Kim, S., Deka, G.C.: Hardware Accelerator Systems for Artificial Intelligence and Machine Learning. Academic Press, United States (2021)
45. Kachris, C., Falsafi, B., Soudris, D. (eds.): Hardware Accelerators in Data Centers. Springer Cham, United States (2019). https://doi.org/10.1007/978-3-319-92792-3
46. Talib, M.A., Majzoub, S., Nasir, Q., Jamal, D.: A systematic literature review on hardware implementation of artificial intelligence algorithms. J. Supercomput. **77**(2), 1897–1938 (2021)
47. Keckler, S., Milojicic, D.: Accelerators. Computer. **55**(1), 108–112 (2022)
48. Bianco, S., Cadene, R., Celona, L., Napoletano, P.: Benchmark analysis of representative deep neural network architectures. IEEE Access. **6**, 64270–64277 (2018)
49. Patterson, D.A., Hennessy, J.L.: Computer Organization and Design ARM Edition: The Hardware Software Interface. Morgan Kaufmann, Cambridge, USA (2016)
50. Park, H., Kim, S.: Hardware accelerator systems for artificial intelligence and machine learning. Adv. Comput. **122**, 51–95 (2021)

51. Park, H., Kim, D., Kim, S.: TMA: Tera-MACs/W neural hardware inference accelerator with a multiplier-less massive parallel processor. Int. J. Circuit Theory Appl. **49**(5), 1399–1409 (2021)
52. WTF is a SIMD, SMT, SIMT: https://medium.com/@valarauca/wtf-is-a-simd-smt-simt-f9fb749f89f1
53. Blake, G., Dreslinski, R.G., Mudge, T.: A survey of multicore processors. IEEE Signal Process. Mag. **26**(6), 26–37 (2009)
54. Computer hardware engineering: https://www.kth.se/social/files/54fdb2c5f276546b06f9acfb/lecture10-spp2.pdf
55. Simultaneous multithreading: https://www.ibm.com/docs/en/sdse/6.4.0?topic=planning-simultaneous-multithreading
56. Computer architecture: SIMD and GPUs (Part I): https://course.ece.cmu.edu/~ece740/f13/lib/exe/fetch.php?media=onur-740-fall13-module5.1.1-simd-and-gpus-part1.pdf
57. Duncan, R.: A survey of parallel computer architectures. Computer. **23**(2), 5–16 (1990)
58. Tino, A., Collange, C., Seznec, A.: SIMT-X: Extending single-instruction multi-threading to out-of-order cores. ACM Trans. Archit. Code Optim. (TACO). **17**(2), 1–23 (2020)
59. Aamodt, T.M., Fung, W.W.L., Rogers, T.G.: General-purpose graphics processor architectures. Synth. Lect. Comput. Archit. **13**(2), 1–140 (2018)
60. Whitepaper-NVIDIA's Next Generation CUDATM Compute Architecture: Fermi. https://www.nvidia.com/content/PDF/fermi_white_papers/NVIDIA_Fermi_Compute_Architecture_Whitepaper.pdf
61. Lindholm, E., Nickolls, J., Oberman, S., Montrym, J.: NVIDIA Tesla: A unified graphics and computing architecture. IEEE Micro. **28**(2), 39–55 (2008)
62. Sze, V., Chen, Y.H., Yang, T.J., Emer, J.S.: Efficient processing of deep neural networks: A tutorial and survey. Proc. IEEE. **105**(12), 2295–2329 (2017)
63. Mao, W., Xiao, Z., Xu, P., Ren, H., Liu, D., Zhao, S., An, F., Yu, H.: Energy-efficient machine learning accelerator for binary neural networks. In: Proceedings of the 2020 on Great Lakes Symposium on VLSI, pp. 77–82 (2020)
64. System Architecture: https://cloud.google.com/tpu/docs/system-architecture-tpu-vm#device
65. Chen, Y., Xie, Y., Song, L., Chen, F., Tang, T.: A survey of accelerator architectures for deep neural networks. Engineering. **6**(3), 264–274 (2020)
66. Esmaeilzadeh, H., Sampson, A., Ceze, L., Burger, D.: Neural acceleration for general-purpose approximate programs. In: 2012 45th Annual IEEE/ACM International Symposium on Microarchitecture, pp. 449–460. IEEE (2012)
67. Rocki, K., Van Essendelft, D., Sharapov, I., Schreiber, R., Morrison, M., Kibardin, V., Portnoy, A., Dietiker, J.F., Syamlal, M., James, M.: Fast stencil-code computation on a wafer-scale processor. In: SC20: International Conference for High Performance Computing, Networking, Storage and Analysis, pp. 1–14. IEEE (2020)
68. Lauterbach, G.: The path to successful wafer-scale integration: The Cerebras story. IEEE Micro. **41**(6), 52–57 (2021)
69. Prezioso, M., Merrikh-Bayat, F., Hoskins, B., Adam, G., Likharev, K.K., Strukov, D.B.: Training and operation of an integrated neuromorphic network based on metal-oxide memristors. Nature. **521**(7550), 61–64 (2015)
70. Schuller, I.K., Stevens, R., Pino, R., Pechan, M.: Neuromorphic Computing–from Materials Research to Systems Architecture Roundtable. USDOE Office of Science (SC), United States (2015)
71. Pehle, C., Billaudelle, S., Cramer, B., Kaiser, J., Schreiber, K., Stradmann, Y., Weis, J., Leibfried, A., Müller, E., Schemmel, J.: The BrainScaleS-2 accelerated neuromorphic system with hybrid plasticity. Front. Neurosci. **16** (2022). https://doi.org/10.3389/fnins.2022.795876
72. McDonough, I.M., Haber, S., Bischof, G.N., Park, D.C.: The Synapse project: Engagement in mentally challenging activities enhances neural efficiency. Restor. Neurol. Neurosci. **33**(6), 865–882 (2015)

73. Ambrogio, S., Narayanan, P., Tsai, H., Shelby, R.M., Boybat, I., Di Nolfo, C., Sidler, S., Giordano, M., Bodini, M., Farinha, N.C., Killeen, B.: Equivalent-accuracy accelerated neural-network training using analogue memory. Nature. **558**(7708), 60–67 (2018)

74. Cho, K., Lee, I., Lim, H., Kang, S.: Efficient systolic-array redundancy architecture for offline/online repair. Electronics. **9**(2), 338 (2020)

75. Capra, M., Bussolino, B., Marchisio, A., Shafique, M., Masera, G., Martina, M.: An updated survey of efficient hardware architectures for accelerating deep convolutional neural networks. Future Internet. **12**(7), 113 (2020)

76. Skliarova, I., Sklyarov, V.: FPGA-Based Hardware Accelerators. Springer Cham, Switzerland (2019). https://doi.org/10.1007/978-3-030-20721-2

77. Karras, K., Pallis, E., Mastorakis, G., Nikoloudakis, Y., Batalla, J.M., Mavromoustakis, C.X., Markakis, E.: A hardware acceleration platform for AI-based inference at the edge. Circuits Syst Signal Process. **39**, 1059–1070 (2020)

78. Mowla, N.I., Doh, I., Chae, K.: A hardware acceleration platform for AI-based inference at the edge. On-device AI-based cognitive detection of bio-modality spoofing in medical cyber physical system. IEEE Access. **7**, 2126–2137 (2018)

79. Dhar, S., Guo, J., Liu, J., Tripathi, S., Kurup, U., Shah, M.: A hardware acceleration platform for AI-based inference at the edge. A survey of on-device machine learning: An algorithms and learning theory perspective. ACM Trans. Internet Things. **2**(3), 1–49 (2021)

80. Architecture Day 2021 Presentation: https://download.intel.com/newsroom/2021/client-computing/intel-architecture-day-2021-presentation.pdf

81. White Paper on AI Chip Technologies: https://www.080910t.com/downloads/AI%20Chip%202018%20EN.pdf

82. Hamdioui, S., Xie, L., Du Nguyen, H.A., Taouil, M., Bertels, K., Corporaal, H., Jiao, H., Catthoor, F., Wouters, D., Eike, L., Van Lunteren, J.: Memristor based computation-in-memory architecture for data-intensive applications. In: 2015 Design, Automation & Test in Europe Conference & Exhibition (DATE), pp. 1718–1725. IEEE (2015)

83. Singh, G., Chelini, L., Corda, S., Awan, A.J., Stuijk, S., Jordans, R., Corporaal, H., Boonstra, A.J.: Near-memory computing: Past, present, and future. Microprocess. Microsyst. **71**, 102868 (2019)

84. Lightspeeur® 5801S Neural Accelerator: https://www.gyrfalcontech.ai/solutions/lightspeeur-5801/

85. AWS Trainium: https://aws.amazon.com/machine-learning/trainium/

86. AWS Inferentia: https://aws.amazon.com/machine-learning/inferentia/

87. Hickmann, B., Chen, J., Rotzin, M., Yang, A., Urbanski, M., Avancha, S.: Intel nervana neural network processor-T (NNP-T) fused floating point many-term dot product. In: 2020 IEEE 27th Symposium on Computer Arithmetic (ARITH), pp. 133–136, Portland, OR, USA (2020)

88. Gaudi® Training Platform White Paper: https://habana.ai/wp-content/uploads/pdf/2020/Habana%20GAUDI%20Training%20Whitepaper%20v1.2.pdf

89. Goya Inference Platform White Paper: https://habana.ai/wp-content/uploads/pdf/2020/Habana%20GOYA%20Inference%20Performance%20Whitepaper%20Nov'20.pdf

90. Introducing the Colossus™ MK2 GC200 IPU: https://www.graphcore.ai/products/ipu

91. NVIDIA A100 Tensor Core GPU Architecture: https://images.nvidia.com/aem-dam/en-zz/Solutions/data-center/nvidia-ampere-architecture-whitepaper.pdf

92. NVIDIA Tesla V100 GPU Architecture: https://images.nvidia.com/content/volta-architecture/pdf/volta-architecture-whitepaper.pdf

93. Investor Presentation Q1 FY2022: https://s22.q4cdn.com/364334381/files/doc_financials/2022/q1/NVDA-F1Q22-Investor-Presentation-FINAL.pdf

94. AI Accelerator Card: https://e.huawei.com/en/products/cloud-computing-dc/atlas

AI Accelerators for Standalone Computer

Taewoo Kim, Junyong Lee, Hyeonseong Jung, and Shiho Kim

1 Introduction to Standalone Compute

Machine learning (ML) has progressed dramatically over the past few decades, and it is having a substantial effect on developing practical artificial intelligence (AI) systems, such as speech recognition, robot control, and so forth [1]. In these applications, deep learning (DL) significantly outperforms its predecessors; ML, widely used recently [2], has the advantage of investigating various knowledge and inferring new facts when a data set is provided.

DNN has grown over time as hardware and software for deep learning have improved [3]. Recently, DNN has often required millions of data and several days of computation time for training [4]. Therefore, training the DNN on platforms with extensive computing capabilities, such as cloud services, has the advantage of economy of scale [5]. This data-aware ML has become a driving force of recent achievement in marvelous results in AI research.

However, there is an approach to implement the DNN domestically on personal devices instead of sending the data to a central server. The rapid increase in the computing power of personal hardware and the prevalence of GPGPU (General Purpose Graphic Processing Unit) for DNN accelerators have made it possible to run AI functions on personal computers entirely [6]. It also addresses considerations associated with privacy, latency, and security [7]. Transferring private data through the Internet has many vulnerabilities in revealing its secrets through network threats such as eavesdropping [8]. And even excluding traditional network attack methods, some notable indirect attacks on the DNN model should reveal confidential information from exposed results [9].

T. Kim · J. Lee · H. Jung · S. Kim (✉)
School of Integrated Technology, Yonsei University, Incheon, South Korea
e-mail: boratw@yonsei.ac.kr; jjunilee@yonsei.ac.kr; jhshseong@yonsei.ac.kr;
shiho@yonsei.ac.kr

© The Author(s), under exclusive license to Springer Nature Switzerland AG 2023
A. Mishra et al. (eds.), *Artificial Intelligence and Hardware Accelerators*,
https://doi.org/10.1007/978-3-031-22170-5_2

53

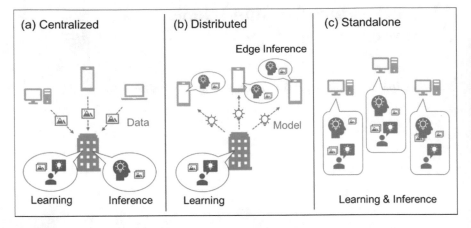

Fig. 1 Overview of learning and inference scheme in (**a**) centralized, (**b**) distributed, and (**c**) standalone computing

Another reason for adapting standalone computing is the data transferring latency between user devices and computing platforms. While deploying the neural network in real-time applications, such as autonomous vehicles, the capacity of the Internet connection restricts the immediate response to harmful situations.

Since GPUs are designed for parallelism in areas where maximally parallel operations are essential, a GPU can perform better for DNN accelerator than a CPU [10]. Nowadays, GPUs are widely used in various AI fields, and many manufacturers provide hardware [11] and software [12] support for computing neural networks using GPUs. On the other hand, specially configured hardware for accelerating AI, called neural processing units (NPU), is gradually expanding the market in application-specific fields (Fig. 1).

This chapter discusses various architectures and strategies to accelerate DNN on standalone computers. Section 2 presents a summary of the architectures and considerations of DNN acceleration. Section 3 looks at the overall GPU architectures for AI and some techniques to accelerate the computation. Section 4 focuses on NPU hardware accelerators, emerging due to low hardware utilization and energy consumption. In the last section, we summarize the chapter and discuss the future of AI in standalone computers.

2   Hardware Accelerators for Standalone Compute

2.1   Inference and Training of DNNs

The neural network (NN) contains multiple layers of nodes called neurons, which are inspired by the neurons and synapses in the human brain [13]. Each node is

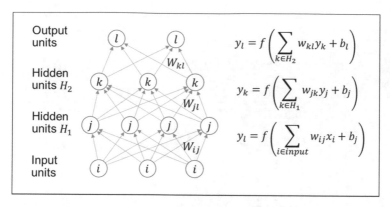

$$y_l = f\left(\sum_{k \in H_2} w_{kl} y_k + b_l\right)$$

$$y_k = f\left(\sum_{k \in H_1} w_{jk} y_j + b_j\right)$$

$$y_l = f\left(\sum_{i \in input} w_{ij} x_i + b_j\right)$$

Fig. 2 A typical Feedforward structure of the neural network

connected to the nodes in the adjacent layer with a specific weighted connection. In feed-forward neural network (FFNN), the information flows from the input layer to the output layer during the feedforwarding progress. The final output contains extracted information from the input, which is the expected result of the network. For training purposes, the feedback or loss of the result starts from the output layer and updates the hidden units backward. This process is called back-propagation (Fig. 2).

In contrast to feedforward neural networks, if the connections between layers form a cycle and the output becomes an input of the layer itself, the network is called a Recurrent Neural Network (RNN) [14]. RNN is typically used for understanding data sequences such as voices. And there is a convolutional neural network (CNN), a particular form of FFNN. In CNN, the nodes of layers are not connected to all nodes in the adjacent layers; instead, they get inputs from the corresponding small part of the previous layer [15] (Fig. 3).

CNN has the advantage of having fewer connections, and it still has enough capability to understand the features of an image. The other major part of a neural network is the activation function, and modern neural networks have activation functions associated with the connections between each layer. To learn the complex knowledge mapping beyond simple linear functions, we need to apply non-linear activation functions to the network. Rectified linear unit (ReLU) has been widely used recently [15], but many improved variants of ReLU and other activation functions exist. And every activation function has its advantages and disadvantages [16].

2.2 Accelerating DNN Computation

The CPU is an essential component for running applications in standalone computers, and it is optimized for executing sequential and various operations. However,

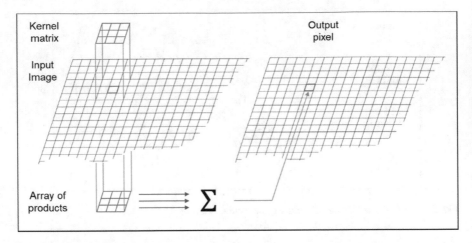

Fig. 3 Details of convolution layer

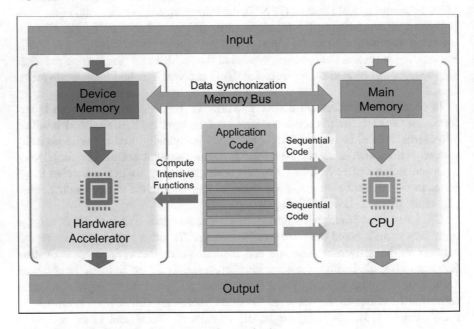

Fig. 4 Schematic diagram of the coprocessing mechanism

some parts of code could be more efficiently computed on dedicated hardware rather than the CPU. By distributing such a compute-intensive code to the coprocessor, the utilization rate of the CPU can be significantly increased (Fig. 4).

Learning and inferring a neural network involve millions of iterations of matrix operations, vector addition and multiplication, and activation functions. Therefore, hardware that supports these operations at the circuit level can accelerate deep learning operations. To demonstrate, GPUs have large parallel operation units for computing matrices and vectors and are highly efficient for deep learning [10]. Also, FPGA and ASIC hardware accelerators could achieve significant performance gains while lowering power consumption by customizing memory hierarchy and allocating dedicated resources [17].

Hardware accelerator vendors provide dedicated APIs for loading applications and utilizing resources on their devices. And many deep learning libraries embed these APIs to accelerate the computation by utilizing available hardware. Later we will look more closely at the software support of hardware accelerators.

One of the problems when using a coprocessor for acceleration is the delay due to memory communication, called the Von-Neumann Bottleneck. It is an operation stall inside the program because memory access operations are significantly slower than computations. It is a bigger problem for coprocessors farther from the main memory than the CPU. Thus, many techniques reduce the traffic burden between the computing and memory units [18].

First, Novel memory management systems like memory prefetching reduce the actual amount of memory operation [19]. Hardware accelerators often implement new memory hierarchies optimized for deep learning to reduce memory movement overhead [20]. Also, the processing-in-memory (PiM) technique proposed that the memory could be processed in place [18].

Likewise, data quantization is frequently used to reduce the size of data. 32-bit floating-point is the standard representation of the neural networks. Replacing it with less precise representations such as FP16, BF16, and INT8 could reduce the data traffic. However, reduced precision causes significant quantization error. Practically, this kind of error may hurt the model accuracy of neural networks (Fig. 5).

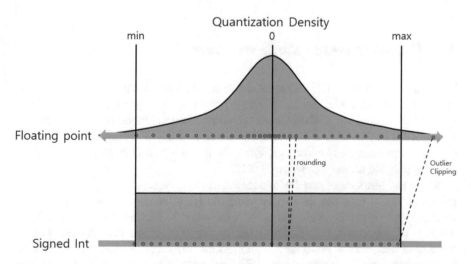

Fig. 5 Floating-point precision and quantization of integer value

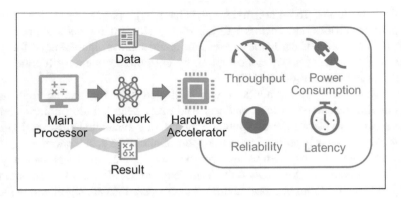

Fig. 6 The DNN accelerators' operation and the hardware design considerations

Nvidia and Google introduced hardware-accelerated mixed precision and numeric stability, using 16-bit and 32-bit data types while training the network. 'Numeric stability' means that the calculation result changes as the precision changes. It could cause by overflow, underflow, and rounding errors. Fortunately, due to the computational structure of DNN, only underflow might cause a severe effect on back-propagation gradient descent. Back-propagation is computed through the losses of the model during training. Suppose an underflow occurs while calculating the difference, losses, and gradient turn to zero. As a result, losses cannot propagate backward. To avoid the zero-gradient problem, multiply many losses and calculate the loss scale value. After final gradients are computed, return to the correct value using the value. It is a major technique to avoid numeric underflow called loss scaling.

2.3 Considerations in Hardware Design

Due to the vast usage of DNNs, recent hardware platforms have a unique feature that mainly targets processing DNNs. Figure 6 shows an analogy of dataflows within a typical DNN accelerator. The process by which the accelerator employs a network is analogous to that of general-purpose processors. Like the compiler translates the program into binary codes for execution, the mapper translates the network shape into mapping for efficient processing [21].

Since the purpose of the DNN accelerator is to compute the neural network more efficiently than the main processor, the DNN accelerator is designed with several emphases. The main factor of efficiency in accelerating is throughput, the rate of successful results delivered. In the application domain, the size of the neural networks is considerably increasing and requires more and more data for learning complex representations. For example, AlexNet [22] model has over 60 million model parameters and 1.4 billion floating-point operations [23]. Many

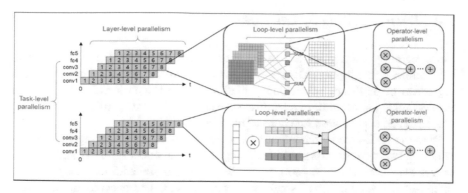

Fig. 7 Four levels of parallelism

hardware accelerators have parallelization strategies to increase throughput by simultaneous work on tasks, layers, loops, and operators [24]. Figure 7 describes four parallelizable parts of network computation.

In addition to parallelization, analysis of data movement could enhance throughput, such as reusing the data, neighboring local memory, and optimizing memory access. Within CNN, a large amount of data is frequently used, such as weights of the convolution operation. On-chip buffers [25] and memory prefetching [26] can increase computation speed. Also, loop-reordering and tiling are proposed to run the loop computation efficiently with limited resources [25, 27]. On the other hand, dataflows can be optimized for CNN shape by reconfiguring the spatial architecture to map the row-stationary computation [28].

Another important part of an accelerator is power consumption. We cannot increase the hardware resources adequate for the inferencing network in embedded edge devices due to the limitation of their power budgets. As mentioned above, the amount of calculation for utilizing DNN is dramatically increasing with the complexity of the operation. Reducing the precision of some computations while feedforwarding can decrease the circuit complexity and power consumption [29–31]. However, the reduction of precision impacts the accuracy inevitably, and it is crucial to reduce the precision of calculation without affecting its operational reliability.

And since the hardware acceleration technology is commonly employed on embedded AI systems that specialize in specific purposes, the DNN processor also needs to consider the related service requirements at the design level. If the system prioritizes guaranteeing the real-time reaction, the processor should be designed to reduce the actual processing time. Otherwise, some systems could demand accurate handling of a considerable scale of data at once, regardless of processing time. In conclusion, there is no universal solution for hardware design. And choosing an acceleration strategy for a DNN depends on the available resources and the target application.

2.4   Deep Learning Frameworks

Deep learning frameworks are being widely used in applications that want to offer artificial intelligence. They provide interfaces to implement AI functionalities regardless of hardware or computing platforms. Most deep learning frameworks also optimize the network in a hardware-friendly manner and utilize the hardware resources and acceleration functions as much as possible.

Because GPUs are commonly used in deep learning, deep learning frameworks widely support constructing and deploying neural networks on GPU and actively reflect the acceleration support from GPU vendors. In addition, they provide optimizations on standalone accelerators such as TPU and specific environments such as cloud servers. In this section, we introduce some popularly used open-source deep learning frameworks.

TensorFlow TensorFlow is an open-source platform for machine learning maintained by the Google Brain team [32]. TensorFlow can run on various platforms, including CPU, GPU, and Google TPU [33]. In 2017, Google announced TensorFlow Lite, a lightweight version for mobile and embedded devices [34].

The biggest strength of TensorFlow is the vast ecosystem founded by numerous developers. Many libraries and toolkits for deep learning developers are based on the TensorFlow platform or emphasize TensorFlow. A famous deep learning high-level wrapper, Keras has used only TensorFlow as its backend since version 2.4 [35].

PyTorch PyTorch is an optimized tensor library for deep learning primarily developed by Facebook [36]. It is based on the Torch library, which was actively used, but development ceased after 2018 [37]. Caffe2, the successor of Caffe, has also been merged into a submodule of Pytorch.

PyTorch is becoming popular incredibly fast for its simplicity and ease of use, and its ecosystem is growing fast enough to threaten TensorFlow. It has the Pytorch Lightning library, which provides high-level interface abstractions for researchers like Keras [38]. It practically works on the CPU or GPU platform and supports only PC environments. Pytorch Mobile has been released to support mobile platforms but is not stable yet [39].

Theano Theano is an open-source python library for optimizing mathematical computations originally developed by Montreal Institute for Learning Algorithms (MILA) [40]. It had existed since the emergence of a neural network and was widely used until the advent of python-based platforms. The development of Theano has been stopped since 2017; however, numerous libraries designed on top of Theano still influence deep learning. It is being continued as Aesera [41], developed by the maintainers of the PyMC library.

MXNet Apache MXNet is an open-source framework for deep learning applications maintained by Apache Software Foundation (ASF) [42]. It emphasizes computational scalability and flexibility in environments where many machines exist, such as Cloud Platform. One feature is that it supports binding for eight different languages.

ONNX To enable collaboration among developers locked on fragmented frameworks, Facebook and Microsoft announced the Open Neural Network Exchange (ONNX) in 2017 [43]. ONNX provides a standard graph model representation and its computation data flow. Any framework that supports ONNX can share its studies and models in the ONNX format. And Hardware vendors can support various frameworks by supporting optimization in the ONNX environment. Most major hardware manufacturers, including Intel, NVIDIA, AMD, Qualcomm, ARM, and Huawei, support the ONNX project as a partner [44].

3 Hardware Accelerators in GPU

3.1 History and Overview

GPU was developed initially in the 1990s to support real-time rendering by sharing graphics-related operations of PCs [45]. Since graphics computation contains many repeated floating-point operations, GPU employs many parallelized computing units. In contrast, the CPU allocates a substantial portion to the control unit and cache memory to support various complicated operations. Figure 8 shows the comparison between CPU and GPU.

In the 2000s, researchers began considering using the GPU's numerous resources for purposes other than graphics operations. Thus, general-purpose computing on GPU (GPGPU) was proposed to utilize high-performance parallelization of GPUs to perform computations traditionally handled by CPU [47]. Early GPGPU was implemented by formulating data and operations in graphics primitives and translated to pass through a fixed graphics pipeline. The left of Fig. 9 represents the data flow in the graphic API model. Blue squares are buffers from which data can

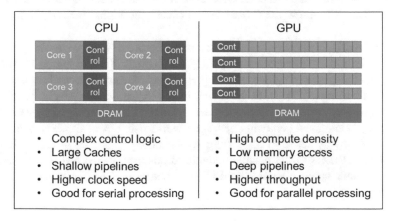

Fig. 8 Comparison between multi-core CPU and GPU [46]

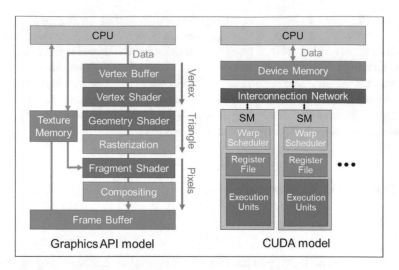

Fig. 9 Graphics API programming model and CUDA programming model [50]

be read and written, and orange squares are programmable stages. And gray squares are fixed pipeline functions useful for image synthesis but may not be necessary for general programming.

In compliance with this inconvenience, GPGPU programming APIs are proposed to perform computation independent of the graphics pipeline model. NVIDIA introduced CUDA [48], which is supported by its products. And Khronos Group releases OpenCL [49], a vendor-independent GPGPU platform. These APIs provide programmable SIMD instructions instead of pixels and shaders. Also, the data stream in the graphics pipeline is bound to the image handling process, and parts of the global memory declared as a texture are exclusive between processors [50]. On the other hand, the GPGPU model supports data communication and synchronization to allow the processors to access the global memory unhindered [46].

With GPGPU technology, tasks consist of numerous repetitive simple operations that can easily be parallelized and greatly accelerated. Since a neural network is essentially a large set of simple neuron operations, they are suitable for computing on GPU platforms [51]. Therefore, AI technologies have actively used GPU hardware resources from the beginning. Many public deep learning libraries support optimizing training and deploying neural networks using these GPGPU APIs.

NVIDIA is very active in supporting AI technology using its products. In most AI studies, NVIDIA GPUs occupy a standard position. In the rest of this section, we mainly talk about NVIDIA's GPU architecture and CUDA-related software libraries.

3.2 GPU Architecture

The GPU's architecture is optimized for graphic operations that perform a huge amount of repetitive and similar tasks. A recent GPU contains thousands of lightweight processors in parallel. These computing units support different data formats. Typical GPUs have 32-bit floating-point units (FP32), 64-bit floating-point units (FP64), and 32-bit integer units (INT32) for general graphic processing.

Because these massive processors can process multiple threads simultaneously, the programming methodology for GPUs is different from CPUs. In contrast to the single instruction multiple data (SIMD) model, the modern GPU implements the single instruction multiple thread (SIMT) architecture. In an SIMT architecture, multiple parallel threads follow divergent control flow and access different addresses, rather than the SIMD model that issues vector data in a single thread [52].

NVIDIA coined 'warp' to describe this bundle of threads. NVIDIA GPUs group 32 threads into a warp, and a warp is executed on a streaming multiprocessor (SM). The SM features hundreds of CUDA cores in general, and each CUDA core has fully pipelined arithmetic logic units [53]. AMD uses 'wavefront' in the same sense, and a wavefront consists of 64 threads (Fig. 10).

SIMT technology allows programmers to accelerate a sequence of operations by thousands of concurrent threads. But instructions with long delays, such as memory access, have the potential to stall the pipeline in starvation and lead to lower utilization of resources [55]. To avoid this underutilization, GPUs have a scheduler to determine the processing order and tolerate the latency. In NVIDIA GPUs, parallelized operations in multiple threads are translated into warp instructions and delivered to the warp scheduler. Then the warp scheduler dispatches the warp instructions to CUDA pipelines in optimal order to maximize pipeline utilization. For Keplar architecture, each SM includes four warp schedulers and eight instruction dispatch units. Each SM selects four warps for each cycle, and each warp allows two independent instructions at once [53].

Another significant element of NVIDIA GPU is the tensor core. NVIDIA presented a new computing unit distinguished from regular CUDA cores, named Tensor Core, since the Volta architecture (announced in 2017). Each tensor core executes MMA (matrix multiplication and accumulation, $D = A \times B + C$) on a 4×4 floating-point matrix. They are built to enhance deep learning by boosting matrix arithmetics, which are often used in deep learning (Fig. 11).

Tensor core supports low-precision operations because it has been shown that reduced precision can significantly increase energy efficiency and throughput with little loss of accuracy [56]. Since the first release of the tensor core, it has supported 16-bit floating-point (FP16, also known as half-precision) data as operands to accelerate inference throughput [54]. Turing architecture adds 8-bit integer (INT8) and 4-bit integer (INT4) operation modes for developers looking for even higher throughput [57].

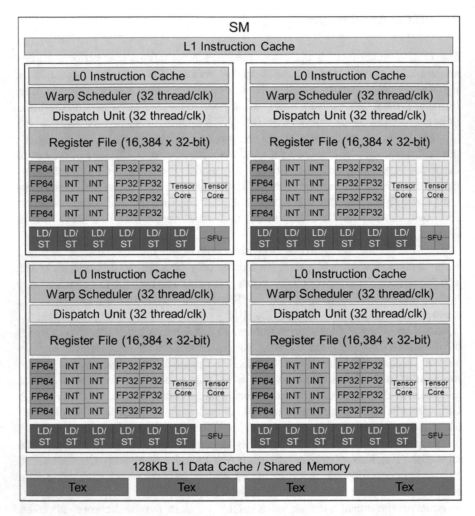

Fig. 10 NVIDIA Volta architecture [54]

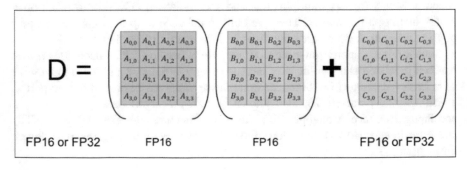

Fig. 11 Matrix multiplication and accumulation [54]

Table 1 shows the features and performances of NVIDIA GPU architectures since Pascal Architecture.

3.3 GPU Acceleration Techniques

AI developers tend to own individual machines, which enable experiments with larger and higher-quality datasets. NVIDIA launched the DGX station to support this, and GPU-based standalone computers include special computing techniques (Fig. 12).

DGX station is a total AI developing solution that includes powerful CPUs and GPUs with high bandwidth communication links. They build PCIe 4.0 for CPU-GPU connection and NVlink for GPU interconnection. The GPU interconnection bandwidth is 3X faster than the PCIe Gen4 ×16. It enables a special memory utilization technique. Also, the machine supports fine-grained structured sparsity, third-generation tensor cores, and a multi-instance GPU (MIG). Due to these effective techniques, it became a standard for high-performance standalone computers.

Memory Utilization Memory utilization is crucial for deep learning, as with all other graphics operations. Since a typical deep learning model is huge, it is impossible to put it all in the GPU's memory. Thus, we need to slice the network into fragments and load and store all the fragments for each data input. Because communication between GPU and CPU's memory takes much more time than the computation within the GPU, memory operations become a bottleneck in deep learning [60]. Therefore, several solutions have been proposed to reduce the number of memory operations and the time taken for data transmission.

A simple and most powerful solution is to increase the memory capacity of the GPU. Coincidentally, 3D applications are also evolving and demanding large amounts of GPU memory. Therefore, the memory capacity of GPUs is increasing. NVIDIA Titan Z, released in 2014, combines two GPUs to achieve 12 GB of memory [61]. However, in December 2018, NVIDIA launched Titan RTX, which includes 24 GB of memory on a single GPU [62].

Another efficient solution is increasing the data bandwidth between memory components. GPU and CPU communication and the communication inside the GPU, such as the connection with the cache memory, dramatically affect the performance. Generally, the external connection of GPU uses PCI-express, and manufacturers have recently started to support PCI-e 4.0, which has twice the bandwidth of 3.0 [63, 64]. Also, GDDR graphic memory increases its bandwidth as the version increases. NVIDIA Turing architecture includes GDDR6 SDRAM, which has approximately 2–3 times higher transfer rate compared to GDDR5 [57].

Finally, GPU vendors offer several memory management technologies to reduce the demand for memory transfer. NVIDIA Turing architecture includes a compression engine that reduces the amount of data transferred from memory to the

Table 1 The features of NVIDIA Tesla GPUs [58]

Tesla GPU	Fermi	Fermi	Kepler	Kepler	Maxwell	Pascal	Volta	Turing	Ampere
	GF100	GF104	GK104	GK110	GM200	GP100	GV100	TU104	GA100
Compute capability	2.0	2.1	3.0	3.5	5.3	6.0	7.0	7.0	8.0
Streaming multiprocessors	16	16	8	15	24	56	84	72	128
FP32 CUDA cores/SM	32	32	192	192	128	64	64	64	64
FP32 CUDA cores	512	512	1536	2880	3072	3584	5376	4608	8192
FP64 units	–	–	512	960	96	1792	2688	–	4096
Tensor Core units	–	–	–	–	–	–	672	576	512
Threads/warp	32	32	32	32	32	32	32	32	32
Max warps/SM	48	48	64	64	64	64	64	64	64
Max threads/SM	1536	1536	2048	2048	2048	2048	2048	2048	2048
32-bit registers/SM	32,768	32,768	65,536	65,536	65,536	65,536	65,536	65,536	65,536
Max registers/thread	63	63	63	255	255	255	255	255	255
Max threads/thread block	1024	1024	1024	1024	1024	1024	1024	1024	1024

Shared memory size Configs	16 KB,48 KB	16 KB, 48 KB	16 KB, 32 KB, 48 KB	16 KB, 32 KB, 48 KB	96 KB	64 KB	Config Up to 96 KB	Config Up to 96 KB	Config Up to 164 KB
Hyper-Q	No	No	No	Yes	Yes	Yes	Yes	Yes	Yes
Dynamic parallelism	No	No	No	Yes	Yes	Yes	Yes	Yes	Yes
Unified memory	No	No	No	No	No	Yes	Yes	Yes	Yes
Pre-emption	No	No	No	No	No	Yes	Yes	Yes	Yes
Sparse matrix	No	No	No	No	No	No	No	No	Yes

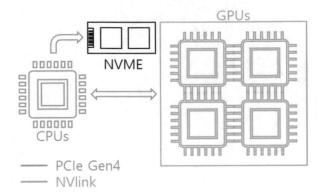

Fig. 12 Diagram of Nvidia DGX station [59]

L2 cache. It was reported that Turing has a 50% increase in effective bandwidth compared to Pascal architecture [57].

Low Precision Arithmetic Low-precision arithmetic, also frequently referred to as quantization, is a powerful tool to accelerate neural network inference. Using low-precision types in deep learning can significantly reduce the size of data with little loss of accuracy [31]. Reducing the data size has the same effect as increasing the memory utilization mentioned above, and it also reduces the usage of resources such as memory footprints [65].

The most frequently used floating-point type is FP32, of which representation is standardized in IEEE 754 [66]. However, in deep learning, it is known that FP32 is too large and can sufficiently represent necessary information with lower precision. Many GPUs support FP16 at the hardware level, and, recently, several other data types are chosen by programmers according to their needs.

Brain Floating-Point Format (BF16) is a widely used type among various types, introduced by Google Brain [67]. BF16 is a 16-bit data format like FP16 but can express a broader range by allocating more bits to express the exponent. BF16 can show almost the same performance as FP32 while inferencing [68]. From the Tensor Core installed in the NVIDIA GA10× GPU with the Ampere architecture, it started to support the BF16 data type [69].

Another valuable data expression in deep learning is the integer type. Currently, a lot of research has been conducted to reduce the amount of data transmission and computation by expressing weights and data in an integer type [70]. The statement mentioned above tells us why modern GPUs are starting to incorporate integer-type computational cores that do not match their original purpose, graphics operation. NVIDIA has installed INT32 computational cores in its tensor cores since the first launch [11] and has supported INT8 and INT4 types from the Turing architecture [57] (Fig. 13).

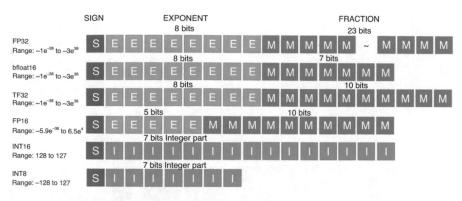

Fig. 13 Various data types being used in GPUs

Multi-GPU System Because deep learning and inference are highly parallelizable operations, multiple GPUs can be used to perform operations simultaneously. The concept of connecting multiple GPUs and using them as one GPU was presented first in the 3D games and applications market. NVIDIA SLI [71] and AMD Crossfire [72] are the multi-instance connectivity technologies provided by GPU manufacturers to meet this demand.

Most deep learning frameworks support distributed training of networks automatically or through simple commands when multiple GPUs are found in a PC. In addition, TensorFlow and Pytorch provide packages named distribution (tf. distribute and torch. distributed) for distributing networks across multiple instances and managing distributed training strategies [73, 74]. Also, NVIDIA provides NVIDIA Collective Communication Library (NCCL) to support deep learning with multiple NVIDIA GPUs natively [75].

The problem with using multiple GPUs is the overhead of sharing data between GPUs [76]. Communication using wires is inherently slower than communication on a chip. In a large-scale environment such as the cloud, hardware acceleration technologies such as NVLink and NVSwitch are used to solve this problem [77]. However, in the PC environment, NVIDIA currently supports NVLink only in its top-tier products and announced that it would gradually stop supporting it [78].

Computing with Sparsity Pruning is a well-known technique to reduce the deep learning model by eliminating unnecessary values. In a trained deep learning model, some weights significantly affect the result, while others have insignificant effects [79]. Weights with values close to 0 generally do not significantly affect accuracy. By omitting calculations by considering these values as 0, the amount of computation in deep learning can be significantly reduced.

However, pruning has not been utilized when using a GPU because the computational pipeline of GPUs is rigid, so it is difficult to skip some computations directly. NVIDIA provides cuSPARSE to support accelerated operations on matrices where most elements are zero [80]. However, unless more than 99% of the elements are filled with zeros, the performance of cuSPARSE is worse than calculating the whole

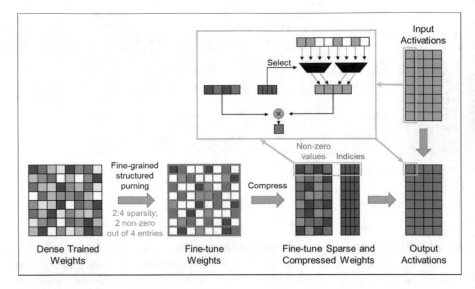

Fig. 14 Fine-grained structured sparsity

values [81]. This performance dependency on the sparsity of the matrix makes it challenging to use meaningfully in deep learning.

GPU manufacturers have started supporting pruning in 2020. NVIDIA introduced fine-grained structured sparsity technology with Ampere architecture [69]. Fine-grained structured sparsity classifies trained weights into several 2-out-of-4 non-zero patterns by region and compresses them to half the size. It allows the amount of computation of the sparse tensor core to be reduced in half (Fig. 14).

3.4 CUDA-Related Libraries

NVIDIA provides several software libraries to support deep learning on its hardware, and they are provided as part of NVIDIA CUDA-X. CUDA-X is a collection of libraries and tools optimized for the GPU environment on top of NVIDIA CUDA [82]. In this section, we introduce some valuable libraries included in CUDA-X.

Math Libraries There are several math libraries related to CUDA. They provide acceleration of mathematical operations using the GPU's CUDA Core. Most of them are provided as part of the CUDA toolkit without needing to install them separately.

- cuBLAS: Basic linear algebra (BLAS) library
- cuFFT: Library for fast Fourier transforms
- cuSOLVER: Dense and sparse direct solvers
- cuSPARSE: BLAS for sparse matrices

- cuTENSOR: Tensor linear algebra library

cuDNN CUDA deep neural network library (cuDNN) is a GPU-accelerated library of primitives for deep neural networks. cuDNN provides highly optimized implementations of mathematical functions widely used in DNN, such as convolution, pooling, normalization, and activation layers. cuDNN acceleration has been widely used in popular deep learning frameworks [83].

TensorRT NVIDIA TensorRT is an SDK for optimizing deep learning inference. TensorRT optimizes trained neural networks to achieve low latency and high throughput. TensorRT provides several performance-boosting techniques, including quantizing models to FP16 or INT8 precision. It also optimizes GPU memory usage by reusing tensors and fusing nodes in a kernel. With Ampere architecture GPUs, TensorRT also uses sparse tensor cores for an additional performance boost [84].

NCCL NVIDIA collective communication library (NCCL) supports deep learning on multiple GPUs. NCCL provides multi-node communication routines optimized for NVIDIA GPUs, such as broadcasting, gathering, and reducing. NCCL is compatible with virtually any multi-GPU parallelization model [75].

DALI NVIDIA data loading library (DALI) is an open-source library for accelerating the input data stream of deep learning applications. It can build a data processing pipeline that includes loading, decoding, and several data augmentation operators. It can mitigate training bottlenecks by duplicating training and pre-processing time [85].

4 Hardware Accelerators in NPU

4.1 History and Overview: Hardware

In the beginning, Intel proposed the electrically trainable artificial neural network (ETANN) around 1989 [86], which was the first implementation of artificial neural in analog circuits. It was an electronically trainable parallel data processor composed of 64 neurons and 10,240 [87] synapses to calculate the dot product between weight array and input vector. It demonstrated the potential of separate accelerators to achieve high efficiency and throughput [87, 88]. After that, Intel released the Ni1000 fully digital chip, including 1024 neurons, to implement a multi-layer perceptron [89].

The demand for real-time digital signal processing operations has increased as technology advances. Therefore, the coprocessor concept has been proposed to share the complex calculation tasks heavily allocated to the CPU. And it is now widely used due to its high efficiency and performance. Digital Signal Processor was used to accelerate these operations, such as Optical Character Recognition (OCR) [90] and audio signal processing [91]. Also, graphics processing unit (GPU)

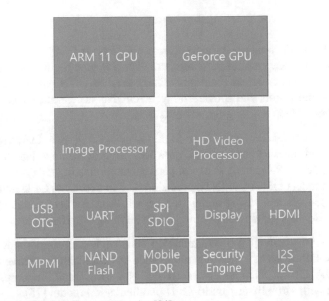

Fig. 15 Diagram of Nvidia Tegra 6XX series [94]

supported by general-purpose computing on GPU (GPGPU) technologies such as Nvidia CUDA is used to accelerate image manipulation and local analysis.

According to AI technology development, efficient real-time AI inferencing is the crucial challenge of edge computing [92]. In times past, data was processed on the server using the network, but the paradigm changed to 'edge-computing' due to various reasons such as privacy, cost, and responsiveness. Edge computing needs a proper device that satisfies both performance and efficiency.

However, general-purpose computing units like CPUs and GPUs are less efficient than specialized processors, FPGA, and NPU [93]. Thus, NVIDIA designed GPU-based system-on-chip (SoC) solutions, including unique computing cores. Finally, NVIDIA launched NVIDIA Tegra and NVIDIA Jetson series (Fig. 15).

Nvidia Tegra is an SoC solution that includes NVIDIA deep learning accelerator (NVDLA) and vision programming interface (VPI). NVDLA consists of the following components to accelerate deep learning inference operations: convolution core, single data processor, planar data processor, channel data processor, and dedicated memory and data reshape engines. VPI is an optimized computer vision and image processing software library. It provides a unified interface to regulate Jetson embedded devices and ×86 devices with discrete GPUs [95].

Nowadays, many flagship mobile SoCs include a neural processing unit [96, 97]. NPU increases the system's overall efficiency with the acceleration of repeated AI functions. It provides a good AI inferencing performance with reasonable power consumption on edge devices [98, 99]. Samsung develops Exynos 2200 with its own NPU [100]. And Qualcomm develops Snapdragon 888 with Hexagon 780. Also, Huawei develops Kirin 9000 with Da Vinch Architecture 2.0 [101]. Also, various artificial intelligence companies such as Google, NVIDIA, Qualcomm, etc., have announced accelerators using NPUs to the market. These particular chips can dramatically improve the processing speed by significantly reducing the required time to read large amounts of data.

4.2 Standalone Accelerating System Characteristics

In mobile environments, the CPU was a major device for calculating DNNs even though the GPU performs better [102]. In many SoCs, the number of memory chips is limited because each additional part increases power consumption and package size. Therefore, there is a limit to increasing the available memory size and transfer bandwidth. Unfortunately, reserving the required memory for both CPU and GPU is impossible. Therefore, the transmission and loading of data cause bottlenecks in the computation [103]. For these and other reasons, most inference workloads ran on CPUs with sufficient memory space.

Acceleration of deep learning on a CPU starts by utilizing the CPU's parallel computation instructions such as single instruction multiple data (SIMD). The deep learning frameworks implement main deep learning functions in a form that can maximize CPU utilization.

Despite advances in semiconductor technology, there was a limitation on CPU performance and efficiency. Thus, many designers choose to mount either more GPUs or additional NPUs for AI computation. For edge devices that value efficiency rather than versatility, NPU is selected.

NPU and CPU programming schemes are different due to their distinguished features. Because NPUs are specially designed devices, it cannot be assumed that every user device necessarily constitutes a system with a specific hardware accelerator. So, programmers use dedicated instruction sets for different hardware. CPU's software stack and programming instructions are well standardized and documented, and therefore programmers can work identically on different hardware [104–106].

Besides, the hardware vendors also provide optimized mathematical APIs for AI application developers. Intel oneAPI [107] and ARM-CL [108] are hardware-dependent libraries for enhancing the performance of mathematical computations. Also, they offer deep learning functions that provide mid-level toolkits to support acceleration in representative deep learning frameworks. Examples are Intel Open-VINO [109] and ARM-NN [110] (Fig. 16).

4.3 Architectures of Hardware Accelerator in NPU

As technology advances, 'Efficiency' and 'Flexibility' are the main hardware development topics. Thus, there are several techniques to achieve both objectives

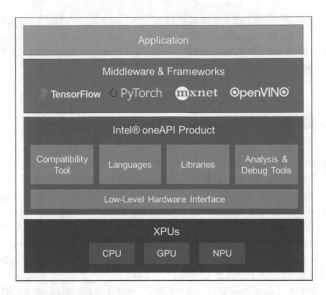

Fig. 16 Overview of Intel oneAPI toolkit

at the hardware level while designing NPU. We will discuss two main perspectives to optimize the accelerator: computation unit and memory architecture.

Optimized computation unit From the perspective of the computation unit, several design strategies achieve flexibility and efficiency in the NPU. First, design an energy-efficient elementary computation unit. Matrix multiplication unit (MMU) consists of thousands of multiply-accumulator (MAC) units to achieve lower energy consumption by parallel processing. Second, embed the sparsity computation unit in NPU. Many AI models have fine-grained sparsity in their weights. However, software utilization is impossible due to sparsity's inherent complexity and randomness. Thus, the hardware-level acceleration of sparse operands is the key factor in the new generation of NPU. Third, re-configurable MAC array. To maintain high utilization in various types of layers, NPU needs to develop a reconfigurable data path to adapt to the array shape. Lastly, support the mixed precision. Unlike the training phase, inferencing adopts reduced precision for efficiency. However, some of the important parts of layers require a more precise expression, such as int16 or float32. Thus, supporting more accurate precision within a few more cycles could achieve precision flexibility more easily compared to using accumulators dedicated to another precision [111, 112].

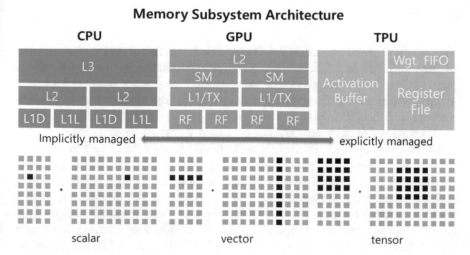

Fig. 17 Memory subsystem architecture

In contrast with GPU, NPU has less memory access during matrix computation. Modern NPU consists of a 'systolic array,' a physically connected MAC circuit structure. At the beginning of matrix computation, NPU loads operands from the main memory into Matrix eXecution Unit (MXU). Then, each result of the calculation is moved to the next MAC. Lastly, a summary of the calculations returns to the main memory. This mechanism does not access the memory while the matrix computation is in progress [113] (Fig. 17).

Specialized memory subsystem architecture From the perspective of memory architecture, there are also strategies to increase performance. First, optimization of the memory subsystem architecture focuses on data structure alignment. Unlike general-purpose processors such as CPU and GPU, NPU has a specific purpose and predictable memory access patterns. Therefore, NPU can utilize a special memory managing strategy.

CPU is designed to calculate all kinds of programming problems. Thus, a multi-layer cache and memory system with scalar memory access is essential. Meanwhile, GPU is specialized to solve rendering problems; in this case, vector-wise memory access is the optimal solution.

However, since neural network architecture is based on matrix multiplication, increasing the data unit dimension is efficient for task acceleration. This kind of 2-d computation primitive is called Tensor. Recently, increasing the number of memory access units in the NPU has been emphasized to increase efficiency, so various methods for memory slicing have been proposed [114–117].

4.4   SOTA Architectures

This section describes the recent developments in the design of the DNN accelerator. In general, DNN is a function that takes a high-dimensional input for various purposes, such as creating a classification label. This process is called inference, and parameters are optimized to minimize a specific loss function through various methods such as SGD [118].

A neural processing unit (NPU) is a chip designed to accelerate program segments using on-chip NN instead of running on the CPU. The NPU consists of several processing engines, and each PE acts as neuron. NPU frequently executes the mathematical operations and deduces the optimal value of DNN, and the segment can be accelerated. Based on this, it can dramatically reduce the time required for general inferencing, which is a differentiated strength from conventional accelerators. A typical NPU is the Nvidia deep learning accelerator (NVDLA) in JETSON Xavier of NVIDIA [119]. The growth of DNN accelerator applications makes many companies participate in the NPU market and sparks competition between high performance and low price (Fig. 18).

Figure 19 above is a picture of the structure of TPU announced by Google. Each square describes a unit constituting the TPU. The blue units are operational components, and the orange units take and store the values required for the operation predictively from the data buffer and store the operation results. Control units play a role in adjusting each unit, and the light green unit is the presented memory in the off-chip [121].

Figure 20 is an overview of the template-based machine learning accelerator used for architecture. The template accelerator is comprised of a 2D array of processing elements. Each processing element performs various arithmetic computations using single instruction multiple data (SIMD) instructions. The main components of the processing engine are several processing cores, each with multiple compute lanes for performing tensor operations in an SIMD manner. In terms of memory, there is

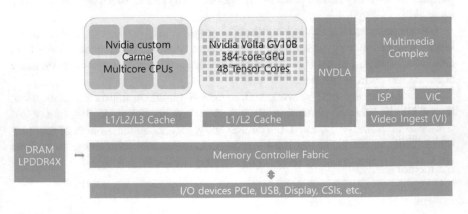

Fig. 18 NVIDIA Jetson Xavier NX [120]

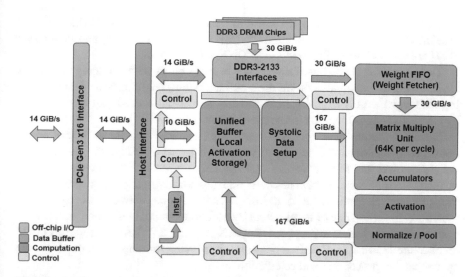

Fig. 19 TPU Block Diagram

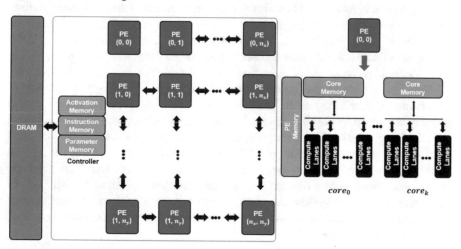

Fig. 20 The architecture of Edge TPU accelerators [121]

a processing engine memory shared across the cores, and each core has a dedicated core memory equipped with it. Therefore, parallel operations can be performed quickly using Coral Dev Board, an embedded board including Edge TPU with the above architecture [122].

On the other hand, ReRAM [123] and Hybrid Memory Cube (HMC) [124] are memory structures that implement processing in memory (PIM) technology. PIM can significantly reduce data movement in computing platforms because data movement between the CPU and off-chip memory consumes two digits more energy than floating-point operations. The PIM is a structure that can greatly benefit from data movement on computing platforms, and the use of ReRAM and HMC in DNN accelerators can accelerate execution. PRIME [125], ISAAC [126], and Pipe Layer [127] are widely known ReRAM-based DNN accelerators.

There is also an HMC-based accelerator in accelerators with emerging memories. HMC vertically integrates DRAM dies, and logic dies, enabling near-data processing with high memory capacity, high memory bandwidth, and low latency provided by HMC. Examples include Neurocube [128] and Tetris [129].

Finally, there are accelerators for emerging applications. First, Sparse neural network accelerators aim to accelerate the calculation of the NN model using sparse weight matrices and feature maps. However, a Sparse network has the disadvantage of incurring additional hardware costs, and research is being conducted to solve this problem. Second, there is a low-precision neural network. A method of increasing the efficiency of accelerators by co-optimizing data accuracy and overall system efficiency has been proposed. ReGAN proposes a ReRAM-based PIM GAM architecture [130]. It is a method of further improving training efficiency by using spatial parallel processing and calculation sharing.

NPU research plays a vital role in the AI semiconductor market because of deep learning research. Many NPUs with various specifications, such as Samsung Electronics' Exynos series [131], Qualcomm's Snapdragon series [132], and Google's Coral series [133], continue to appear in the market. Samsung's Exynos 2200 implemented ARMv9, the latest architecture of ARM. Compared to its predecessor, ARMv9 doubles the speed of AI operations and provides an enhanced on-device AI function, which is the device's own AI operation. Qualcomm announced Snapdragon 888 at Snapdragon Tech Summit 2021 with twice the size of shared memory along with twice the performance of tensor accelerators. It is also equipped with the 7th generation Qualcomm engine, which improves inference performance up to 4 times and power efficiency by 1.7 times compared to the previous generation. There is also a fast-growing area where various companies test their products and measure performance through an AI accelerator benchmark called MLPerf. In MLPerf, every contest verifies the product performance of various companies such as Google and Qualcomm [134]. Table 2 shows the newest benchmark result of MLPerf Inference: Edge v2.0.

5 Summary

As the level of deep learning rises, the demand for deep learning in general user-level applications increases. Therefore, the technology for performing deep learning on the devices of general users is developing widely. This chapter summarizes hardware accelerators and assistant software technologies for performing deep learning on standalone platforms.

Since the GPU is one of the main components that consists of the PC and most of the operations of deep learning can be easily parallelized and accelerated on the GPU, most PC applications that utilize deep learning use the GPU as the primary

Table 2 2022.05 MLPerf inference: Edge v2.0 benchmark [135]

Submitter	System	Image classification			Object detection (large)			Natural language processing		
		ImageNet ResNet			COCO SSD-large			SQuAD v1.1 BERT		
		Single stream	Multi-stream	Offline	Single stream	Multi-stream	Offline	Single stream	Offline	
		Latency in ms	Latency in ms	Samples/s	Latency in ms	Latency in ms	Samples/s	Latency in ms	Samples/s	
Alibaba	Alibaba Cloud Sinian Platform (Haishen)	0.75	2.52	7047.83	26.97	120.24	150.6	14.42	220.61	[136]
Azure	NC24_A100_v4 (1× A100-PCIe-80 GB, TensorRT)	0.47	0.72	35788.4	1.86	11.28	875.48	1.56	3073.03	[137]
Azure	NC24_A100_v4 (1× A100-PCIe-80 GB, TensorRT, Triton)	0.45	0.84	35310.5	1.83	11.62	870.9	1.61	3108.65	[137]
Dell	Dell PowerEdge XE2420 (1× A30, TensorRT)	0.52	0.92	17418.2	3.16	19.7	492.31	2.29	1717.98	[138]

(continued)

Table 2 (continued)

Submitter	System	Task	Image classification ImageNet ResNet			Object detection (large) COCO SSD-large			Natural language processing SQuAD v1.1 BERT		
			Single stream	Multi-stream	Offline	Single stream	Multi-stream	Offline	Single stream	Offline	
		Data model									
		Scenario									
		Units	Latency in ms	Latency in ms	Samples/s	Latency in ms	Latency in ms	Samples/s	Latency in ms	Samples/s	
Dell	Dell PowerEdge XE2420 (1 × A30, MaxQ, TensorRT)		0.77	1.18	16,520				2.31	1670.45	[138]
Dell	Dell PowerEdge XE8545 (1× A100-SXM-80 GB-MIG-1 × 1g.10 gb, TensorRT)		0.66	1.99	5224.5	8.18	60.12	134.84	5.38	483.38	[137]
Dell	Dell PowerEdge XE8545 (1× A100-SXM-80 GB, TensorRT)		0.46	0.67	40,471				1.59	3777.6	[137]
Dell	Dell PowerEdge XR12 (1× A2, TensorRT)		1.1	3.54	3010.97				9.01	244.47	[139]

GIGABYTE	Gigabyte R282-Z93 (2× QAIC100) QUALCOMM Cloud AI 100 PCIe/HHHL Pro	0.34		46361.4	8.73	28.03	885.04	10.25	1437.71	[136]
Inspur	NE5260M5 (2× A100-PCIe-80 GB, TensorRT)	0.46	0.64	75802.6	2.03	6.17	1868.3	1.54	6356.91	[137]
Inspur	NE5260M5 (2× A100-PCIe-80 GB, TensorRT, MaxQ)	0.51	0.76	70351.6	2.09	6.53	1723	1.59	5795.9	[137]
Inspur	NE5260M5 (2× A100-PCIe-80 GB, TensorRT, Triton)	0.45	0.68	71880.4	2.01	6.23	1860.4	1.59	6348.94	[137]
Inspur	NF5488A5 (1× A100-SXM-80 GB, TensorRT)	0.48	0.7	40991.3	1.76	8.81	1023.6	1.62	3749.63	[137]
Inspur	NF5688M6 (1× A100-SXM-80 GB, TensorRT)	0.46	0.66	41271.3	1.69	8.68	1025.7	1.54	3778.76	[137]
Inspur	NF5688M6 (1× A100-SXM-80 GB, TensorRT, Triton)	0.43	0.75	39,340	1.65	8.8	1023.9			[137]
Krai	AI development kit Qualcomm Snapdragon 865 QUALCOMM Cloud AI 100 DM.2									
Krai	Odroid N2+ (odroid)	339.09	2738.34	2.96	21.54	91.86	224.65	17.38	361.09	[136]

(continued)

Table 2 (continued)

Submitter	System / Units	Image classification ImageNet ResNet			Object detection (large) COCO SSD-large			Natural language processing SQuAD v1.1 BERT		
		Single stream Latency in ms	Multi-stream Latency in ms	Offline Samples/s	Single stream Latency in ms	Multi-stream Latency in ms	Offline Samples/s	Single stream Latency in ms	Offline Samples/s	
Krai	Odroid N2+ (odroid) Arm Mali-G52 MP6 GPU	246.29	2023.21	4.12						[140]
Krai	Gigabyte R282-Z93 (8× QAIC100)	0.33	0.42	179,618	9.02	12.65	3620	10.41	5783.39	[136]
Krai	Raspberry Pi 3 model B rev. 1.2 (rpi3)	2311.1	22167.54	0.47						
Krai	Raspberry Pi 4 model B rev. 1.4 (rpi4)	349.11	3455.7	3						
Krai	NVIDIA Jetson AGX Xavier	76.74	740.61	13.19						
Lenovo	Lenovo ThinkEdge SE450 NVIDIA A100-PCIe-80 GB	0.45	3.66	72835.4	1.84	14.82	1805.9	1.59	6310.58	[137]

Lenovo	Lenovo ThinkEdge SE450 NVIDIA A30	0.47	3.86	36199.4	2.74	22.04	957.21	2.17	3340.66	[138]
NVIDIA	Gigabyte G482-Z54 (1× A30-MIG-1×1g.6 gb, TensorRT)	0.71	2.27	5088.67	8.48	61.53	128.92	5.85	502.02	[138]
NVIDIA	Gigabyte G482-Z54 (1× A30-MIG-1×1g.6 gb, TensorRT, Triton)	0.74	2.68	5078.34	8.41	62.45	133.41	6.03	502.17	[138]
NVIDIA	NVIDIA DGX A100 (1× A100-SXM-80 GB-MIG-1×1g.10 gb, TensorRT)	0.68	2	5231.07	8.18	63.91	134.91	5.35	484.02	[137]
NVIDIA	Gigabyte G482-Z54 (1× A100-PCIe-80 GB, TensorRT)	0.5	0.76	36864.6	1.93	10.96	913.68	1.57	3206.7	[137]

(continued)

Table 2 (continued)

		Task								
		Image classification			Object detection (large)			Natural language processing		
		ImageNet ResNet			COCO SSD-large			SQuAD v1.1 BERT		
	Data model	Single stream	Multi-stream	Offline	Single stream	Multi-stream	Offline	Single stream	Offline	
Submitter	System	Units								
	Scenario	Latency in ms	Latency in ms	Samples/s	Latency in ms	Latency in ms	Samples/s	Latency in ms	Samples/s	
NVIDIA	Supermicro AS-1114S-WTRT (1× A2, TensorRT)	0.71	3.21	3036.57	14.64	113.33	72.27	8.9	246.64	[139]
NVIDIA	Gigabyte G482-Z54 (1× A30, TensorRT)	0.52	0.96	18072.9	3.03	20.04	480.91	2.2	1692.41	[138]
NVIDIA	NVIDIA DGX A100 (1× A100-SXM-80 GB, TensorRT)	0.47	0.69	39649.2	1.7	8.78	1001.8	1.55	3560.73	[137]
NVIDIA	NVIDIA DGX A100 (1× A100-SXM-80 GB, TensorRT, Triton)	0.43	0.8	39,649	1.66	8.94	993.55	1.59	3549.38	[137]
NVIDIA	NVIDIA drive-A100-PCIe (1× drive-A100-PCIe, TensorRT)	0.77	1.27	31431.5				1.81	3128.81	[137]

NVIDIA	NVIDIA Jetson Xavier NX (TensorRT)	2.46	10.53	1243.14	42.37	323.34	36.69	46.44	61.4	[141]
Nettrix	Nettrix X640 G40 (A100-PCIe-80 GB×1, TensorRT)	0.45	0.75	37826.5	2	11.8	937.41	1.54	3275.23	[137]
Nettrix	Nettrix X640 G40 (1× A30, TensorRT)	0.48	0.8	19169.9	2.66	17.45	481.64	2.16	1689.63	[138]
Nettrix	Nettrix X660 G45 (1× A100-SXM-80 GB, TensorRT)	0.45	0.65	39556.9	1.69	8.89	991.83	1.54	3542.5	[137]
Qualcomm	AI development kit QUALCOMM Cloud AI 100 DM.2e	0.84	3.91	5849.3	25.68	129.49	128.77	16.38	179.47	[136]
Qualcomm	AI development kit QUALCOMM Cloud AI 100 DM.2	0.89	3.3	9780.43	22.21	96.91	213.59	15.41	315.89	[136]
Qualcomm	Foxconn Gloria Qualcomm Snapdragon 865 QUALCOMM Cloud AI 100 DM.2e	0.82	3.86	7309.24	27.07	114.14	163.42	13.73	230.3	[136]

(continued)

Table 2 (continued)

Task	Image classification			Object detection (large)			Natural language processing		
Data model	ImageNet ResNet			COCO SSD-large			SQuAD v1.1 BERT		
Scenario	Single stream	Multi-stream	Offline	Single stream	Multi-stream	Offline	Single stream	Offline	
Submitter / System / Units	Latency in ms	Latency in ms	Samples/s	Latency in ms	Latency in ms	Samples/s	Latency in ms	Samples/s	
Qualcomm / Gigabyte R282-Z93 (1× QAIC100) QUALCOMM Cloud AI 100 PCIe/HHHL Pro	0.35	1.02	23807.5	8.94	43.45	443.16	10.27	724.08	[136]
Qualcomm / Gigabyte R282-Z93 (5× QAIC100) QUALCOMM Cloud AI 100 PCIe/HHHL Pro	0.34	0.42	114,869	8.61	25.94	2224.5	10.25	3606.5	[136]
Supermicro SYS-220HE-FTNR_1×A2_TRT	0.7	3.41	2735.6	16.01	125.02	66.92	9.88	243.12	[139]
Supermicro SYS-220HE-FTNR_2×A2_TRT	0.74	3.17	5777.52	14.26	112.59	138.66	8.46	481.86	[139]
NVIDIA Orin (TensorRT)	0.69	3.96	6138.84	8.03	60.68	207.66	7.64	476.34	[142]
FuriosaAI Supermicro SYS-420GP-TNR 1x Warboy	0.71	3.97	2758.44	13.43	107.87	79.92			[143]

accelerator. NVIDIA supports accelerating deep learning operations at the hardware level of GPUs, such as Tensor Core, as well as computation acceleration libraries using its products.

There is also an increasing demand for deep learning on low-power, low-cost devices such as mobile platforms. NPUs specialized only in deep learning are being proposed for use in these platforms. NPUs consist of architectures that are more optimized in matrix operations frequently used in deep learning, and their memory structures are optimized for repetitive operations. In addition, various techniques are being applied to increase throughput, such as lowering precision or skipping calculations.

Acknowledgments This work was supported by Institute of Information & Communications Technology Planning & Evaluation (IITP) grant funded by the Korea government (MSIT) (2022-0-00966, Development of AI processor with a Deep Reinforcement Learning Accelerator adaptable to Dynamic Environment).

References

1. Jordan, M.I., Mitchell, T.M.: Machine learning: Trends, perspectives, and prospects. Science (1979). **349**, 255–260 (2015)
2. Pouyanfar, S., Sadiq, S., Yan, Y., Tian, H., Tao, Y., Reyes, M.P., Shyu, M.-L., Chen, S.-C., Iyengar, S.S.: A survey on deep learning: Algorithms, techniques, and applications. ACM Comput. Surv. (CSUR). **51**(1–36) (2018)
3. Goodfellow, I., Bengio, Y., Courville, A.: Deep Learning. The MIT Press, Cambridge, MA, USA (2016)
4. Kim, J.H., Grady, B., Lian, R., Brothers, J., Anderson, J.H.: FPGA-based CNN inference accelerator synthesized from multi-threaded C software. In: 2017 30th IEEE International System-on-Chip Conference (SOCC), pp. 268–273. IEEE (2017)
5. Saiyeda, A., Mir, M.A.: Cloud computing for deep learning analytics: Asurvey of current trends and challenges. Int. J. Adv. Res. Comput. Sci. **8**(2), 68–72 (2017)
6. Kim, S., Deka, G.C.: Hardware Accelerator Systems for Artificial Intelligence and Machine Learning. Academic Press, Elsevier Science (2021)
7. Gupta, N.: Introduction to hardware accelerator systems for artificial intelligence and machine learning. In: Advances in Computers, pp. 1–21. Elsevier (2021)
8. Zlatanov, N.: Computer security and mobile security challenges. In: Conference: Tech Security Conference At, (2015)
9. Mireshghallah, F., Taram, M., Vepakomma, P., Singh, A., Raskar, R., Esmaeilzadeh, H.: Privacy in deep learning: A survey. arXiv preprint arXiv:2004.12254. (2020)
10. Oh, K.-S., Jung, K.: GPU implementation of neural networks. Pattern Recogn. **37**, 1311–1314 (2004)
11. Durant, L., Giroux, O., Harris, M., Stam, N.: Inside Volta: The World's Most Advanced Data Center GPU. https://developer.nvidia.com/blog/inside-volta/
12. About CUDA: https://developer.nvidia.com/about-cuda
13. Rosenblatt, F.: The perceptron: A probabilistic model for information storage and organization in the brain. Psychol. Rev. **65**, 386 (1958)
14. Sak, H., Senior, A., Beaufays, F.: Long short-term memory recurrent neural network architectures for large scale acoustic modeling. In: Proceedings of the Annual Conference of the International Speech Communication Association. INTERSPEECH (2014)

15. Albawi, S., Mohammed, T.A., Al-Zawi, S.: Understanding of a convolutional neural network. In: 2017 International Conference on Engineering and Technology (ICET), pp. 1–6. IEEE (2017)
16. Sharma, S., Sharma, S., Athaiya, A.: Activation functions in neural networks. Towards Data Sci. **6**, 310–316 (2017)
17. Shawahna, A., Sait, S.M., El-Maleh, A.: FPGA-based accelerators of deep learning networks for learning and classification: A review. IEEE Access. **7**, 7823–7859 (2018)
18. Lasserre, D.: Breaking the Von Neumann Bottleneck: A Key to Powering Next-Gen AI Apps. https://www.electronicdesign.com/technologies/embedded-revolution/article/21156009/gsi-technology-breaking-the-von-neumann-bottleneck-a-key-to-powering-nextgen-ai-apps (2021)
19. Shriram, S.B., Garg, A., Kulkarni, P.: Dynamic memory management for gpu-based training of deep neural networks. In: 2019 IEEE International Parallel and Distributed Processing Symposium (IPDPS), pp. 200–209. IEEE (2019)
20. Bang, S., Wang, J., Li, Z., Gao, C., Kim, Y., Dong, Q., Chen, Y.-P., Fick, L., Sun, X., Dreslinski, R.: 14.7 a 288 μw programmable deep-learning processor with 270 kb on-chip weight storage using non-uniform memory hierarchy for mobile intelligence. In: 2017 IEEE International Solid-State Circuits Conference (ISSCC), pp. 250–251. IEEE (2017)
21. Sze, V., Chen, Y.-H., Yang, T.-J., Emer, J.S.: Efficient processing of deep neural networks: A tutorial and survey. Proc. IEEE. **105**, 2295–2329 (2017)
22. Krizhevsky, A., Sutskever, I., Hinton, G.E.: Imagenet classification with deep convolutional neural networks. Commun. ACM, ACM New York, NY, USA **60**(6), 84–90 (2017)
23. Suda, N., Chandra, V., Dasika, G., Mohanty, A., Ma, Y., Vrudhula, S., Seo, J., Cao, Y.: Throughput-optimized OpenCL-based FPGA accelerator for large-scale convolutional neural networks. In: Proceedings of the 2016 ACM/SIGDA International Symposium on Field-Programmable Gate Arrays, pp. 16–25 (2016)
24. Liu, Z., Dou, Y., Jiang, J., Xu, J., Li, S., Zhou, Y., Xu, Y.: Throughput-optimized FPGA accelerator for deep convolutional neural networks. ACM Trans. Reconfigurable Technol. Syst. (TRETS). **10**(1–23), 1 (2017)
25. Sun, F., Wang, C., Gong, L., Xu, C., Zhang, Y., Lu, Y., Li, X., Zhou, X.: A high-performance accelerator for large-scale convolutional neural networks. In: 2017 IEEE International Symposium on Parallel and Distributed Processing with Applications and 2017 IEEE International Conference on Ubiquitous Computing and Communications (ISPA/IUCC), pp. 622–629. IEEE (2017)
26. Nguyen, D.T., Nguyen, T.N., Kim, H., Lee, H.-J.: A high-throughput and power-efficient FPGA implementation of YOLO CNN for object detection. IEEE Trans. Very Large Scale Integr. (VLSI) Syst. **27**, 1861–1873 (2019)
27. Alwani, M., Chen, H., Ferdman, M., Milder, P.: Fused-layer CNN accelerators. In: 2016 49th Annual IEEE/ACM International Symposium on Microarchitecture (MICRO), pp. 1–12. IEEE (2016)
28. Chen, Y.-H., Krishna, T., Emer, J.S., Sze, V.: Eyeriss: An energy-efficient reconfigurable accelerator for deep convolutional neural networks. IEEE J. Solid State Circuits. **52**, 127–138 (2016)
29. Lai, L., Suda, N., Chandra, V.: Deep convolutional neural network inference with floating-point weights and fixed-point activations. arXiv preprint arXiv:1703.03073. (2017)
30. Judd, P., Albericio, J., Hetherington, T., Aamodt, T., Jerger, N.E., Urtasun, R., Moshovos, A.: Reduced-precision strategies for bounded memory in deep neural nets. arXiv preprint arXiv:1511.05236. (2015)
31. Hubara, I., Courbariaux, M., Soudry, D., El-Yaniv, R., Bengio, Y.: Quantized neural networks: Training neural networks with low precision weights and activations. J. Mach. Learn. Res. **18**, 6869–6898 (2017)
32. TensorFlow: https://www.tensorflow.org/
33. Cloud Tensor Processing Units (TPUs): https://cloud.google.com/tpu/docs/tpus

34. Vincent, J.: Google's new machine learning framework is going to put more AI on your phone. https://www.theverge.com/2017/5/17/15645908/google-ai-tensorflowlite-machine-learning-announcement-io-2017
35. Keras 2.4.0: https://github.com/keras-team/keras/releases/tag/2.4.0
36. PyTorch: https://pytorch.org/
37. Torch7: https://github.com/torch/torch7
38. PyTorch Lightning: https://www.pytorchlightning.ai/
39. PyTorch Mobile: https://pytorch.org/mobile/home/
40. Theano: https://github.com/Theano/Theano
41. Aesera: https://github.com/aesara-devs/aesara
42. MXNet: https://mxnet.apache.org/versions/1.9.0/
43. Boyd, E.: Microsoft and Facebook create open ecosystem for AI model interoperability. https://azure.microsoft.com/en-us/blog/microsoft-and-facebook-create-open-ecosystem-for-ai-model-interoperability/
44. ONNX: https://onnx.ai/about.html
45. NVIDIA Launches the World's First Graphics Processing Unit: Geforce 256. https://pressreleases.responsesource.com/news/3992/nvidia-launches-the-world-s-first-graphics-processing-unit-geforce-256/ (1999)
46. Tao, B.: Understand the mobile graphics processing unit. https://embeddedcomputing.com/technology/processing/understand-the-mobile-graphics-processing-unit (2014)
47. Fung, J., Mann, S.: Computer vision signal processing on graphics processing units. In: 2004 IEEE International Conference on Acoustics, Speech, and Signal Processing, pp. V–93. IEEE (2004)
48. CUDA Zone: https://developer.nvidia.com/cuda-zone
49. OpenCL: https://www.khronos.org/opencl/
50. Khan, M., Anisiu, M.-C., Domoszali, L., Iványi, A., Kasa, Z., Pirzada, S., Szécsi, L., Szidarovszky, F., Szirmay-Kalos, L., Vizvári, B.: Algorithms of informatics, vol. III. AnTonCom, Budapest, Hungary (electronic), Mondat Kft. Budapest, Hungary (print) (2013)
51. Li, X., Zhang, G., Huang, H.H., Wang, Z., Zheng, W., Performance analysis of GPU-based convolutional neural networks. In: Proceedings of the International Conference on Parallel Processing, (2016)
52. Lindholm, E., Nickolls, J., Oberman, S., Montrym, J.: NVIDIA Tesla: A unified graphics and computing architecture. IEEE Micro 28, 39–55 (2008)
53. NVIDIA's Next Generation CUDA Compute Architecture: Kepler TM GK110/210. https://www.nvidia.com/content/dam/en-zz/Solutions/Data-Center/tesla-product-literature/NVIDIA-Kepler-GK110-GK210-Architecture-Whitepaper.pdf
54. NVIDIA Tesla V100 GPU Architecture: https://images.nvidia.com/content/volta-architecture/pdf/volta-architecture-whitepaper.pdf
55. Narasiman, V., Shebanow, M., Lee, C.J., Miftakhutdinov, R., Mutlu, O., Patt, Y.N.: Improving GPU performance via large warps and two-level warp scheduling. In: Proceedings of the Annual International Symposium on Microarchitecture. MICRO (2011)
56. Gupta, S., Agrawal, A., Gopalakrishnan, K., Narayanan, P.: Deep learning with limited numerical precision. In: 32nd International Conference on Machine Learning, ICML 2015 (2015)
57. NVIDIA Turing GPU Architecture
58. Morgan, T.P.: Diving Deep into the NVIDIA Ampere GPU Architecture. https://www.nextplatform.com/2020/05/28/diving-deep-into-the-nvidia-ampere-gpu-architecture/ (2020)
59. NVIDIA DGX Station A100 System Architecture: https://images.nvidia.com/aem-dam/Solutions/Data-Center/nvidia-dgx-station-a100-system-architecture-white-paper.pdf
60. Xu, Q., Jeon, H., Annavaram, M.: Graph processing on GPUs: Where are the bottlenecks? In: IISWC 2014 – IEEE International Symposium on Workload Characterization, (2014)
61. NVIDIA GeForce GTX TITAN Z: https://www.techpowerup.com/gpu-specs/geforce-gtx-titan-z.c2575

62. NVIDIA TITAN RTX: https://www.nvidia.com/en-us/deep-learning-ai/products/titan-rtx/
63. Cutress, I.: Intel's 11th Gen Core Tiger Lake SoC Detailed: SuperFin, Willow Cove and Xe-LP. https://www.anandtech.com/show/15971/intels-11th-gen-core-tiger-lake-soc-detailed-superfin-willow-cove-and-xelp/5
64. Mujtaba, H.: AMD Ryzen 3rd Generation 'Mattise' AM4 Desktop CPUs Based on Zen 2 Launching in Mid of 2019 – X570 Platform, 8 Core/16 Thread SKU Demoed and PCIe Gen 4.0 Support. https://wccftech.com/amd-ryzen-3000-zen-2-desktop-am4-processors-launching-mid-2019/ (2019)
65. Wu, H.: Low Precision Inference on GPU. https://developer.download.nvidia.com/video/gputechconf/gtc/2019/presentation/s9659-inference-at-reduced-precision-on-gpus.pdf
66. IEEE Standard for Floating-Point Arithmetic. IEEE Std 754–2019 (Revision of IEEE 754–2008). 1–84 (2019). https://doi.org/10.1109/IEEESTD.2019.8766229
67. Wang, S., Kanwar, P.: BFloat16: The secret to high performance on Cloud TPUs. https://cloud.google.com/blog/products/ai-machine-learning/bfloat16-the-secret-to-high-performance-on-cloud-tpus, (2019)
68. Kalamkar, D., Mudigere, D., Mellempudi, N., Das, D., Banerjee, K., Avancha, S., Vooturi, D.T., Jammalamadaka, N., Huang, J., Yuen, H.: A study of BFLOAT16 for deep learning training. arXiv preprint arXiv:1905.12322. (2019)
69. NVIDIA Ampere Architecture: https://www.nvidia.com/en-us/data-center/ampere-architecture/
70. Wu, S., Li, G., Chen, F., Shi, L.: Training and inference with integers in deep neural networks. arXiv preprint arXiv:1802.04680. (2018)
71. SLI: https://www.nvidia.com/en-gb/geforce/technologies/sli/
72. AMD Crossfire Technology: https://www.amd.com/en/technologies/crossfire
73. Li, S.: Pytorch Distributed Overview. https://pytorch.org/tutorials/beginner/dist_overview.html
74. Distributed training with TensorFlow: https://www.tensorflow.org/guide/distributed_training
75. NVIDIA NCCL: https://developer.nvidia.com/nccl
76. Pal, S., Ebrahimi, E., Zulfiqar, A., Fu, Y., Zhang, V., Migacz, S., Nellans, D., Gupta, P.: Optimizing multi-GPU parallelization strategies for deep learning training. IEEE Micro. **39**, 91 (2019)
77. NVLink and NVSwitch: https://www.nvidia.com/en-us/data-center/nvlink/
78. Lilly, P.: Multi-GPU technology is not quite dead but Nvidia is close to pulling the trigger. https://www.pcgamer.com/multi-gpu-technology-is-not-quite-dead-but-nvidia-is-close-to-pulling-the-trigger/ (2020)
79. Han, S., Mao, H., Dally, W.J.: Deep compression: Compressing deep neural networks with pruning, trained quantization and Huffman coding. In: 4th International Conference on Learning Representations, ICLR 2016 – Conference Track Proceedings, (2016)
80. cuSPARSE: https://docs.nvidia.com/cuda/cusparse/index.html
81. Shi, S., Wang, Q., Chu, X.: Efficient sparse-dense matrix-matrix multiplication on GPUs using the customized sparse storage format. In: Proceedings of the International Conference on Parallel and Distributed Systems – ICPADS, (2020)
82. NVIDIA CUDA-X GPU-Accelerated Libraries: https://developer.nvidia.com/gpu-accelerated-libraries
83. NVIDIA cuDNN: https://developer.nvidia.com/cudnn
84. NVIDIA TensorRT: https://developer.nvidia.com/tensorrt
85. NVIDIA Data Loading Library: https://developer.nvidia.com/dali
86. Holler, M., Tam, S., Castro, H., Benson, R.: Electrically trainable artificial neural network (ETANN) with 10240 "floating gate" synapses. In: IJCNN International Joint Conference on Neural Networks, (1989)
87. Calvin, J., Rogers, S.K., Zahirniak, D.R., Ruck, D.W., Oxley, M.E.: Characterization of the 80170NX (ETANN) chip sigmoidal transfer function for a device Vgain = 3.3 V. In: Applications of Artificial Neural Networks IV, pp. 654–661. International Society for Optics and Photonics (1993)

88. Kern, L.R.: Design and development of a real-time neural processor using the intel 80170nx etann. In: [Proceedings 1992] IJCNN International Joint Conference on Neural Networks, pp. 684–689. IEEE (1992)

89. Perrone, M., Cooper, L.: The Ni1000: high speed parallel VLSI for implementing multilayer perceptrons. Adv. Neural Inf. Proces. Syst. **7**, 747–754 (1994)

90. Almohri, H., Gray, J.S., Alnajjar, H.: A real-time DSP-based optical character recognition system for Isolated Arabic characters using the TI TMS320C6416T. In: The 2008 IAJC-IJME International Conference, (2008)

91. Georgiev, P., Lane, N.D., Rachuri, K.K., Mascolo, C.: Dsp. ear: Leveraging co-processor support for continuous audio sensing on smartphones. In: Proceedings of the 12th ACM Conference on Embedded Network Sensor Systems, pp. 295–309 (2014)

92. Zhou, Z., Chen, X., Li, E., Zeng, L., Luo, K., Zhang, J.: Edge intelligence: Paving the last mile of artificial intelligence with edge computing. Proc. IEEE. **107**, 1738–1762 (2019)

93. Li, Y., Hao, C., Zhang, X., Liu, X., Chen, Y., Xiong, J., Hwu, W., Chen, D.: Edd: Efficient differentiable dnn architecture and implementation co-search for embedded ai solutions. In: 2020 57th ACM/IEEE Design Automation Conference (DAC), pp. 1–6. IEEE (2020)

94. Nvidia Unleashes Tegra System-on-Chip for Handheld Devices: http://piefae.blogspot.com/2012/01/nvidia-unleashes-tegra-system-on-chip.html

95. NVDLA Primer: http://nvdla.org/primer.html

96. Song, J., Cho, Y., Park, J.-S., Jang, J.-W., Lee, S., Song, J.-H., Lee, J.-G., Kang, I.: 7.1 An 11.5 TOPS/W 1024-MAC butterfly structure dual-core sparsity-aware neural processing unit in 8 nm flagship mobile SoC. In: 2019 IEEE International Solid-State Circuits Conference-(ISSCC), pp. 130–132. IEEE (2019)

97. Park, J.-S., Jang, J.-W., Lee, H., Lee, D., Lee, S., Jung, H., Lee, S., Kwon, S., Jeong, K., Song, J.-H.: 9.5 a 6k-mac feature-map-sparsity-aware neural processing unit in 5 nm flagship mobile soc. In: 2021 IEEE International Solid-State Circuits Conference (ISSCC), pp. 152–154. IEEE (2021)

98. Ignatov, A., Timofte, R., Chou, W., Wang, K., Wu, M., Hartley, T., van Gool, L.: Ai benchmark: Running deep neural networks on android smartphones. In: Proceedings of the European Conference on Computer Vision (ECCV) Workshops, p. 0 (2018)

99. Kim, Y.D., Jeong, W., Jung, L., Shin, D., Song, J.G., Song, J., Kwon, H., Lee, J., Jung, J., Kang, M.: 2.4 a 7 nm high-performance and energy-efficient mobile application processor with tri-cluster CPUs and a sparsity-aware NPU. In: 2020 IEEE International Solid-State Circuits Conference-(ISSCC), pp. 48–50. IEEE (2020)

100. Samsung Introduces Game Changing Exynos 2200 Processor With Xclipse GPU Powered by AMD RDNA 2 Architecture: https://news.samsung.com/global/samsung-introduces-game-changing-exynos-2200-processor-with-xclipse-gpu-powered-by-amd-rdna-2-architecture (2022)

101. Kirin 9000: https://www.hisilicon.com/en/products/Kirin/Kirin-flagship-chips/Kirin-9000

102. Park, J., Naumov, M., Basu, P., Deng, S., Kalaiah, A., Khudia, D., Law, J., Malani, P., Malevich, A., Nadathur, S.: Deep learning inference in Facebook data centers: Characterization, performance optimizations and hardware implications. arXiv preprint arXiv:1811.09886 (2018)

103. Wu, Y., Cao, W., Sahin, S., Liu, L.: Experimental characterizations and analysis of deep learning frameworks. In: 2018 IEEE International Conference on Big Data (Big Data), pp. 372–377. IEEE (2018)

104. Wu, C.-J., Brooks, D., Chen, K., Chen, D., Choudhury, S., Dukhan, M., Hazelwood, K., Isaac, E., Jia, Y., Jia, B.: Machine learning at Facebook: Understanding inference at the edge. In: 2019 IEEE International Symposium on High Performance Computer Architecture (HPCA), pp. 331–344. IEEE (2019)

105. Dean, J., Corrado, G., Monga, R., Chen, K., Devin, M., Mao, M., Ranzato, M., Senior, A., Tucker, P., Yang, K.: Large scale distributed deep networks. Adv. Neural Inf. Proces. Syst. **25**, 1223–1231 (2012)

106. Zhang, M., Rajbhandari, S., Wang, W., He, Y.: DeepCPU: Serving RNN-based deep learning models 10x faster. In: 2018 USENIX Annual Technical Conference (USENIX ATC 18), pp. 951–965 (2018)
107. Intel oneAPI Deep Neural Network Library: https://www.intel.com/content/www/us/en/developer/tools/oneapi/onednn.html
108. Arm Compute Library: https://github.com/ARM-software/ComputeLibrary
109. Intel distribution of openVINO toolkit: https://www.intel.com/content/www/us/en/developer/tools/openvino-toolkit/overview.html
110. ARM NN SDK: https://www.arm.com/products/silicon-ip-cpu/ethos/arm-nn
111. Jang, J.-W., Lee, S., Kim, D., Park, H., Ardestani, A.S., Choi, Y., Kim, C., Kim, Y., Yu, H., Abdel-Aziz, H.: Sparsity-aware and re-configurable NPU architecture for Samsung flagship mobile SoC. In: 2021 ACM/IEEE 48th Annual International Symposium on Computer Architecture (ISCA), pp. 15–28. IEEE (2021)
112. Salvator, D.: How Sparsity Adds Umph to AI Inference. https://blogs.nvidia.com/blog/2020/05/14/sparsity-ai-inference/
113. Esmaeilzadeh, H., Sampson, A., Ceze, L., Burger, D.: Neural acceleration for general-purpose approximate programs. In: 2012 45th Annual IEEE/ACM International Symposium on Microarchitecture, pp. 449–460. IEEE (2012)
114. Zhu, M., Zhang, T., Gu, Z., Xie, Y.: Sparse tensor core: Algorithm and hardware co-design for vector-wise sparse neural networks on modern gpus. In: Proceedings of the 52nd Annual IEEE/ACM International Symposium on Microarchitecture, pp. 359–371 (2019)
115. Mei, X., Chu, X.: Dissecting GPU memory hierarchy through microbenchmarking. IEEE Trans. Parallel Distrib. Syst. **28**, 72–86 (2016)
116. Sousa, R., Jung, B., Kwak, J., Frank, M., Araujo, G.: Efficient tensor slicing for multicore NPUs using memory burst modeling. In: 2021 IEEE 33rd International Symposium on Computer Architecture and High Performance Computing (SBAC-PAD), pp. 84–93. IEEE (2021)
117. Kwon, Y., Rhu, M.: A disaggregated memory system for deep learning. IEEE Micro. **39**, 82–90 (2019)
118. Chen, Y., Xie, Y., Song, L., Chen, F., Tang, T.: A survey of accelerator architectures for deep neural networks. Engineering. **6**, 264–274 (2020)
119. Marie, L.: NVIDIA Announces Jetson Xavier NX, World's Smallest Supercomputer for AI at the Edge. https://nvidianews.nvidia.com/news/nvidia-announces-jetson-xavier-nx-worlds-smallest-supercomputer-for-ai-at-the-edge
120. Jetson Xavier NX: https://developer.nvidia.com/embedded/jetson-xavier-nx
121. Jouppi, N.P., Young, C., Patil, N., Patterson, D., Agrawal, G., Bajwa, R., Bates, S., Bhatia, S., Boden, N., Borchers, A.: In-datacenter performance analysis of a tensor processing unit. In: Proceedings of the 44th Annual International Symposium on Computer Architecture, pp. 1–12 (2017)
122. Yazdanbakhsh, A., Seshadri, K., Akin, B., Laudon, J., Narayanaswami, R.: An evaluation of edge tpu accelerators for convolutional neural networks. arXiv preprint arXiv:2102.10423. (2021)
123. Strukov, D.B., Snider, G.S., Stewart, D.R., Williams, R.S.: The missing memristor found. Nature. **453**, 80–83 (2008)
124. Pawlowski, J.T.: Hybrid memory cube (HMC). In: 2011 IEEE Hot chips 23 symposium (HCS), pp. 1–24. IEEE (2011)
125. Chi, P., Li, S., Xu, C., Zhang, T., Zhao, J., Liu, Y., Wang, Y., Xie, Y.: Prime: A novel processing-in-memory architecture for neural network computation in reram-based main memory. ACM SIGARCH Comput. Archit. News. **44**, 27–39 (2016)
126. Shafiee, A., Nag, A., Muralimanohar, N., Balasubramonian, R., Strachan, J.P., Hu, M., Williams, R.S., Srikumar, V.: ISAAC: A convolutional neural network accelerator with in-situ analog arithmetic in crossbars. ACM SIGARCH Comput. Archit. News. **44**, 14–26 (2016)

127. Song, L., Qian, X., Li, H., Chen, Y.: Pipelayer: A pipelined reram-based accelerator for deep learning. In: 2017 IEEE International Symposium on High Performance Computer Architecture (HPCA), pp. 541–552. IEEE (2017)
128. Kim, D., Kung, J., Chai, S., Yalamanchili, S., Mukhopadhyay, S.: Neurocube: A programmable digital neuromorphic architecture with high-density 3D memory. ACM SIGARCH Comput. Archit. News. **44**, 380–392 (2016)
129. Lu, H., Wei, X., Lin, N., Yan, G., Li, X.: Tetris: Re-architecting convolutional neural network computation for machine learning accelerators. In: 2018 IEEE/ACM International Conference on Computer-Aided Design (ICCAD), pp. 1–8. IEEE (2018)
130. Chen, F., Song, L., Chen, Y.: Regan: A pipelined reram-based accelerator for generative adversarial networks. In: 2018 23rd Asia and South Pacific Design Automation Conference (ASP-DAC), pp. 178–183. IEEE (2018)
131. Mobile performance redefined: https://semiconductor.samsung.com/us/processor/mobile-processor/
132. Snapdragon Platforms Your devices deserve Snapdragon: https://www.qualcomm.com/snapdragon
133. Coral: https://coral.ai/
134. MLPerf Benchmarks: https://www.nvidia.com/en-us/data-center/resources/mlperf-benchmarks/
135. v2.0 Results: https://mlcommons.org/en/inference-edge-20/
136. Cloud AI 100: https://www.qualcomm.com/products/technology/processors/cloud-artificial-intelligence/cloud-ai-100
137. NVIDIA A100 Tensor Core GPU: https://www.nvidia.com/en-us/data-center/a100/
138. NVIDIA A30 Tensor Core GPU: https://www.nvidia.com/en-us/data-center/products/a30-gpu/
139. NVIDIA A2 Tensor Core GPU: https://www.nvidia.com/en-us/data-center/products/a2/
140. Mali-G52: https://developer.arm.com/Processors/Mali-G52
141. Jetson Xavier NX Series: https://www.nvidia.com/en-us/autonomous-machines/embedded-systems/jetson-xavier-nx/
142. Jetson AGX Orin: https://www.nvidia.com/en-us/autonomous-machines/embedded-systems/jetson-orin/
143. FuriosaAI's first silicon Warboy marks its successful debut at MLPerf 2021 with top performance: https://www.furiosa.ai/

AI Accelerators for Cloud and Server Applications

Rakesh Shrestha, Rojeena Bajracharya, Ashutosh Mishra, and Shiho Kim

1 Introduction

Artificial intelligence (AI) accelerator is a type of particular hardware accelerator developed to speed the use of computers, in particular artificial neural networks, machine visualization, and machine learning (ML). It has a heterogeneous architecture because this technique enables a given system to accommodate numerous specialized processors to serve specific activities, delivering the computational efficiency required by AI applications. The AI accelerators are based on a design-by optimization concept where the data engineers use an in-built parallelized computing scheme that consumes a massive amount of data and trains themselves through iterative optimization like neural networks. AI accelerators' advantages are energy efficiency, low latency, high computational speed, high scalability, and heterogeneous architecture. The AI accelerator performs hundreds to thousands more efficiently than general CPU machines. It is suitable in the cloud, data centers, and servers, as well as for edge devices, Internet of things (IoT), and numerous sensors consuming low power and inexpensive operation cost. In various modern intelligent applications, ML is commonly used. To facilitate such applications, several hardware platforms are imposed, such as graphics process unit (GPU), field programmable gate arrays (FPGAs), and application-specific integrated circuits

R. Shrestha
RISE Research Institute of Sweden, Västerås, Sweden
e-mail: rakez.shrestha@ri.se

R. Bajracharya
Department of Electronic Engineering, Kyung Hee University, Yongin, South Korea
e-mail: rojeena@khu.ac.kr

A. Mishra (✉) · S. Kim
School of Integrated Technology, Yonsei University, Incheon, South Korea
e-mail: ashutoshmishra@yonsei.ac.kr; shiho@yonsei.ac.kr

© The Author(s), under exclusive license to Springer Nature Switzerland AG 2023
A. Mishra et al. (eds.), *Artificial Intelligence and Hardware Accelerators*,
https://doi.org/10.1007/978-3-031-22170-5_3

95

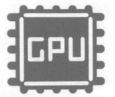

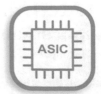

- Domain programmable
- Partially general-purpose
- Parallel workloads
- Power-hungry

- Hardware programmable
- Partially general-purpose
- Any workloads
- High performance
- Low power

- Hard wired
- Special-purpose
- Dedicated workloads
- High performance
- Low power

Fig. 1 Comparison of different hardware architectures

(ASICs). In this chapter, we focus on the AI accelerator used for cloud and servers with an illustration. When using the ML model as a part of a software application, such as filters for pictures or text translations, the inference step will consume lots of time. If the model consumes hundreds of seconds resulting in slow processing outcomes, it will directly affect the user experience; as a result, the users will be reluctant to use such applications. Thus, the total application latency can be minimized, and the inference step can be accelerated by speeding up inference and delivering a smooth application experience. It can be carried out by outsourcing ML model prediction computation to an AI accelerator.

Figure 1 shows a comparison of popular hardware architectures. The GPU is the most commonly adopted accelerator due to its fast calculation speed and compatibility with various techniques. The GPUs have been used as accelerators in hyper-scale data centers for operations such as encryption, networking, gaming, graphics, and machine learning. One example of GPU is NVIDIA GPUs containing compute unified device architecture (CUDA) cores in their chips. They are used for parallel workloads, however, they are power-hungry. The FPGAs are integrated circuit programmable arrays that consist of programmable logic blocks. FPGAs can be general-purpose accelerators that customers can customize for a specific job after production. They necessitate a more flexible development environment, and, as a result, they are often more expensive than GPUs. Furthermore, unlike GPUs, FPGAs do not allow for production-level customization. FPGAs are more powerful than GPUs for computing ML algorithms at a low speed, low power, and cost with high performance. FPGAs have grown in popularity because of their flexibility in hardware acceleration and parallel computing capabilities. FPGAs are often deployed in areas such as autonomous driving systems that utilize image data from cameras mounted on automobiles, parallel computing, and visual recognition/reconstruction techniques such as deep learning networks.

On the other hand, ASIC is considered a dedicated AI accelerator built for a particular application or function and cannot be reprogrammed, unlike FPGAs and GPUs. While compared with FPGAs and GPUs, ASICs are much more efficient because they can be readily customized for specific activities. They are suitable for use cases like trade, gaming, and crypto mining. ASIC design achieves the most efficient energy capacity at the lowest reconfiguration value, making it suitable for specific ML algorithms such as deep convolution neural networks. Some of the custom ASICS are vision processing unit (VPU) and tensor processing units (TPU). The VPUs are custom ASICs used in image processing and computer vision. They consume ultra-low power and are suitable for trained models. TPU is also customized ASICs created by Google to speed up machine learning applications. It is built to run on TensorFlow using several processing primitives known as tensors. Tensors are expansions of vectors and matrices that can have many dimensions. Another excellent example of a specialized processor is Amazon Web Service (AWS) Inferentia, which is also a custom-designed ASIC by AWS for deep-learning inference applications.

Similarly, the Tesla full self-driving (FSD) computer chip is a new System on Chip (SoC) used for very high demanding autonomous vehicle driving workloads. It consists of two neural processing units (NPUs) that act as neural network accelerators and are fully custom-designed that were produced by Tesla engineers. Each NPU chip has a 96 × 96 MAC array with 8-bit multiply and 32-bit add units. Moreover, FSD has a set of 12 ARM Cortex-A72 CPUs and a GPU; however, most of the computational capability is delivered by the NPUs.

The cloud is an ideal location for training as it gathers a massive amount of data from various sources, while edge computing is the ideal option for inference. In the case of cloud computing and data centers, FPGA allows the adoption of customized hardware that may be easily modified based on the requirements of customers and users. This adaptability to requirements qualifies them for cloud computing systems where the architecture may be changed based on requirements at any moment without bearing additional expenses or a lengthy development schedule. FPGAs enable cloud-based ML providers such as Meta's Facebook to construct their data center architecture and train models much quicker while maintaining the same latency limits for end users. The two leading manufacturers of FPGAs are Xilinx and Altera.

ASICs have lower latency and better electrical properties than FPGAs; they may outperform FPGAs in terms of performance. They are used for mining Proof of Work (PoW)–based cryptocurrencies like Bitcoin, Ethereum, etc. The cloud computing providers give incentives to the users that use their platform for efficiently mining cryptocurrency, which is one reason ASICS is popular. In contrast with GPUs and FPGAs, ASICS also give the highest levels of security, power efficiency, and versatility since they can be molded to perform any task that matches their design criteria. They also provided a virtualized environment with enormous flexibility and scalability enabled by a vast infrastructure based on homogeneous and parallel computing platforms.

This chapter will focus on ASIC and FPGA as they are supposed to accelerate the ML/AI tasks in the cloud. Several parts of the content discussed are equally applicable to other forms of acceleration. We also provide various cloud, server, and edge computing AI accelerators. We also discuss the staggering cost of training AI accelerators to boost the performance of cutting-edge models.

2 Background

The AI accelerator is a heterogeneous architecture because it enables a given system to accommodate numerous specialized processors to serve specific activities, delivering the computational efficiency required by AI applications. AI accelerators provide a high computational speed that helps to decrease the latency, which is very useful in life-threatening applications such as in-vehicle advanced driver assistance systems (ADAS). Scalability is a significant concern in parallel processing. However, in the realm of neural networks, AI accelerators increase performance speed almost comparable to the multiple cores used. The AI accelerators are different from the traditional software design, where the software engineers focus on developing an algorithmic approach to solve specific problems and then implementing them in a high-level language. Various AI accelerator designs may have several performance tradeoffs, but they always need a software stack to allow system-level performance; else, the hardware may go unused. As intelligence reaches the edge in many applications, AI accelerators are becoming more distinct. The edge supports a wide range of applications, demanding AI accelerators tailored to particular features like latency, efficient energy, and memory capacity, depending on the demands of the end application. The existing standardized instruction set architectures (ISA) for executing software is unsuitable for AI accelerators. The AI accelerator architecture utilizes and computes big data from ubiquitous connectivity to provide powerful and energy-efficient processing.

There are two different AI accelerator spaces: cloud data centers and edge computing. The data center AI accelerator needs massively scalable computing architectures and requires a large number of memory chips. For such demands, Cerebras developed the wafer-scale engine (WSE) that can produce a large chip for deep-learning networks. The WSE is one type of methodology for accelerating AI applications; however, there are other types of hardware AI accelerators for applications that do not need one large chip. On the contrary, edge computing AI accelerator spaces are located at the edge devices and are more stringent to energy efficiency than centralized computing. The AI accelerators are incorporated into edge SoC devices that provide near-instantaneous results, for instance, interactive smartphone applications or industrial robots. Some manufacturers include Intel Gen 3 Intel® Movidius VPU, NVIDIA Jetson AGX Xavier Series, and other FPGA-based AI accelerator for edge computing. Edge computing can potentially drastically minimize the cost and energy consumption per inference channel. Inference at the edge is appropriate for object identification, automatic license plate recognition, behavior monitoring, facial recognition, and visual inspection applications. However, the google edge TPU performs better than the above-mentioned H/W edge accelerators.

Several other hardware AI accelerators are independent of the WSE approach for a single large chip, such as GPUs, massive multicore scalar processors, and spatial accelerators (e.g., Google TPU). Most of them are individual chips that may be coupled in order of hundreds to form more extensive systems capable of processing substantial neural networks. Coarse-grain reconfigurable architectures (CGRA) is gaining traction in this sector because they can provide compelling tradeoffs among performance and energy efficiency and program flexibility for various networks.

3 Hardware Accelerators in Clouds

Cloud computing service is a competitive market as it provides a complete solution from storage, hardware, networks, interfaces, and services at a reasonable cost. Several commercial cloud service providers (CSPs) provide various services to the customers according to their requirements. These cloud service providers use various AI accelerators, as discussed before, for ML to enhance their performance. Some of the most popular CSPs based on their market share are Amazon Web Service (AWS) by Amazon, Azure by Microsoft, Google Cloud by Google, Aliyun by Alibaba, IBM SmartCloud by IBM, and Tencent Cloud by Tencent. The comparison between these cloud service providers is given in Table 1. The Amazon Elastic Compute Cloud (EC2) web service enables application programs to operate on its native website, that is, Amazon.com. Microsoft Azure was the first CSP introduced in March of 2006.

It enables customers to operate enterprise-level applications with a computing platform, that is, platform as a service (PaaS). Microsoft Azure and Google cloud were established around the same time in 2008, and Google compute engine executes large-scale workloads on virtual servers hosted by Google's infrastructure. Alibaba, established in September 2009, is a CSP that has been gaining popularity recently, delivering cloud computing services to online commerce and its e-commerce ecosystem. Oracle started providing cloud services in 2016. Oracle is an autonomous database that can repair itself. It offers its customers a whole cloud service package from storage and networks to application services and can be interoperable with Azure.

Similarly, IBM started its SmartCloud in October 2017. It is a hybrid cloud whose AI techniques focus on social business tools in the cloud, such as email, calendar, filesharing, etc. Like other cloud services, Tencent is a new and emerging CPS established in January 2018. It provides a wide range of services, including a gaming ecosystem and live-video streaming in the cloud at an affordable cost. Table 1 provides a detailed comparison between these various CSPs based on features such as attributes, market share values, average cost, advantages, and disadvantages.

The CSP, as mentioned above, uses different types of hardware technology as an AI accelerator from several hardware manufacturers. Some of them use more than one type of hardware accelerator from the respective manufacturers so that the customers of the CPS can choose them according to their needs and requirements. In general, the GPUs are manufactured by NVIDIA and AMD, and FPGAs are manufactured by Xilinx and Intel, while Google, Intel, and Samsung manufacture ASICs. In the case of Microsoft Azure, Amazon AWS, or Alibaba Aliyun, the customers can choose between GPUs and FPGAs in their cloud computing systems.

Table 1 Comparison between various cloud service providers

	Amazon AWS	Microsoft Azure	Google Cloud	Alibaba Aliyun	Oracle	IBM	Tencent
Date	March 2006	October 2008	April 2008	September 2009	October 2016	October 2017	January 2018
Attributes	Over 175 cloud services	Over 600 cloud services	Over 100 cloud services	About 200 cloud services	65 cloud services	174 cloud services with AI	Offers 70 cloud services
MarketShare	Approx.32% Largest share in IaaS & Paas	Almost 20% Huge market share in PaaS	Almost 7% Market share in IaaS	Increased to 6% Market share in IaaS and PaaS	Around 2% Low market share in PaaS	Less than 5% Hybrid cloud and AI	Around 2.8% Higher share in IaaS than Oracle.
Location	275 countries	140 countries/54 regions	200 countries/25 regions	23 regions	25 regions	More than 6 regions	27 regions
Cost	Comparatively Expensive	Price similar to AWS	Provides price calculator	25% less expensive than other clouds	Complex cloud pricing models	Dynamic price range	Affordable
Advantages	Suitable for all sizes of business Some services are free for testing Secure	Suitable for Microsoft services Strong player in all use cases including edge computing	Suitable for the data science field and ML Open source products	Suitable for hosting huge application tasks and scalable in terms of users. More reliable than public ISPs	Autonomous database that repairs itself Interoperable with Azure Can migrate existing workloads	Hybrid cloud service Provides more than 20 free cloud products and several free trials 24/7 support service	Performance better than Alibaba cloud Good for the gaming ecosystem and live video streaming in the cloud
Disadvantages	Difficult to switch to other cloud service providers	Low ratio of availability No capacity guarantee	Not yet reached enterprise-level maturity	Limited acceptance outside Asia Need international exposure	Slow adoption by users	Complex cloud platform due to legacy products	Less global exposure despite its popularity as a gaming platform

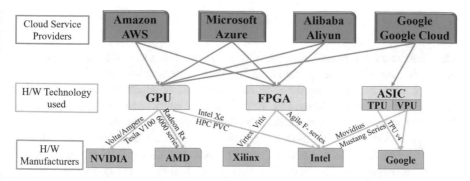

Fig. 2 Mapping of popular hardware accelerators used in cloud service providers

On the other hand, Google Cloud customers have options to choose between GPUs and ASICs. Microsoft Azure and AWS allow customers to choose between GPUs and FPGAs in their cloud computing systems. In recent years, ASICs have increased in popularity, with large technology firms such as Google and Intel. They utilize an ASIC-based system to drive their cloud-computing platform. Because ASICs have lower latency and better electrical properties than FPGAs, they can outperform FPGAs in terms of performance. The mapping of popular hardware accelerators used in cloud service providers and the corresponding hardware manufacturers is given in Fig. 2.

NVIDIA and AMD are the front runners for manufacturing GPUs used as cloud and server accelerators, as shown in Fig. 2. The NVIDIA Volta, Ampere architecture, and Tesla V100 series are computational hardware cards aimed for inferencing and training. AMD's Radeon Rx 6000 series are powerful graphic cards used for gaming that support hardware (HW)-accelerated raytracing. Variable-rate shading was used to achieve realistic graphics and great speed. Xilinx is one of the largest manufacturers of FPGA. It is redefining its fundamental technology by improving its devices to make them ideal for neural network deployment with the help of the adaptive compute acceleration platform (ACAP). An ACAP combines an FPGA's distributed memory, DSP blocks, and logic with a multiprocessor SoC, all connected through a network on chip (NoC) to one or more hardware-adaptable compute engines. However, Google believes that FPGAs are less power efficient than ASICs considering their programming nature. Google's TPU has an instruction set that allows it to work when the TensorFlow program changes or a new type of program is created. Intel manufactures both FPGA and ASIC, that is, VPU. Intel's Agile F-Series with advanced DSP capabilities provides a complete design environment within the PCI Express (PCI-e) form factor. The Agile F-series are optimized for cloud, data center, and edge computing applications. The Mustang-V100 is PCI-e based that uses Intel's Movidius VPU, accelerating the workload burdens of computer vision and AI applications.

4   Hardware Accelerators in Data Centers

Data centers are pretty different from cloud servers. The data centers are on-premise computing facilities, that is, hardware installed at or near the organization's local network sites. At the same time, cloud computing is an off-premise and storing facility accessed via an Internet connection. In other words, the data center is a hardware setup for the organization, while the cloud is a service provider. Three significant attributes determine organizations' decisions to adopt either the in-house data center or cloud computing: their business demands, data security, and equipment expenses. The dedicated data centers provide complete organization control over their hardware and internal data for running several applications and their workloads. The data centers provide more security to the organizational data as they are physically connected with the 'Intranet'. The data centers are provided with the latest security updates and ensure that only organizational personnel with approved credentials can access the data. On the other hand, cloud servers are available to anybody with the appropriate credentials from any location with an 'Internet' connection. They might be sharing the cloud resources with other cloud users. This connection opens up many entry points, all of which must be protected to ensure the security of data transferred to and from these locations. For small organizations, data centers might become very costly as compared with the clouds because data centers require significant time to set up, and maintenance costs will be very high. They might opt for cloud servers as a cheap and best solution for their organizations. Accelerators are utilized in data centers nowadays in a variety of ways. Embedded accelerators (ISA extensions) and general-purpose graphics processing units are frequently utilized. Hardware design has evolved into a critical enabler of AI advancement. Simultaneously, it presents a new set of difficulties to its early adopters for both cloud and edge sectors testing conventional chip technology's performance, power, and space limitations. The designs of AI data center are characterized by huge dimensions, various physical hierarchy levels, and architectures that are synchronous locally and asynchronously worldwide. Several design angles, tremendous variability, ultra-low energy requirements, and heterogeneous integration are necessary for the AI designs such as sensors.

4.1   Design of HW Accelerator for Data Centers

FPGAs are increasingly being utilized in clusters and data centers, and they stand out as one of the best candidates for hardware accelerators in data centers to speed applications in various ways because it provides micro-architectural flexibility, parallel computing, high data bandwidth, and are inexpensive. In data centers, servers can be equipped with FPGAs, either as a dedicated PCI adapter with a potential Ethernet connection, as FPGA directly attached to the CPU, or as an FPGA mounted directly on the motherboard, allowing for customizable hardware acceleration across a pool of server farms. Moreover, sophisticated orchestration can dynamically load the FPGA with the required acceleration, tailor-made for the server workloads. It is possible to speed up jobs, reducing execution time and resource consumption. Examples of different types of FPGAs and reconfigurable logic accelerators currently used are Catapult [1], DySER [2], VEAL [3], CHARM [4], etc. FPGA is also used to accelerate applications such as PageRank, neural networks, belief propagation, etc. A custom accelerator requires various design tools and languages for fast response time. C/C++, Python, CUDA, Java, OpenCL, and other high-level languages are used for custom accelerators. A comprehensive and in-depth understanding of the application is necessary to build effective hardware accelerators for data centers. The applications can be divided into two classes based on the workload characteristics, performance metrics for each application, and resource requirements in terms of Memory, CPU, and network input/output [5]. They are as follows:

4.1.1   Batch Processing Applications

This offline batch processing works with large amounts of data collected and stored in the cloud data centers. The performance metrics in this class are throughput and data completion. Some of the batch cloud applications, along with their examples, are data analytics (e.g., Mahout), in-memory analytics (e.g., Spark Mlib), and graph analytics (e.g., GraphX).

4.1.2   Streaming Processing Applications

This online processing application processes a high volume of streaming data comparably simpler than batch processing applications. The performance metric in this class is delay (i.e., N-th percentile delay). Some batch cloud applications, along with their examples, are data caching (e.g., Memcached), Data Serving (e.g., Cassandra), Media Streaming (e.g., Nginx), Web Search (e.g., PageRank), and Web Serving (e.g., Nginx-Memcached-MySql)

4.2   Design Consideration for HW Accelerators in the Data Center

Hardware accelerators such as FPGA perform specific tasks in the data centers and provide workload acceleration to the CPU in servers at the AI-centric data centers. However, specific design considerations exist for the hardware accelerators used in the data centers. Some of the design considerations are discussed below:

4.2.1 HW Accelerator Architecture

The HW accelerator architecture is one of the crucial considerations in data centers. AI chip engineers should figure out what to accelerate, how to accelerate, and how to operate accelerators on various neural networks, including convolutional neural network (CNN), deep neural network (DNN), and recurrent neural network (RNN), as well as various data sources. AI designers can architect multiple HW accelerators by optimizing the algorithms on a single AI chip.

4.2.2 Programmable HW Accelerators

The technology is evolving quickly, and new types of HW and software algorithms are generated daily. This emergence of new products causes a requirement for the HW accelerators to be used for longer. Thus, flexible HW architectures with programmability capabilities allow AI designers to manage a variety of workloads, neural network topologies, new standards, updated algorithms, etc.

4.2.3 AI Design Ecosystem

The AI designers must carefully examine the accessibility of developer tools for model generation, chip assessment, and proof-of-concept designs. It is also important to investigate which AI frameworks are supported by the hardware accelerators, such as Caffe, PyTorch, TensorFlow, etc. Some companies like Intel and Nvidia attempt to deliver AI processor and software toolkits.

4.2.4 Hardware Accelerator IPs

AI accelerators embedded on chips are also accessible as hardware IP core or intellectual property (IP) core. The IP core is a reusable logic, cell, or IC layout design unit used by ASICs and FPGA designers as building blocks. Several semiconductor companies employ an IP license approach to supply AI accelerators for usage in customized chips.

4.2.5 Energy and Power Efficiency

Recently, energy consumption in data center and cloud computing has been a big issue due to environmental concerns. Since AI chips handle energy-consuming workloads for data center applications like multimedia streaming and massive simulations, HW accelerators in these AI chips must maintain energy efficiency while performing complicated workloads.

5   Heterogeneous Parallel Architectures in Data Centers and Cloud

Key requirements of AI compute are high compute density, high bandwidth access to memory, and chip-to-chip connectivity with high bandwidth, low latency, and low energy. Individual architecture at data centers and cloud applications fail to deliver the critical requirements of AI computing. Heterogeneous integration provides high-performance computing to meet these requirements. High-performance computing requires a heterogeneous architecture to accelerate the computation efficiency of the computer [6]. Such architectures utilize multiple hardware working in parallel for accelerating computation performance. For server and cloud applications, recent AI accelerators adopt popular architectures such as heterogeneous computing architectures and/or hybrid computing architectures [7].

5.1   Heterogeneous Computing Architectures in Data Centers and Cloud

Emerging and future computing systems are heading toward 'extreme heterogeneity' [8]. Figure 3 represents the ongoing trends toward the extreme heterogeneity of computing architectures. They incorporate more than one hardware architecture on a single chip, which is why such architectures are called heterogeneous computing architectures. Heterogeneous integration combinations of popular architectures such as CPUs, GPUs, FPGAs, ASICs, and AI accelerators based on emerging memories are needed to enable an upward trajectory for AI system performance [9].

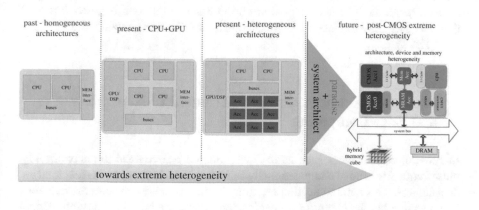

Fig. 3 Trends toward extreme heterogeneity of computing architectures

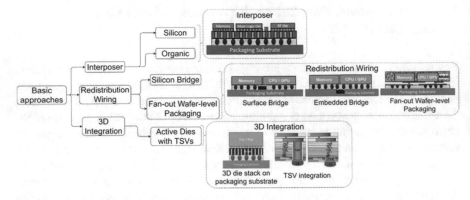

Fig. 4 Heterogeneous integration options

Advantages of such integration are:

- Increase performance
- Decrease data communication
- Different devices for different roles
- Increase flexibility and reliability
- Increase energy efficiency
- Cheaper flop

Basic heterogeneous integration options are represented in Fig. 4. There are three primary options: interposer, redistribution wiring, and three-dimensional (3D) integration [9]. The interposer option has two further divisions; the silicon interposer and the organic interposer. The redistribution wiring option can be either a silicon bridge with back end of line (BEOL) wiring or a fan-out wafer-level packaging (FOWLP) option. The 3D integration option has several subcategories, but in all these, the key common factor is that there are active dies that contain through-silicon vias (TSVs).

6 Hardware Accelerators for Distributed In-Network and Edge Computing

In general, edge devices such as IoT with limited resources at the network edge must frequently offload computationally intensive tasks to more powerful cloud and data centers. This communication requirement incurs delays because the data must be transmitted to the cloud for offloading, preprocessing, training, and inferencing. Then the results are returned to the IoT devices. A viable solution to this issue is to offload data closer to the edge platforms, such as edge devices with fast processing capabilities (e.g., cloudlets, fog, or in-network computing devices). These intermediate and small-scale data centers deployed closer to the data sources,

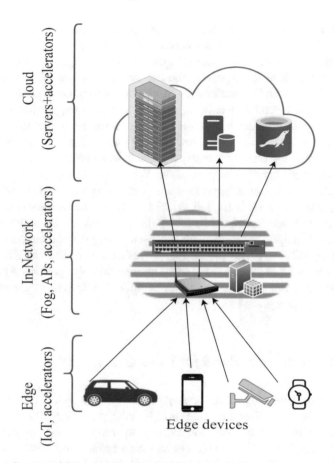

Fig. 5 Hierarchy among cloud, in-network, and edge computing

or edge devices provide cloud-like services with low latencies. Figure 5 shows the hierarchy among cloud, in-network, and edge computing [10].

However, this solution is still subjected to communication delays due to software network stack, PCIe interconnection, limited resources, etc. Recently, IoT evolved from available devices that send data to the cloud for analysis to smart gadgets that can do complex training and inference independently. This section discusses how the HW accelerators can be used on distributed in-network and near-edge computing.

6.1 HW Accelerator Model for In-Network Computing

The cloudlets and in-network infrastructures (like bridges, switches, and gateways) connect the edge devices and perform data processing. They are computationally capable of training and inferencing near-the-edge devices. In-network computing is an emerging example of conventional centralized cloud computing where the computation is distributed in the networking infrastructure. As such, AI architects use AI workloads on various input data types such as images, audio, texts, multimedia, and sensor data for improving the decision process based on deep learning frameworks (like Torch, TensorFlow, Caffe, etc.) and neural networks (like recurrent neural networks). Several AI workloads need a huge memory, parallel computation, and low latency. An efficient and optimized AI platform is required to overcome the computational resources to provide low latency and accurate decisions. Thus, the in-network computing platforms satisfy these requirements considering the weight, size, and power consumption requirements. When developing an AI platform, the architects should employ a heterogeneous computing architecture with various core types, such as CPU, GPU, FPGA, and ASIC, in the in-network computing devices that can boost the performance of the applications. These heterogeneous core devices are helpful for in-network processing because of their flexibility, high processing capabilities, packet processing, and other network services. Moreover, it can reduce the data and execution latency. The intention is to perform AI workloads on the best-suited core, resulting in quicker computation, lower power consumption, and lower delay for a given task.

6.2 HW Accelerator Model for Edge Computing

IoT applications frequently utilize intelligent gateways to collect information from sensor nodes over Wi-Fi or Bluetooth and execute processing. The advancement in hardware lowers the computation latency by bringing the computation near the edge of the devices. This edge computation reduces network traverse time; however, the data transmission latency becomes a significant issue and a possible constraint for IoT applications that use offloaded computing. The computations near the edge devices will introduce slight communication delays that will eventually increase the total delay of the application. One of the possible solutions is to use edge computing. Edge computing refers to data processing at a node or data source directly connected to it, such as a cluster head, IoT gateways, or IoT devices [11]. The main reason for using HW accelerators in edge computing is that contemporary IoT devices can independently do complex inferencing and pattern matching. Several advantages of executing AI algorithms locally on an intelligent device include low latency, improved security, better mobility, fewer hops, and low communication cost. The edge node acceleration can be achieved with ASIC accelerators using TPU/VPU that are optimized for AI inference tasks based on computer vision and deep learning. One example is Google's Pixel 6 pro that uses TPU as an AI accelerator. Similarly, FPGAs use SoC platforms like the Xilinx Zynq, which tightly connect an FPGA with an Arm CPU. Software operating on the processing system controls and communicates with the FPGA fabric used in industrial applications. One such example is ADLINK [12]. ADLINK offers flexible heterogeneous embedded computing solutions (such as GPU and

VPU acceleration boards) and deep learning profiling to optimize AI-powered edge devices' performance in AI training and inferencing applications. In addition, ADLINK provides a high-performance capability for processing data for deep learning inferencing, autonomous learning, and pattern-matching.

7 Infrastructure for Deploying FPGAs

FPGA chips are the emerging cloud power-efficient acceleration trend [13]. Accelerators have also been used to improve and optimize latency-sensitive applications. Several commercial vendors offer FPGAs in their cloud platforms. Recently, several areas, including memory system optimization, management of network-accessible FPGAs, FPGA-IP cores for data center acceleration, etc., are the key research topics of FPGAs in data centers. These research topics also focus on the latency and performance tradeoffs in FPGA-based acceleration. In recent progress, AWS has initiated the elastic compute cloud (EC2) F1 instances to deploy FPGAs on a small scale in the cloud [14]. EC2 F1 instances are custom hardware accelerators using FPGA. They are easy to program, simulate, debug, and compile, including an FPGA developer, and support hardware-level development on the cloud. The F1 instance accelerations were successfully applied to genomics, search, data analytics, network security, electronic design automation (EDA), image, video processing, etc. Recently, Xilinx, acquired by advanced micro devices (AMD), developed the Xilinx® Alveo™ U55C data center accelerator card [15]. It is an FPGA accelerator card for data centers and high-performance computing (HPC) systems. It uses the 16 nm XCU55 UltraScale+ FPGA system-on-chip with 16GBs of high-speed, high bandwidth memory (HBM2). Likewise, there are several other industrial deployments of FPGA-based accelerators in the cloud and data centers. Bobda et al. have presented a classification of FPGA acceleration in data centers and cloud [16]. Major players are Amazon, Microsoft, Baidu, Alibaba, Huawei, Tencent, etc., as given in Table 2.

Bobda et al. have introduced the commonly used taxonomies in FPGA architectures, as shown in Fig. 6.

(a) Bump-in-the-wire (large-scale compute, network, and storage acceleration)
(b) Coprocessor (local compute acceleration)
(c) Storage attached (local storage acceleration)
(d) Back-end (ultra-low latency, rack-scale FPGA-FPGA communication)
(e) Smart network interface card (NIC) (local network acceleration)
(f) Network HW (flexible routing/switching protocols)
(g) Local cluster (multi-accelerator system)
(h) Shared memory (cache-coherent acceleration)

According to Bobda et al. [16], multiple factors govern the trends in the FPGA-based architectures in the cloud. These are modularity, power-usage-effectiveness (PUE), performance, resilience, scalability, security, total-cost-of-ownership (TCO),

Table 2 Production group *vs.* research group cloud architecture based on different categories

	Company	Board type	Network connectivity	Intra-node connectivity	Use case
Production group	Amazon AWS F1	Custom	Nome	FPGA, CPU Storage	Customer
	Microsoft Catapult v2	Custom	Main	CPU, ASIC Storage	AAAS, Provider
	Alibaba	Custom	No	FPGA. CPU. Storage	Customer, Provider
	Baidu	Custom	No	CPU Storage	Customer, Provider
	Tencent	Off- the shelf	No	CPL Storage	Customer
	Huawei	Custom	No	FPGA, CPU, Storage	Customer, Provider
Research group	IBM CloudFPGA	Custom	Main	Storage	Research
	Microsoft Catapult	Custom	Secondary	CPU, Storage	Research
	Enzian	Custom	Secondary	CPU, Storage	Research
	Cygnus	Out of the box	Secondary	FPGA, CPU, GPU, Storage	Research
	Novo-C	Out of the box	No	FPGA, CPU, Storage	Research
	Novo-G#	Out of the box	Secondary	FPGA, CPU, Storage	Research
	IBM Power8+ CAPI	Out of the box	No	FPGA, CPU, GPU, Storage	Research
	IBM SuperVessel	Out of the box	No	CPU, Storage	Research

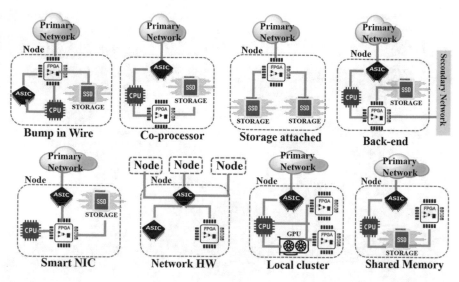

Fig. 6 Common FPGA Architectures. (Figures adapted from Ref. [16])

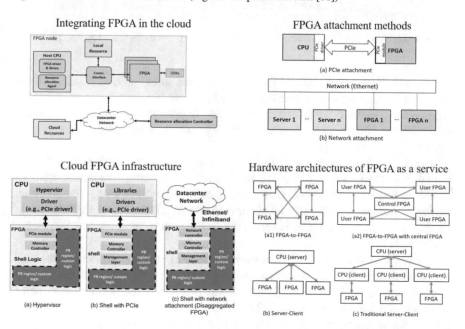

Fig. 7 Infrastructures of FPGA as a service in the cloud. (Figures imported from Ref. [17])

etc. Qassem et al. have surveyed various infrastructures and frameworks using FPGA-based hardware acceleration in the cloud for infrastructures of FPGA as a service in the cloud, as shown in Fig. 7 [17].

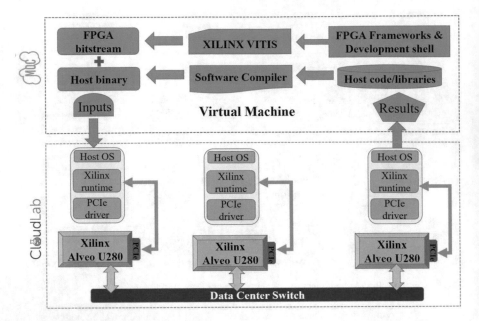

Fig. 8 Open cloud testbed for research in cloud platforms. (Figure adapted from Ref. [20])

There requires a coexistence of the hardware and software in an efficient FPGA-based accelerator in cloud applications. The hardware framework is required for the high-performance and flexible usage of FPGA. In contrast, the software framework facilitates the utilization of other and FPGA accelerators in a heterogeneous cloud computing architecture [18].

The open cloud testbed (OCT) is a project funded by US National Science Foundation (NSF) [19]. It provides a testbed for cloud platform research, including hardware and software [19–21]. It aims to bring together cloud researchers and hardware innovators. The CloudLab and the Mass Open Cloud (MOC) are the leading research labs of OCT. Figure 8 depicts the open cloud testbed for research in cloud platforms.

Asiatici et al. present an efficient on-chip memory system to minimize the number of pipeline stalls [22]. It leads to a bandwidth advantage that can significantly speed up latency-insensitive applications. The increase in the non-blocking caches increases the DRAM bandwidth. Therefore, a miss-optimized memory system elevates the speedup and reduces the area for a data center FPGA system. The use of FPGA in high-performance computing (HPC) and computational flow dynamics has been investigated by Hogervorst et al. in [23]. They have studied the usage of FPGA in the context of reconfigurable hardware with the increase in on-chip memory size, an increment in the number of logic cells, and the integration of high-bandwidth memories onboard. Alonso et al. have presented the scaling performance of DNN inference in FPGA Clouds [24]. They have involved an automatic partitioning technique to maximize the performance and scalability of FPGA-based pipeline data flow of DNN inference accelerators. They have distinguished the two DNN inference acceleration architectures on FPGA and they are matrix of processing engines (MPE) architecture and a feed-forward data flow (DF) architecture. The MPE architectures are systolic array matrix multipliers. It executes instructions on fixed precision weights and activations. In dataflow accelerator (DFA) architecture, the DNN computational graph is converted into a pipeline of FPGA-accelerated operations. It stores parameters such as weights and biases needed for computation in on-chip memory (OCM).

8   Infrastructure for Deploying ASIC

This section will discuss a specific AI accelerator, that is, Google TPU accelerators, which is an ASIC. It is designed to operate on TensorFlow and uses numerous processing primitives known as tensors. The tensor is a mathematical term, a generalized matrix that maps geometric vectors, scalars, and similar objects in a multilinear fashion to a subsequent tensor. It is a multidimensional expansion of vectors and matrices. In the neural network (NN) inference, each neuron undergoes calculations like multiplying the input data (x) by the weights (w) to express the signal strength, adding the result to represent the condition of the neuron, and using an activation function (f) to adjust artificial neuron activity. These matrix multiplications require a trained model's most intensive and computationally expensive tasks. Each prediction involves a series of complex processing. As a result, multiplying and extracting slices from arrays consumes a significant amount of CPU time and memory. Thus, TPUs were created to alleviate these particular workloads. One of the most famous ASICs revolutionary innovations is Google's custom-developed ASICs called the Google TPU. TPU is an ASIC specifically designed to accelerate machine learning workloads very efficiently.

8.1   Tensor Processing Unit (TPU) Accelerators

TPU is Google's exclusive product and was developed by Google to accelerate ML applications. One TPU board consists of four TPU chips, and each chip consists of two TPU cores. Each TPU core has vector, scalar, and matrix multiplication units. The resources offered by a TPU core differ depending on the versions. There are three versions of Google TPU processing. The TPU1 architecture is designed to support inference only, while the TPU2 architecture is a better version than TPU v1, allowing both training and inference. Google's second-generation Maxwell TPU2 supports 64-bit, 80 teraflops, and provides high-energy efficiency, that is, more than nine times compared to common CPUs. It consists of 8 GiB of HBM for each TPU core, a single matrix multiplier unit (MXU) for each TPU core, and up to 512 total TPU cores. It provides data throughput 2–3 times greater than the first generation of TPUs employed in Google's data centers. The range also provides particular support for deep learning inference. Google claimed that their second-generation TPU can do inferences at 4500 images/second (for ResNet-50), a task requiring 16 high-end Nvidia K80 GPUs. Google also produced an advanced version of TPU2 called TPU3 [25]. TPU3 configurations give significant performance benefits per core for compute-bound models than TPU2. TPU3 configurations can run new models with batch sizes that were impossible with TPU2 configurations. TPU3 may, for example, enable deeper ResNets and bigger images using RetinaNet. It consists of 16 GiB of HBM for each TPU core, doubles MXU for each TPU core, and up to 2048 total TPU cores. Google also claimed that its 32 teraflops variation of the new TPU design outperforms first-generation TPUs by six times. Google TPU may be rented through Google Cloud Services for approx. $1.35 per machine per hour. Several companies are using TPU in their data centers, such as Google's Alphabet, Baidu, and Alibaba. The TPU was designed to function as a coprocessor on the PCIe I/O bus, enabling it to be plugged into servers at data centers in the same way that a GPU would. Furthermore, rather than collecting TPU instructions, the host server delivers them to it for execution to simplify hardware architecture and debugging. The objective was to execute whole inference models on the TPU to minimize interaction time with the host CPU while being flexible enough to meet the demands of the neural networks [26]. TPUs solve the issue of neural network training and prediction in various ways to make it easier, such as quantization, focusing on inference mathematics, parallel processing, and a systolic array. The main component of TPU is the MXU that consists of 256 by 256 (i.e., 65,536) MACs to perform 8-bit mathematical operations. The MXU gets its input from the weighted FIFO and unified buffer (UB). The UB works as registers or local activation storage with 24 MB of SRAM and generates 16-bit output aggregated in the 4 MiB of 32-bit accumulators. The activation units, which are hardwired activation functions, execute a non-linear function on the accumulators, and the output goes to the UB, as shown in Fig. 9. They are managed by a set of high-level instructions that concentrate on the essential calculations necessary for NN inference. TensorFlow graph API requests are translated into TPU instructions using a custom compiler and software stack. A custom compiler and software layer transform TensorFlow graph API calls into TPU operations. The MXU in the TPU is designed to operate hundreds and thousands of matrix operations that assist in parallel processing. The parallel processing makes the NN operation faster. In the TPU, MXU employs systolic array and execution systolic operation to conserve energy by minimizing UB reads and writes because accessing a big SRAM requires far more power than arithmetic operations.

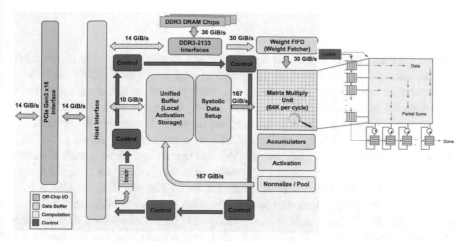

Fig. 9 The architectural view of the TPU accelerator. (Figure imported from Ref. [26])

8.2 Cloud TPU

Google created the TPUs for ML in 2017, and after 1 year, they moved the TPUs to cloud computing and opened them for commercial use. Data processing and immense computing powers are increasingly shifting to the cloud. As mentioned previously, cloud computing is scalable and flexible because it can re-utilize the same standard servers stacked in large quantities and add more servers to increase the computing capability and scalability. However, there are some overheads and hindrances to this flexibility and scalability as it requires orchestration and virtualization of hardware accelerators in cloud data centers to move the task and data smoothly.

Cloud TPU is ideally suited for programs that spend a significant portion of their processing time conducting matrix multiplications. The Cloud TPUs are extremely fast regarding dense vector and matrix calculations. One Cloud TPU device comprises 4 chips with 2 TPU cores each. As a result, a program should leverage each of the 8 cores to get the most out of Cloud TPU. The data transfer between the Cloud TPU and host memory is slow compared to the calculation speed. In contrast, the PCIe bus is significantly slower than the Cloud TPU connection and on-chip HBM. To speed up the process, the Cloud TPU programming model is designed to run most of the training tasks on the TPU. With the help of the TPU programming model, the TensorFlow server, which runs on the host computer (i.e., CPU connected to the Cloud TPU hardware), retrieves and preprocesses data before feeding it to the Cloud TPU hardware. In addition, data parallelism can be achieved when the Cloud TPU cores execute programs residing in their corresponding HBM synchronously.

8.3 Edge TPU

Lately, ML models trained in the clouds require to run inferencing at the edge devices such as IoT. IoT devices like sensors and intelligent devices aggregate instantaneous data, make smart decisions, and provide a response, as well as communicate with other devices or the cloud. Google introduced Edge TPU coprocessor to speed up the ML inferencing on these low-powered IoT devices. Thus, a single Edge TPU can perform 2 trillion operations per second (2 TOPS), using only 1 watt of power. The Edge TPU can run state-of-the-art mobile vision models such as MobileNet V2 with approximately 400 frames per second energy-efficiently. The low-power AI accelerator improves Cloud TPU and Cloud IoT, which ultimately provides an end-to-end platform (i.e., edge-to-cloud, hardware, and software) for AI-based solutions. The Edge TPU has various form factors, such as a single-board machine, a PCIe/M.2 card, and a surface-mounted module for custom prototype and production equipment. The Edge TPU also assists different model architectures built with TensorFlow, comprising models built with Keras. Moreover, it supports simultaneous inferencing for applications that run multiple models on a single Edge TPU by co-compiling the models on the shared Edge TPU scratchpad memory. If multiple Edge TPUs are in the system, one may enhance the performance by allocating and running each model to a specific Edge TPU. Model pipelining can be done in Edge TPUs that enable the execution of various portions of the same model on different Edge TPUs. This Edge TPU is suitable for applications that demand rapid throughput and performance. Despite Edge TPU, which is meant particularly for inferencing, one may use it to speed up transfer learning with a pre-trained model based on Python API that executes the backbone of the model on the Edge TPU during the training process.

9 SOTA Architectures for Cloud and Edge

The cloud is an excellent place to train since it gives access to massive amounts of data from different servers. The more data an AI application considers during training, the better the result. Furthermore, the cloud can save money by allowing expensive hardware such as GPUs to train multiple AI models simultaneously. Capacity is not a problem because training occurs on each model on an as-needed basis. AI algorithms that use inference handle less input but must respond quickly. For instance, when a self-driving car sees an obstacle on the road, it does not have time to transfer pictures to the cloud for processing. Similarly, medical apps that analyze critically ill patients do not have the same flexibility when analyzing brain scans after a hemorrhage. Thus, edge computing or in-device computing is the ideal option for inference. In this section, we will discuss the state-of-the-art architectures for ASIC-based Cloud and Edge computing regarding TPU.

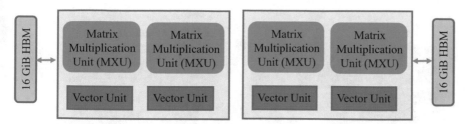

Fig. 10 TPU v3 architecture and its processor layout

9.1 Advances in Cloud and Edge Accelerator

Google has developed three TPU versions since its inception. In this section, we will focus on the latest TPU v3 architecture and its processor layout as given in Fig. 10. There are 4 TPU chipsets, each consisting of 2 cores in the TPU v3 board. Each TPU core has 16 GiB of HBM as well as two MXUs on each core. The TPU v3 Pod has 2048 TPU cores and 32 TiB memory [27]. The improved FLOPS per core and memory capacity in the TPU v3 configuration can help the models perform better. Compared with TPU v2, the TPU v3 configurations can operate on new models with larger batch sizes that TPU v2 cannot handle. TPU v3 enables deeper ResNets and bigger images using RetinaNet.

There are two types of configurations available for both TPU v2 and TPU v3 devices at the Google data center. While creating the TPU node, it is necessary to specify the TPU type. The first configuration type is single-device TPUs that are standalone TPU devices (i.e., connected only to the single device) with no dedicated high-speed network linked to other TPU devices. The chips in single-device TPUs are linked on the device; thus, connectivity between chips does not require the host CPU or host networking resources. The single-device TPU is not part of a TPU Pod configuration and does not occupy any part of a TPU Pod.

Similarly, the second configuration type is the TPU pod configuration, which is a high-speed computer with multiple TPU devices with up to 2048 TPU cores that can distribute the processing workload over multiple TPU boards. TPUs were created to be scaled up to a TPU Pod, which is interconnected through a dedicated high-speed network link that avoids the communication delays of going through the CPU hosts. ML tasks are distributed over all the TPU devices by the hosts in the TPU node. The TPU chips in a TPU Pod are connected to the device, so interaction between chips does not need host CPU or host network devices. This architectural design results in a high bisection bandwidth, the bandwidth obtainable between the two halves of the TPU Pod in the worse split. The TPU v3 Pod may be configured with up to 256 devices for a maximum of 2048 TPU v3 cores and 32 TiB of TPU memory, as shown in Fig. 11. In reality, the Cloud TPU software stack simplifies the process of creating, executing, and feeding TPU Cloud applications [28].

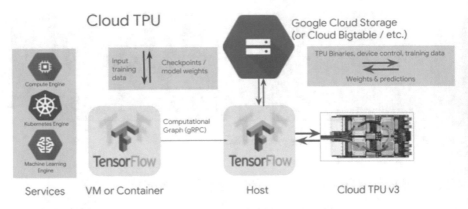

Fig. 11 Cloud TPU V3 configuration. (Figure imported from Ref. [28])

9.1.1 Cloud TPU System Architecture

High-level Tensorflow APIs simplify the execution of replicated network models on Cloud TPU hardware. TensorFlow applications can connect TPU nodes through Google Cloud containers, instances, or services. The software program requires access to the TPU node over the Virtual Private Cloud (VPC) network. A VPC network is a virtual version of a real network created within Google's production network utilizing Andromeda. When users start their applications, TensorFlow builds a computation graph and transmits it to the TPU node through google Remote Procedure Call (gRPC). A gRPC is an open-source, universal RPC framework developed by Google that can operate in any environment. It can seamlessly link services in and across data centers with plug-and-play support for load balancing, tracing, status checking, and authorization. The TPU type users choose for the TPU node defines the number of devices available for the particular workload. The TPU node quickly generates the computation graph and transmits the program binary to one or more TPU devices for processing. Input data to the model are frequently saved in cloud storage. The inputs are streamed to one or more TPU devices for utilization by the TPU node. The Cloud TPU system architecture is depicted in the block diagram in Fig. 12, which includes the NN model, TPU Estimator, TensorFlow client, TensorFlow server, and accelerated linear algebra (XLA) compiler. TPU Estimators are a bundle of high-level APIs necessary while writing an NN model using Cloud TPU. It expands on estimators that simplify NN models for Cloud TPU and extract maximum TPU performance. The TPU Estimator transforms the user programs into TensorFlow operations. A TensorFlow client converts the TensorFlow operations into a computational graph, then used for communication with the TensorFlow server. TensorFlow servers are hosted on Cloud TPU servers. The server executes certain operations when it gets a computational graph from the TensorFlow client. Such operations are loading inputs from cloud storage, partitioning the graphs so they can run on Cloud TPUs, generating XLA operations, and invoking the XLA compiler. The XLA compiler

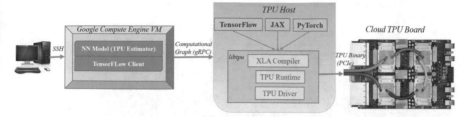

Fig. 12 Cloud TPU system architecture. (Figure adapted from Ref. [27])

is a just-in-time compiler that accepts High-Level Optimizer (HLO) instructions provided by the TensorFlow server as inputs. XLA creates binary code for Cloud TPU processing, comprising data orchestration from on-chip memory to hardware execution units and inter-chip connectivity. The resulting binary is loaded into Cloud TPU through a PCIe connection between the Cloud TPU server and the Cloud TPU and then executed.

9.1.2 Cloud TPU VM Architecture

Based on the method of interaction with the TPU host located at the Google data center, there are two types of Cloud TPU virtual machine (VM) architectures, that is, TPU nodes and TPU VMs. The architecture of conventional TPU node requires users to create one or more VMs that let the users connect with the TPU host machines of the Cloud TPU utilizing gRPC remotely, as shown in Fig. 12. The TPU VM architecture is flexible and simple as compared to the TPU nodes architecture. In Cloud TPU VM architecture, the TPU VMs run on TPU host computers, which are connected directly by Secure Shell (SSH) to Google Compute Engine VM running on the TPU host, which is then connected to TPU accelerators. This feature helps the developers obtain performance gains and cost benefits because the codes no longer require making round trips across data center networks to reach the TPUs. The node TPU architecture with user virtual machine and host is shown in Fig. 13. The user does not have access to the TPU host. Similarly, Cloud TPU virtual machine is shown in Fig. 14. The Cloud TPU VM gives users direct access to TPU host machines, allowing them to develop and deploy TensorFlow, PyTorch, and JAX on Cloud TPUs. Cloud TPU VMs allow users to set their interactive development platform on each TPU host machine rather than accessing Cloud TPUs remotely over the network.

Some of the features of Cloud TPU VMs are as follows:

- With Cloud TPU VMs, the users can process the data directly on the Cloud TPU host by removing the requirement of extra compute engine VMs, as was in the case of Cloud TPU nodes.
- The developer enjoys direct access to debug logs and error messages from the TPU compiler and runtime.

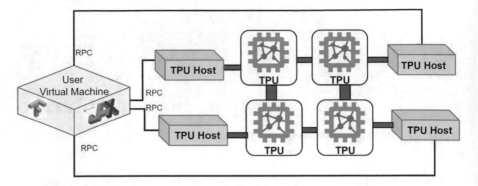

Fig. 13 TPU node architecture. (Figure adapted from Ref. [27])

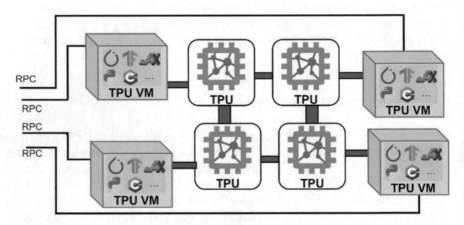

Fig. 14 Cloud TPU VM. (Figure adapted from Ref. [27])

- Ability to perform customized operations in the input pipeline, use local storage space, and integrate Cloud TPUs into user's research/project.
- Use a single TPU VM to build and develop ML models, then extend it on a Cloud TPU to take full advantage of the super-fast TPU connection.
- In addition to supported frameworks like TensorFlow, PyTorch, and JAX integration on Cloud TPU, developers can write their integrations using a new '*libtpu*' shared library on the VM for low-level access to the TPU.

9.2   Staggering Cost of Training SOTA AI Models

While it is exciting to see AI researchers boosting the performance of cutting-edge models to greater levels, the expenses of such procedures are also soaring. It would take 60 hrs to train the XLNet-Large model running on 512 TPU v3 chips (or 128

TPU v3 devices) for 500 K steps with an Adam optimizer, TPU linear learning rate decay with a batch size of 2048, as discussed in [29]. According to the Google researcher, the approximate cost is about $61,440. The detailed calculation is as follows: the cost of running a Cloud TPU v3 'device', which includes four separate integrated chips on the Google cloud, is $8/hour. For a 128 TPU v3 device (i.e., 512 TPU v3 chips), the cost would be 128 (device) × 60 hrs × $8 = $61,440, which seemed pretty expensive at that time. The high cost is due to the huge training model consisting of 24 transformer blocks, 1024 hidden units in each layer, and 16 attention heads [30].

BERT (bidirectional encoder representations of transformers) is a popular bidirectional transformer model that redefines state-of-the-art for eleven natural language processing tasks. The pre-training of BERT-large was carried out on 16 Cloud TPU devices (i.e., 64 total TPU chips because TPU v2 consists of 4 TPU chips each with two cores) that took almost 4 days (i.e., 96 hrs) to complete. According to Google research, the approximate cost for Cloud TPU v2 use for training was 16 TPU v2 devices × 96 hrs × $4.5 = $6912.

Similarly, the generative pre-trained transformer (GPT) series created by OpenAI are huge linguistic models that can generate realistic text paragraphs. Despite lacking task-specific training data, the model performed well across various linguistic tasks. The cost of training the GPT-2 in 128 Google Cloud TPU v3 chips (i.e., 32 TPU v3 devices) per hour was calculated as 32 total TPU chips × $8/hr = $256. However, OpenAI has not mentioned the training time for their model. We can see that AI research has increased along with the computation cost. AI research is a commodity where big companies like Facebook, Google, Baidu, etc., are drawing billions of dollars in revenue gains. Because of this, it is tough to enter this market for new researchers with low computational resources. Moreover, the AI research is centralized and non-collaborative, where the model data are isolated and cannot be shared to speed up the training process. As a result, several researchers cannot contribute to creating massive neural networks because the required experiments are incredibly costly. If AI research is decentralized and collaborative and shares computation and information, then the future of AI is very bright.

With the continued advancement of algorithms, processing devices, and engineering initiatives, the cost of computing used for training models is assumed to decrease considerably in the coming years.

10 Security and Privacy Issues

Hardware accelerator is a new field; however, it also has a vulnerability, and security and privacy issues exist. Some of the vulnerabilities in HW accelerators exist in its software platform. Adversaries may use the accelerator controller's vulnerabilities to conduct a variety of exploits such as SoC architecture. The adversaries might tamper with the system bus to inject malicious instructions that cause the accelerators to run without authorization. It is necessary to provide secure storage of sensitive information with necessary encapsulation; otherwise, the attackers can access internal and sensitive information through an output port or shared memory space. Even though the HW accelerator's design is safe, the physical attacks that take advantage of the implementation's flaws can threaten the accelerator's operation. Side-channel attacks can extract confidential information from high-end cloud servers and embedded devices [31]. The HW accelerators can also be used in autonomous vehicles, but some security and privacy issues need to be considered before implementing HW accelerators [32–34].

Apart from the end-user hardware vulnerabilities, the HW accelerators should be protected from reverse engineering, including the insertion of hardware Trojans and unauthorized copying. Reverse engineering causes all the assets to be accessible to unauthorized entities. In contrast, unauthorized copies might result in financial damage incurring billions of euros for the IP and technology providers [31].

Some of the tentative solutions for securing HW accelerators are as follows:

- Dynamic data flow monitoring: The system can be protected from malicious data injection such as unauthorized code execution or buffer overflow by utilizing hardware logic for dynamic data flow monitoring. It protects by tracking and marking insecure data during the code execution process.
- Side channel prevention: Enforcing constant time computation to defeat timing attacks or adding randomness to prevent the attacker from extracting secret keys using power analysis must be used to protect security protocols from side-channel attacks. One crucial step in a security-aware design pipeline is to identify the security primitives that need to be secured, then automatically construct side-channel countermeasures specific to the target platform.
- Reverse engineering protection: One way to protect the HW accelerator from reverse engineering is to embed a secret key or change the functionality of the HW accelerator by including additional logic and connection during data path. This protection makes it hard for the attackers to complete the reverse engineering, which increases the cost for the attacker and minimizes the probability of inserting the HW trojans.

In the case of an FPGA multi-tenant setting, privacy remains an issue as the information can be leaked through cross-wire talk, or sensors can be used to leak cryptographic keys or machine learning (ML) models secretly [35].

Security and privacy are still problematic since unauthorized individuals can design new sensing circuits using low-level access to FPGA hardware, which could lead to data leaks. Existing FPGA-accelerated cloud infrastructures essentially use current FPGA boards in servers and make them available to the end customer. Furthermore, security remains a challenge because low-level access to FPGA hardware allows suspicious users to create new sensor circuits that may significantly contribute to data leaks. The FPGA can be used in drone monitoring and detection systems based on machine learning methods like CNN [36–38]. In the case of cloud computing, FPGAs cause vulnerability to other cloud users or infrastructure as it provides a way for malicious attackers to spy on the data centers or cloud.

11 Summary

This chapter discussed AI accelerators and their capability for high-performance computing and parallel processing dedicated to efficiently processing AI-based workloads like neural networks. Cognitive systems, which seek to replicate human mental processes, will become more prominent in the future. Compared to today's neural networks, cognitive systems have a better knowledge of interpreting data at a higher degree of abstraction. We presented the main concept of AI accelerators and their role in cloud and server applications. This chapter helps the readers to understand the concept of various HW accelerator types such as cloud, data centers, and edge computing. It also discusses the infrastructure for deploying FPGAs and ASICs in cloud and server applications.

References

1. Putnam, A., Caulfield, A.M., Chung, E.S., Chiou, D., Constantinides, K., Demme, J., Esmaeilzadeh, H., Fowers, J., Gopal, G.P., Gray, J., Haselman, M., Hauck, S., Heil, S., Hormati, A., Kim, J.-Y., Lanka, S., Larus, J., Peterson, E., Pope, S., Smith, A., Thong, J., Xiao, P.Y., Burger, D.: A reconfigurable fabric for accelerating large-scale datacenter services. In: Proceeding of the 41st Annual International Symposium on Computer Architecture, ISCA'14, pp. 13–24. IEEE Press, Piscataway, NJ, USA (2014)
2. Govindaraju, V., Ho, C.-H., Sankaralingam, K.: Dynamically specialized datapaths for energy efficient computing. In: 2011 IEEE 17th International Symposium on High Performance Computer Architecture (HPCA), pp. 503–514 (2011)
3. Clark, N., Hormati, A., Mahlke, S.: Veal: Virtualized execution accelerator for loops. In: ISCA'08. 35th International Symposium on Computer Architecture, 2008, pp. 389–400 (2008)
4. Cong, J., Ghodrat, M.A., Gill, M., Grigorian, B., Reinman, G.: Charm: A composable heterogeneous accelerator-rich microprocessor. In: Proceedings of the 2012 ACM/IEEE International Symposium on Low Power Electronics and Design
5. A Survey on Reconfigurable Accelerators for Cloud Computing
6. Chapter 2: High Performance Computing and Data Centers: Heterogeneous Integration Roadmap, 2021 Version. Available online: https://eps.ieee.org/images/files/HIR_2021/ch02_hpc.pdf. Accessed on 13 Jan 2022
7. The Future of the Data Center: Heterogeneous Computing. Available online: https://www.dataversity.net/future-data-center-heterogeneous-computing/. Accessed on 13 Jan 2022
8. Shalf, J.: The future of computing beyond Moore's law. Phil. Trans. R. Soc. A. **378**(2166), 20190061 (2020)
9. Kumar, A., Farooq, M.: Enabling AI with heterogeneous integration. Available online: https:/ /www.chipscalereview.com/wp-content/uploads/2021/01/Reprint-from-ChipScale_Nov-Dec_2020-IBM-WP.pdf. Accessed on 13 Jan 2022
10. Shrestha, R., Bajracharya, R., Nam, S.Y.: Challenges of future VANET and cloud-based approaches. Wirel. Commun. Mob. Comput. **2018** (2018)
11. Cooke, R.A., Fahmy, S.A.: A model for distributed in-network and near-edge computing with heterogeneous hardware. Futur. Gener. Comput. Syst. **105**, 395–409 (2020)
12. ADLINK Technology, Heterogeneous Computing for Artificial Intelligence at the Edge, (2021). Available online: https://go.adlinktech.com/SB-Heterogeneous-Computing-for-Artificial-Intelligence-at-the-Edge_LP.html. Accessed on 23 Dec 2022

13. Doğan, A., Ebclŏglu, K.: Cloud building block Chip for creating FPGA and ASIC clouds. ACM Trans. Reconfigurable Technol. Syst. **15**(2), 35 (2021). https://doi.org/10.1145/3466822
14. Right Sizing: Provisioning Instances to Match Workloads. Available online: https://d1.awsstatic.com/whitepapers/cost-optimization-right-sizing.pdf. Accessed on 23 May 2022
15. Alveo U55C Data Center Accelerator Cards Data Sheet. Available online: https://www.xilinx.com/content/dam/xilinx/support/documents/data_sheets/ds978-u55c.pdf. Accessed on 23 May 2022
16. Bobda, C., Mbongue, J.M., Chow, P., Ewais, M., Tarafdar, N., Vega, J.C., Eguro, K., Koch, D., Handagala, S., Leeser, M., Herbordt, M.: The future of FPGA acceleration in datacenters and the cloud. ACM Trans. Reconfigurable Technol. Syst. (TRETS). **15**(3), 1–42 (2022)
17. Al Qassem, L.M., Stouraitis, T., Damiani, E., Elfadel, I.A.M.: FPGAaaS: A survey of infrastructures and systems. IEEE Trans. Serv. Comput. (2020)
18. Steinert, F., Kreowsky, P., Wisotzky, E.L., Unger, C., Stabernack, B.: A hardware/software framework for the integration of FPGA-based accelerators into cloud computing infrastructures. In: 2020 IEEE International Conference on Smart Cloud (SmartCloud), pp. 23–28. IEEE (2020)
19. Open Cloud Testbed (OCT) – Mass Open Cloud. Available online: https://massopen.cloud/connected-initiatives/open-cloud-testbed/. Accessed on 23 May 2022
20. Leeser, M., Handagala, S., Zink, M.: FPGAs in the cloud. Comput. Sci. Eng. **23**(6), 72–76 (2021). https://doi.org/10.1109/MCSE.2021.3127288
21. Zink, M., et al.: The open cloud testbed (OCT): A platform for research into new cloud technologies. In: 2021 IEEE 10th International Conference on Cloud Networking (CloudNet), pp. 140–147 (2021). https://doi.org/10.1109/CloudNet53349.2021.9657109
22. Asiatici, M., Ienne, P.: Request, coalesce, serve, and forget: Miss-optimized memory systems for bandwidth-bound cache-unfriendly applications on FPGAs. ACM Trans. Reconfigurable Technol. Syst. (TRETS). **15**(2), 1–33 (2021)
23. Hogervorst, T., Nane, R., Marchiori, G., Qiu, T.D., Blatt, M., Rustad, A.B.: Hardware acceleration of high-performance computational flow dynamics using high-bandwidth memory-enabled field-programmable gate arrays. ACM Trans. Reconfigurable Technol. Syst. (TRETS). **15**(2), 1–35 (2021)
24. Alonso, T., Petrica, L., Ruiz, M., Petri-Koenig, J., Umuroglu, Y., Stamelos, I., Koromilas, E., Blott, M., Vissers, K.: Elastic-DF: Scaling performance of DNN inference in FPGA clouds through automatic partitioning. ACM Trans. Reconfigurable Technol. Syst. (TRETS). **15**(2), 1–34 (2021)
25. Teich, P.: Tearing Apart Google's TPU 3.0 AI Coprocessor (2018)
26. Jouppi, N.P., et al.: In-Datacenter Performance Analysis of a Tensor Processing Unit TM (2017)
27. Google Cloud: System Architecture. https://cloud.google.com/tpu/docs/system-architecture-tpu-vm. Accessed on 15 June
28. Ceshine Lee Tensorflow: Training CV Models on TPU without Using Cloud Storage. https://medium.com/the-artificial-impostor/tensorflow-training-cv-models-on-tpu-without-using-cloud-storage-60b20f0a7cd6 (2020). Accessed on 27 Apr 2022
29. Yang, Z., Dai, Z., Yang, Y., Carbonell, J., Salakhutdinov, R., Quoc, V.L.: XLNet: Generalized Autoregressive Pretraining for Language Understanding. https://arxiv.org/pdf/1906.08237.pdf
30. Synced: The staggering cost of training SOTA AI models 2019. https://syncedreview.com/2019/06/27/the-staggering-cost-of-training-sota-ai-models/. Accessed on 12 June
31. Pilato, C., Garg, S., Wu, K., Karri, R., Regazzoni, F.: Securing hardware accelerators: A new challenge for high-level synthesis. IEEE Embed. Syst. Lett. **10**(3), 77–80 (2017)
32. Kim, S., Shrestha, R.: Security and privacy in intelligent autonomous vehicles. In: Automotive Cyber Security, pp. 35–66. Springer, Singapore (2020)
33. Shrestha, R., Djuraev, S., Nam, S.Y.: Sybil attack detection in vehicular network based on received signal strength. In: 2014 International Conference on Connected Vehicles and Expo (ICCVE), pp. 745–746. IEEE (2014)

34. Kim, S., Shrestha, R.: Internet of vehicles, vehicular social networks, and cybersecurity. In: Automotive Cyber Security, pp. 149–181. Springer, Singapore (2020)
35. Turan, F., Verbauwhede, I.: Trust in FPGA-accelerated cloud computing. ACM Comput. Surv. (CSUR). **53**(6), 1–28 (2020)
36. Hobden, P., Srivastava, S., Nurellari, E.: FPGA-based CNN for real-time UAV tracking and detection. Front. Space Technol. **3**, 878010 (2022)
37. Shrestha, R., Omidkar, A., Roudi, S.A., Abbas, R., Kim, S.: Machine-learning-enabled intrusion detection system for cellular-connected UAV networks. Electronics. **10**(13), 1549 (2021)
38. Bajracharya, R., Shrestha, R., Jung, H.: Wireless infrastructure drone based on NR-U: A perspective. In: 2021 International Conference on Information and Communication Technology Convergence (ICTC), pp. 834–838. IEEE (2021)

Overviewing AI-Dedicated Hardware for On-Device AI in Smartphones

Hyunbin Park and Shiho Kim

1 Introduction

Since the birth of Lenet-5, the first deep neural network, in 1988, deep learning has made remarkable progress [1]. In the early to mid-2010s, a tremendous amount of computation required for deep learning was a barrier to on-device deep learning in a smartphone. Therefore, to process deep-learning using a smartphone's data, one option was to transmit the data to a PC or server equipped with one or more graphics processing units (GPUs) and perform deep-learning processing on the data. However, the data transfer latency between a smartphone and a server disturbs real-time deep learning processing, and the data leakage out of the smartphone may weaken user privacy. In addition, it is costly to build and maintain a server environment. Thus, on-device deep learning in the smartphone was the way to go.

The key strategy that enabled on-device deep-learning under the limited hardware resources of the limited power of a smartphone is to perform only inference in the smartphone. One inference processing requires one forward path computation of a neural network while training a neural network requires both forward and backward path computations of more than millions of data with dozens of epochs [1–3].

Figure 1 shows on-device artificial intelligence (AI) using single instruction multiple data (SIMD) coprocessors in a smartphone's application processor (AP) chip. Various SIMD coprocessors for on-device AI such as central processing unit (CPU), GPU, digital signal processor (DSP), and neural processing unit (NPU) are integrated into AP chips. Each SIMD coprocessor has unique advantages and disadvantages, which this chapter explains in detail. Neural network (NN) models

H. Park · S. Kim (✉)
School of Integrated Technology, Yonsei University, Incheon, South Korea
e-mail: bin9000@yonsei.ac.kr; shiho@yonsei.ac.kr

© The Author(s), under exclusive license to Springer Nature Switzerland AG 2023
A. Mishra et al. (eds.), *Artificial Intelligence and Hardware Accelerators*,
https://doi.org/10.1007/978-3-031-22170-5_4

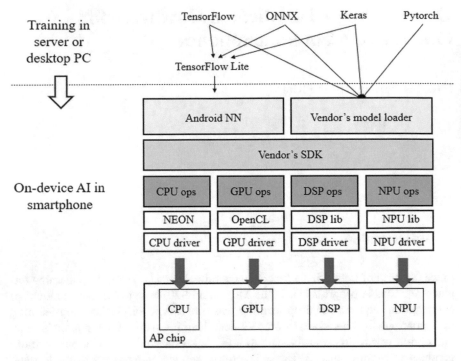

Fig. 1 On-device AI using various SIMD processors in AP chip

trained on a server or a desktop with various machine learning (ML) frameworks such as TensorFlow, Keras, ONNX, and Pytorch are transmitted to a smartphone for on-device inference. The vendor's Software Development Kit (SDK) can choose the SIMD coprocessor to run based on the predetermined criterion, or developers and users can select the coprocessor.

In the case of some deep learning applications that need to guarantee real-time inference, only on-device neural network processing is performed since processing in the cloud cannot guarantee latency. The typical real-time applications are illustrated in Fig. 2. Deep learning applications mainly used in smartphone industries are as follows: voice wakeup, speech recognition, language translation, classification, gesture recognition, face unlock, image enhancement, object detection, semantic segmentation, instance segmentation, depth estimation, frame rate interpolation, pose tracking, super-resolution, and so on. Real-time segmentation can apply different brightness for each object or out-focus on an object with a single camera. Frame rate interpolation and super-resolution can increase the image resolution of recorded videos or increase frame rate, for example, full high definition 30 (FHD 30) fps video to Ultra HD (UHD) 60 fps. Recently, GAN-based applications such as object erasers have been applied [4].

On the other hand, voice assistant service processes highly computational workloads also in the cloud server, as shown in Fig. 3. A smartphone detects a

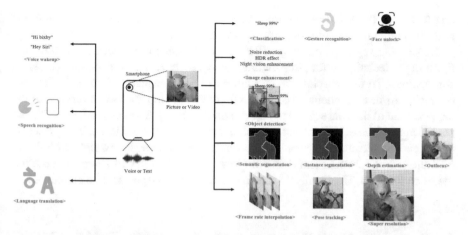

Fig. 2 Real-time on-device ML applications used in a smartphone

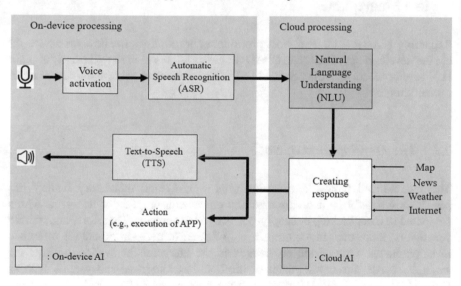

Fig. 3 Example of voice assistant service with on-device processing and cloud processing

keyword that wakes up the smartphone as always-on. If it detects the keyword, it performs automatic speech recognition (ASR) to convert the user's speech into sentences. These sentences are sent to a cloud server, and the user's intention is grasped by running natural language understanding (NLU), which is too compute-intensive for smartphones. The cloud server collects the user's intentions and information on the Internet, such as maps, news, weather, etc., and produces a response, which is transmitted to the smartphone. Depending on the response, the appropriate action is performed, or the response is converted to artificial voice via text-to-speech (TTS). These all processes should be performed within latency that is short enough to guarantee quality of service (QoS).

Developing system on chip (SoC), semiconductor process, SIMD, and lightweight techniques for deep neural networks have allowed on-device AI in smartphones. In this chapter, the authors will briefly explain this overall flow for on-device AI in smartphones. Section 2 will cover a comprehensive overview of the development of hardware (HW) to achieve on-device AI. Section 3 will describe the architecture of the SIMD processor, such as the CPU, GPU, DSP, and NPU. Section 4 will explain techniques for on-device inferences, such as lightweight techniques of deep neural networks and data compression. Section 5 will share the authors' views on future NPUs. Finally, in the last section, this chapter will be concluded.

2   Overview of HW Development to Achieve On-Device AI in a Smartphone

Regarding hardware, the two most prominent contributions to allow on-device AI are the development of SoC and SIMD processors. This section will briefly explain HW developments regarding SoC and SIMD processors to achieve on-device AI in a smartphone.

2.1   The Development of SoC

The standardized typical device size (and weight) and mandatory battery life permit only strictly limited chip size and power consumption of neural networks compared to other desktop or supply-powered environments. Therefore, under these constraints, innovation is required in chip design to perform on-device inference in smartphones. One of the challenges is the hardware innovation of SoC for smartphone AP. The development of the SoC has integrated subsystems such as CPU, GPU, NPU, DSP, 5G modem, camera processing HW, display processing HW, audio processing HW, security processing HW, and other HWs in AP chip [5]. Representative global AP vendors in 2021 include Apple, Qualcomm, Samsung, MediaTek, HiSilicon, and UNISOC, as shown in Fig. 4.

In the case of desktops, CPU cores, GPU, and other processing units on the motherboard are connected with an off-chip connection, increasing power consumption. On the other hand, the internal chip connection of the AP chip allows communication between circuits efficiently in terms of speed and power. In addition, circuits made in the same semiconductor process and optimized interconnection/operation between circuits are also factors that increase efficiency. In 2017, 10 nm AP was launched, and in 2021, 5 nm AP was launched. The semiconductor technology development has led to more computational hardware

Fig. 4 Representative global AP vendors in 2021

units integrated under limited chip area and power. Triple NPU cores integrated in Samsung Exynos 2100 released in 2021 can compute 26 Tera-operations per second (TOPS) [6].

2.2 The Development of SIMD Processor: CPU, GPU, DSP, and NPU

The AP chip of smartphones integrates various SIMD processors, which can be utilized for AI operations. The SIMD processors include CPU, GPU, DSP, and NPU. As shown in Fig. 5, each SIMD processor has pros and cons. From CPU to NPU in Fig. 5, AI computing efficiency increases while supportable operations decrease. Smartphone developers and users can determine hardware to perform AI operations according to the characteristics of each HW.

As shown in Fig. 6, in 2016, Qualcomm's AP, Snapdragon 820/835, integrated the vector processor Hexagon 680 DSP, which was used for low-light photo and video applications [7]. Qualcomm allows Snapdragon APs to process inference of AI in DSP using Snapdragon Neural Processing Engine (SNPE) SDK. Since Huawei released the first NPU [8], global AP vendors have mainly used NPU and DSP for on-device AI inference, while they also use CPU and GPU depending on the situation.

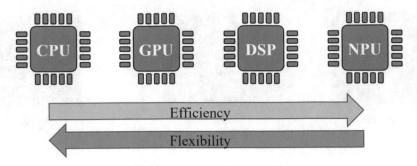

Fig. 5 AI computing efficiency and flexibility according to HW

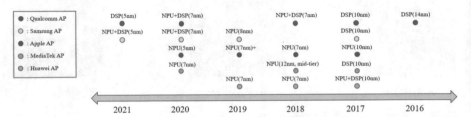

Fig. 6 The semiconductor technology and released year of some remarkable AI-dedicated HWs

Since the 2010s, AP vendors have integrated multicore-based CPUs in their APs [11]. These multicores are mainly intended to process multiple threads in parallel to increase performance, not to process millions of multiplications and additions based on SIMD topology, which is inefficient in terms of TOPs/W when processing the deep learning computation. The increase in the need for efficient SIMD processing in CPUs has led CPU vendors to contain SIMD coprocessors in their CPUs. For example, ARM integrated a NEON SIMD coprocessor in the CPU [12], and Apple also integrated an ML accelerator in their CPU in the A14 AP chip [13]. AP processors include a mobile GPU, which can perform SIMD deep learning processing using its parallel computational units. However, in most cases, it is used for computation for graphics. Mobile GPUs include fewer computational units compared to GPUs equipped with desktops but consume much less power. As shown in Fig. 7, peak AI computational capability using CPU, GPU, NPU, and DSP doubles yearly.

3 Overview of NPU and Review of Commercial NPU

It has become common for APs to integrate AI-dedicated HWs. The AI-dedicated HWs are called different names for each AP vendor, and the authors will collectively refer to them as NPU. This section will briefly explain what NPU is and what features make inference efficient. Also, the architecture and prominent features of

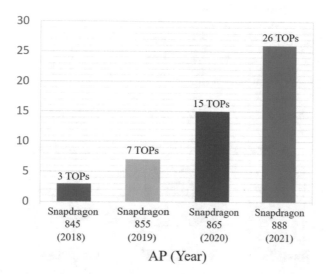

Fig. 7 Qualcomm Snapdragon AP's AI capability. (Data is obtained from Qualcomm's introduction material of Snapdragon 865 5G [9] and Qualcomm Redefines Premium with the Flagship Snapdragon 888 5G Mobile Platform [10])

three commercial NPUs that were opened, Samsung Exynos NPU, Huawei NPU, and MediaTek NPU, will be introduced.

3.1 Overview of NPU

An NPU is an AI-dedicated HW that accelerates deep learning operations. According to Han et al. [14] and Chen et al. [15], energy consumed by DRAM access is much larger than those consumed by computation of multiply-and-accumulation (MAC) in NPU. For this reason, NPU's core strategy is to reduce DRAM memory access by maximizing data reuse. Most vision neural networks have been based on 2D/3D/4D convolution operations. Thus, NPUs focus on computing these convolution operations efficiently. As shown in Fig. 8, in the case of a 3 × 3 convolution operation, 66% input and 100% weight are overlapped between two adjacent operations. The ideal NPU would therefore receive input feature map and weight from the DRAM to the NPU's SRAM and reuse it.

The architecture of APs containing NPUs of the smartphone industry typically has a common feature shown in Fig. 9. The AP and DRAM communicate with each other through an off-chip connection. Subsystems such as CPU, GPU, and NPU are connected through on-chip or off-chip buses such as Advanced eXtensible Interface (AXI) interconnect inside AP. NPU includes a convolution accelerator composed of SRAM and MAC arrays and a circuit that computes non-linear operations such as max-pooling and other activations. A typical MAC unit consists of a multiplier, adder, and registers called accumulators.

$1·1+8·3+0·0+4·4+2·0+0·0+1·1+5·2+2·2=56$

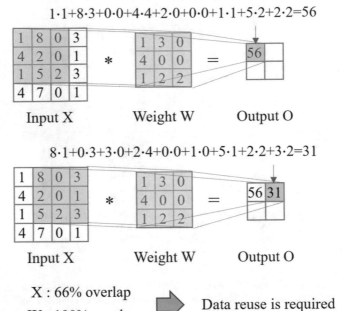

Input X   Weight W   Output O

$8·1+0·3+3·0+2·4+0·0+1·0+5·1+2·2+3·2=31$

Input X   Weight W   Output O

X : 66% overlap
W : 100% overlap   Data reuse is required

Fig. 8 Illustration of 3 × 3 convolution

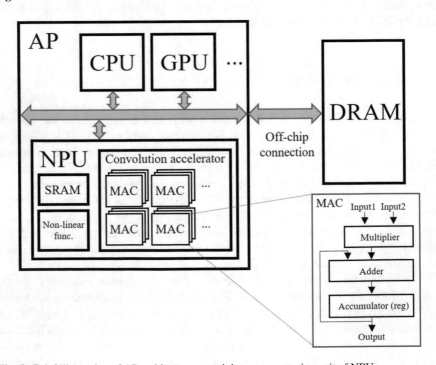

Fig. 9 Brief illustration of AP architecture containing a coprocessing unit of NPUs

	Energy(pJ)	Area(μm^2)
8b Add	0.03	36
16b Add	0.05	67
16b FP Add	0.4	1360
32b Add	0.1	137
32b FP Add	0.9	4184
8b Mult	0.2	282
16b FP Mult	1.1	1640
32b Mult	3.1	3495
32b FP Mult	370	7700

Fig. 10 Energy cost of multiplier and adder for various precision in 45 nm technology. (The data is obtained in Gholami et al. [18])

Traditional von Neumann architecture-based HWs have used Floating-Point 32 (FP32) precision for precise computations [16]. However, the characteristics of the human brain do not require precise calculations, and these characteristics are also applied to deep learning. Various studies have demonstrated that well-designed quantization with fewer bits than 32-bit (e.g., INT8, FP16, and INT4) does not reduce accuracy [17]. This characteristic reduces the bit-width of the NPU's MAC units and SRAM. The reduced bit-width of the HWs results in less energy cost and chip area, as shown in Fig. 10. Therefore, all global AP vendors have adopted NPU architectures with less bit-width than FP32. Qualcomm's NPU, which is called Hexagon, can assign precision to Activation Integer 8-bit with Activation Integer 8-bit (A8W8) and Activation Integer 16-bit with Activation Integer 8-bit (A16W8) when quantizing a model and running the quantized model on the smartphone using SNPE SDK [19]. Therefore, it seems that Qualcomm's NPU can selectively support the precision of both W8A8 and W8A16. Apple officially announced that Apple's NPU, called Neural Engine, supports multi-precision at September Event 2018 [20]. But, details are not opened officially. MediaTek's NPU, called AI (Artificial Intelligence) Processing Unit (APU), supports the precision of Integer 8-bit (INT8), Integer 16-bit (INT16), and FP16 [21]. Huawei's NPU supports INT8 and FP16 [22].

In summary, the NPUs of global AP vendors support FP16 and/or INT8 precision operations. Each of these precisions has its pros and cons. Table 1 summarizes the pros and cons in terms of bit-width. FP16 has an advantage in inference accuracy, and INT8 has advantages in TOPs/W, DRAM bandwidth (BW), and internal memory efficiency. Also, INT8 allows neural networks to have more zero weight and activation since it uses lower bit-width. So, it can achieve better efficiency in NPU with a zero-skipping computation feature.

Table 1 Pros and cons of FP16 and INT8

	FP16	INT8
Inference accuracy	Better	
TOPs/W		Better
DRAM BW		Better
Internal memory efficiency (area and capacity)		Better
Zero weights/activations ratio		Better

3.2   NPU Architectures of Global AP Vendors

In this subsection, NPU architectures in the industry of AP vendors introduced in international conferences will be reviewed. Qualcomm and Apple have not opened their NPU architecture, as the authors know. On the other hand, Samsung, MediaTek, and Huawei unveiled their NPU architectures in ISSCC 2019, ISSCC 2020, and Hot chips, respectively. These are not the latest NPU architectures of the vendors. However, reviewing them provides help to know NPU industry trends.

The first NPU architecture to introduce is Samsung's NPU [23], which was opened at ISSCC 2019. Figure 11 illustrates the NPU architecture. This NPU is integrated with an 8 nm technology node flagship mobile SoC. This NPU has a dual-core, and each core has 8-bit 512 MACs and 512 kB SRAM. Each core also has two data-staging units (DSUs), which have feature maps and weights and delivers them appropriately in MAC. For convolution operations, the dispatcher in the DSU selects non-zero weights for weight-zero skipping, and the feature-map selector selects the activation values to be calculated with the non-zero weights. These selected non-zero weights and corresponding activations are delivered to MAC arrays (MAAs), and dual MACs included in MAAs perform MAC operations. This dual MAC includes two multipliers, three adders, and registers for accumulation. The result processed in MAAs is transmitted to the data-returning unit (DRU). The DRU performs activation operations such as ReLU or PReLU on the partial sums of the output feature map and delivers the results to the SRAM. Peak Giga-Operations per second (GOPS) of this Samsung NPU was achieved at 1910 GOPS at 8 nm CMOS technology and achieved 6937 GOPS when the weight zero-skipping feature is applied. Also, 3.4 TOPS/W was achieved when inference Inception V3.

The Da Vinci Core of Huawei NPU [24], shown in Fig. 12, has computational units such as a 1D scalar unit, 2D vector unit, 3D matrix unit, and internal memories such as L0/L1 buffer, cache, and queue. The 3D matrix unit consists of 4096 FP16 MACs and 8192 INT8 MACs, which compute 3D convolution efficiently. The 2D Vector unit processes a vector computation of 2048 binary INT8/FP16/FP32 vector, activation operations, non-minimum suppression (NMS) operation, and ROI function. This core achieved 8 TOPS at FP16 precision.

The last NPU to be introduced is MediaTek NPU announced at ISSCC 2020 [25], as shown in Fig. 13. This NPU is integrated with 7 nm technology AP. Similar to Huawei's Da Vinci Core, this NPU includes a 1D engine, 2D engine, 3D convolu-

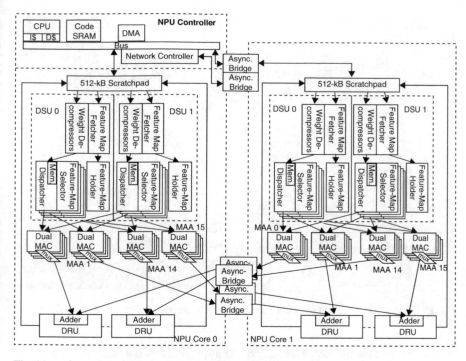

Fig. 11 NPU architecture of Samsung 8 nm flagship AP. (Figure imported from Song et al. [23])

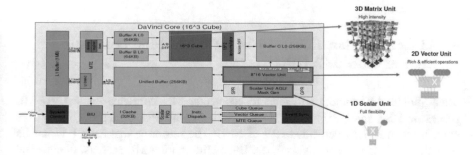

Fig. 12 NPU architecture of Huawei Da Vinci AI Core. (Figure imported from Liao et al. [24])

tional engine (CE), and SRAM. The 1D engine processes normalization operations, activation operations such as ReLU and PReLU, and arithmetic operations such as addition and multiplication. The 2D engine supports average and max pooling operations. The CE can be configurable to process $2 \times 8b$, $1 \times 16b$, or 0.5 FP16 MAC computations per cycle. This NPU performs zero skipping of activation in the fully connected (FC) layer. This zero skipping scheme employs simple control so that if activation is zero, then the row of the weight matrix to be multiplied with the zero activation does not load from DRAM. This reduced DRAM BW by 87% and

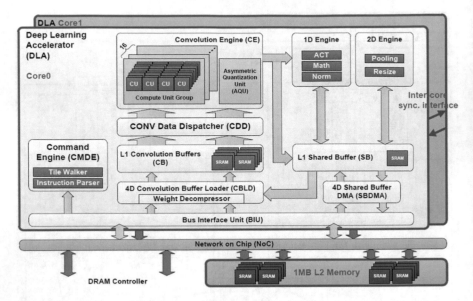

Fig. 13 NPU architecture of MediaTek 7 nm 5G AP. (Figure imported from Lin et al. [25])

operations by 71% in Long Short-Term Memory (LSTM). It achieved 3.6 TOPS @ INT8, 1.8 TOPS @ INT16, and 0.9 TFLOPS @ FP16.

4 AI Acceleration of Non-NPU: CPU Machine Learning Coprocessor, DSP, GPU

Many global AP vendors still tend to accelerate ML inference using CPU ML processor, DSP, and GPU in several cases, rather than NPU, even though their APs contain integrated AI-dedicated HW of NPU. The reason is that each HW has its own unique strengths and weaknesses. Depending on the ML models, the optimal HW to run them varies. This section will briefly introduce the CPU ML coprocessor, DSP, and GPU architecture and discuss in what situations they have advantages over NPU.

Figure 14 shows the architecture of ARM NEON, one of the ML coprocessors of the CPU. Two input vectors contain a series of binary bits, and these vectors are added or multiplied in parallel by multiple adders and multipliers. That is, ARM NEON is a SIMD processor that performs vector processing. Since it is included in the CPU, it has fewer MACs than DSP, GPU, and NPU. Therefore, it takes more time to infer ML models than NPU. However, NPU requires a lot of time to initialize and assign operations and data, sometimes greater than inference time. Therefore, AI acceleration in CPU requires less setup time and can achieve better inference time in small ML models.

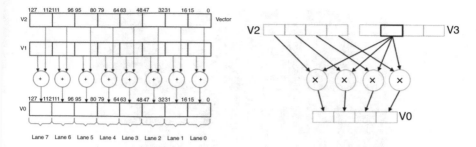

Fig. 14 Vector processing in ARM NEON: Addition and Multiplication. (Figure imported from Introducing Neon for Armv8-A [26])

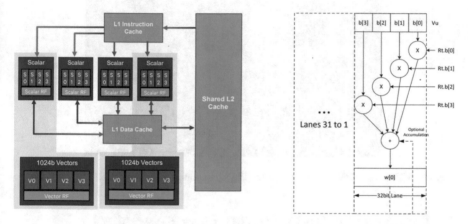

Fig. 15 Qualcomm Hexagon 680 DSP architecture. (Figure imported from Codrescu et al. [27])

DSP is a vector processor, which is usually more powerful than a CPU ML coprocessor. Qualcomm Hexagon 680 DSP contains two 1024-bit vector processing units, four scalar processing units, and L1/L2 caches [27], as shown in Fig. 15. DSP shows ML performance as excellent as NPU for some ML models, although it has fewer MACs than NPU. However, DSP is not optimized for convolution operation, which accounts for most of vision AI. Instead, DSP can compute matrix multiplication more efficiently than NPU, resulting in better performance in non-convolution-based AI models such as Recurrent Neural Networks (RNN), LSTM, and Transformer.

Mobile GPU also has multiple computational cores, which can accelerate inference, although it is optimized for graphic processing. Figure 16 shows the architecture of the ARM Mali-T760 GPU, and it contains a scalable core (SC), cache, and control logic. Mobile GPU typically shows less performance than NPU and DSP. However, since GPU focuses on general-purpose computations, it usually supports AI operations that NPU and DSP do not. Numerous new deep learning models are developed every year, and most of those include new operations. Therefore, the latest ML models would not be run on NPU. In this case, users and developers can run the ML models on mobile GPU.

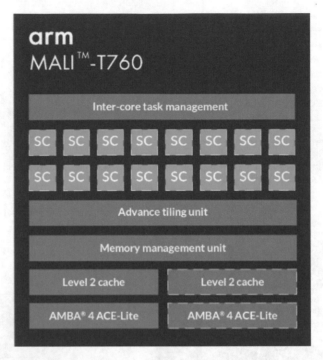

Fig. 16 ARM Mali-T760 GPU architecture. (Figure imported from Arm Mali-T760 GPU [28])

5   Techniques for On-Device Inference: Re-architect Network, Quantization, Pruning, and Data Compression

Sections 3 and 4 reviewed HWs for accelerating ML with on-device. This section will describe lightweight techniques of ML models and data compression techniques to reduce DRAM band width.

Mobilenet V1 proposed a depthwise separable convolution operation that reduces the amount of computation compared to the convolution operation [29].

In standard 3D convolution, after the inputs of the three-dimensional region are filtered, as shown in Fig. 17, and the weights corresponding to them are multiplied by each other, all the multiplication results are summed up. The computational cost of the 3D convolution is arithmetically expressed as Eq. (1):

$$W_H \times W_W \times I_D \times I_H \times I_W \times O_D \tag{1}$$

where I_H, I_W, and I_D are the width, height, and depth of the input feature map, W_H and W_W are the height and width of the filter, O_D is the depth of the output feature map. If W_H, $W_W = 3$, I_D, $O_D = 64$, and I_H, $I_W = 224$, then it requires 1849 million operations.

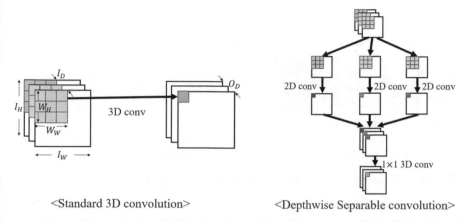

<Standard 3D convolution>   <Depthwise Separable convolution>

Fig. 17 Illustration of standard 3D convolution and depthwise separable convolution

On the other hand, in depthwise separable convolution, 2D convolution is performed on each 2D channel of a filter over each corresponding channel of an input, as shown in Fig. 17. After its 2D outputs are stacked, a 1×1 convolution operation is performed. The computational cost of the depthwise separable convolution is arithmetically expressed as Eq. (2):

$$W_{\mathrm{H}} \times W_{\mathrm{W}} \times I_{\mathrm{H}} \times I_{\mathrm{W}} \times I_{\mathrm{D}} + I_{\mathrm{D}} \times I_{\mathrm{H}} \times I_{\mathrm{W}} \times O_{\mathrm{D}} \tag{2}$$

If W_{H}, $W_{\mathrm{W}} = 3$ and I_{D}, $O_{\mathrm{D}}=64$, I_{H}, $I_{\mathrm{W}} = 224$, then it requires 234 million operations. This number of operations corresponds to 12% of the computational cost of the 3D convolution operation.

The inverted residual block is the second re-architect strategy for on-device inference used in MobilNet v2 [30]. Limited SRAM size in NPU makes memory-efficient computation import. The inverted residual block requires less memory than the residual block. Figure 18 illustrates the efficient memory usage of the inverted residual block, where the inverted residual block has fewer channels in the first and last layers and more channels in the intermediate layers. Well-designed compiler operates inverted residual block as one operation, where only the first and last layers are stored in SRAM. Although the deeper intermediate feature maps are produced in the inverted residual block than the residual block, these can be temporarily held in the buffer, register, or accumulator in MAC. Therefore, minor input and output feature map size in inverted residual blocks lead to memory-efficient computation than the residual block.

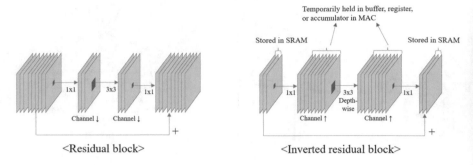

Fig. 18 Illustration of residual block and the inverted residual block

The third re-architect strategy for on-device inference is a neural architecture search (NAS) technique employed in MobileNet v3 [31]. Before NAS was invented, ML models were designed in a way that humans directly design the structure of neural networks with trial-and-error and improve the structure by analyzing the feedback on the trained result. However, such a way to explore infinite possibilities consumes a lot of time and human resources. NAS technique allows machine learning to explore infinite possibilities. It searches the architecture of an ML model to maximize the target goals such as accuracy, parameters, computational cost, the latency of target HW, etc. It helps create an optimal network much faster than humans.

According to MobileNet V3 [31], MobileNet V3-Large 1.0 is designed with the target of better accuracy. The accuracy was improved by 3.2% compared to MobileNet V2 1.0, while the number of parameters increased from 3.4 million to 5.4 million, and the MAC count was decreased. MobileNet V3-Small 1.0 is designed with the target of fewer parameters or latency. MAC count and parameters decreased from 300 million to 66 million and 3.4 million to 2.9 million, respectively. However, instead, the accuracy also decreased by 4.6%.

The second technique for on-device inference to be introduced is quantization. The quantization is to convert an ML model trained with FP32 precision into a model with fewer bit widths, such as INT8 or FP16. According to the neuron's mathematical model [32], a single neuron receives multiple input signals and classifies them approximately on and off. Therefore, using well-designed quantization, the computational cost of the inference can be reduced while the accuracy is maintained. As shown in Table 2, various quantization techniques have been studied in academia and industry to reduce the bit-width of weight and activation: Dynamic Fixed Point [33], BinaryConnect [34], Binary Weight Networks (BWN) [35], Ternary Weight Networks (TWN) [36], XNOR-NET [35], Quantized Neural Networks (QNN) [37], DoReFa-Net [38], Incremental Network Quantization (INQ) [39], and TensorRT [40].

The BWN network quantized AlexNet into 1-bit weight and 32-bit activation and achieved only a 0.8% accuracy drop in the ImageNet dataset. However, XNOR-NET and BNN networks quantized weights and activations into 1-bit, showing 11% and 29.8% accuracy drop in the ImageNet dataset, respectively. QNN and

Table 2 Accuracy drop in terms of quantization techniques

| Quantization method | Model | Bitwidth | | Accuracy drop compared to FP32 (%) |
		Weight	Activation	
Dynamic Fixed Point	AlexNet	8	8	0.6
BinaryConnect		1	32	19.2
Binary Weight Network (BWN)		1	32	0.8
Ternary Weight Networks (TWN)		2	32	3.7
XNOR-NET		1	1	11
Binarized Neural Networks (BNN)		1	1	29.8
Quantized Neural Networks (QNN)		1	2	6.5
DoReFa-Net		1	2	7.63
Incremental Network Quantization (INQ)		5	32	−0.2
TensorRT	AlexNet	8	8	−0.01
	ResNet-50			0.12
	GoogleNet			0.16

DoReFa-Net reduce the quantized weights and activations of AlexNet into 1-bit and 2-bit, respectively, resulting in an accuracy drop between 6% and 8%. Excessive reduction of bit-width causes a significant accuracy drop. NVIDIA TensorRT quantized weights and activations of AlexNet, ResNet50, and GoogleNet into 8-bit and achieved only a 0.2% accuracy drop. Therefore, AP vendors have employed more than 8-bit precision in their NPUs [23–25].

TMA accelerator [41] computes the multiplication of an INT8 input and an INT5 weight by adding two partial sub integers (PSIs) as expressed in Eq. (3):

$$w \cdot X = s1 \cdot 2^{n1} \cdot X + s2 \cdot 2^{n2} \cdot X = \text{PSI1} + \text{PSI2} \qquad (s1, s2 \in {-1, 0, 1}) \tag{3}$$

where $n1$ and $n2$ are integers from 0 to 4. This quantization allows the multiplication to be computed with only two barrel-shifters and two multiplexers. Although this quantization causes an error in the multiplication when the weight is 13, 11, −11, or −13, it only causes an accuracy drop of 3.9% on the ImageNet dataset.

As shown in Fig. 19, there are two representative quantization schemes: symmetric and asymmetric quantization. Symmetric quantization maps FP32 numbers in the range of maximum absolute value into a signed integer range, and it just requires scaling the FP numbers. Therefore, it is easy to implement the technique. However, it does not fully utilize the quantized range. Asymmetric quantization maps FP32 numbers into unsigned integer range. In this case, all FP32 numbers are mapped into the unsigned integer range. Therefore, the full range of the quantized integer is

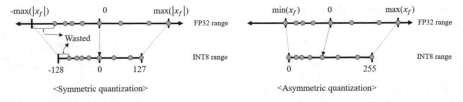

Fig. 19 Number mapping of symmetric and asymmetric quantization

Model	Top-1 Accuracy (Original)	Top-1 Accuracy (Post Training Quantized)	Latency * (Original) (ms)	Latency * (Post Training Quantized) (ms)	Size (Original) (MB)	Size (Optimized) (MB)
Inception_v3	0.78	0.772	1130	845	95.7	23.9
Resnet_v2_101	0.770	0.768	3973	2868	178.3	44.9

Fig. 20 INT8 weight and activation quantization of TensorFlow Lite [42]

Model	Accuracy metric type	Accuracy (float32 activations)	Accuracy (int8 activations)	Accuracy (int16 activations)
MobileBert	F1(Exact match)	88.81(81.23)	2.08(0)	88.73(81.15)

Fig. 21 INT8 weight and INT16 activation quantization of TensorFlow Lite. (Table is adapted from MediaTek Dimensity 800 introduction [42])

utilized. Thus, the accuracy drop caused by quantization can be reduced generally compared to the symmetric one.

ML SDKs of AP vendors such as TensorFlow Lite, Qualcomm SNPE, and Apple Core ML uses post-training quantization, which quantizes pre-trained models to make quantized models. It does not re-train a model for quantization but just uses a pre-trained model. So, the quantization process of post-training quantization takes just a few seconds or minutes. Figure 20 shows a comparison of the original FP32 model and quantized model based on Int8 quantized with TensorFlow Lite. The accuracy error caused by post-training quantization of Inception V3 and ResNet 101 was below 0.01%. Latency and model size of the quantized model is reduced, where the latency is measured on a large CPU core of Pixel2.

However, several neural networks still reduce the inference accuracy after the quantization with INT8. Figure 21 compares the accuracy of the original MobileBert and quantized MobileBerts. The quantization with INT8 reduced the accuracy, resulting in 2.08% of accuracy. In this case, the bit-width of weight and activation should be increased. The quantization with INT16 permitted only a 0.08% of accuracy drop.

Figure 22 shows an example of a quantized graph of post-training quantization. Suppose the data distribution to be quantized is known. In that case, we can quantize neural networks with just mapping, as illustrated in Fig. 19. However, if the input is

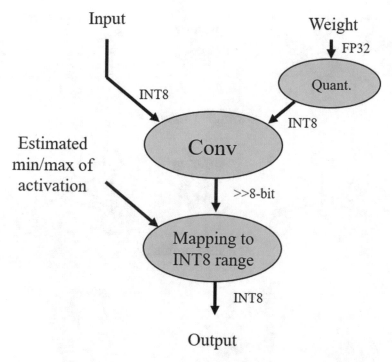

Fig. 22 Example of a quantized graph of post-training quantization

unknown, the distribution of convolution layer output is also unknown. So, a well-designed estimation technique of the distribution is required. Qualcomm's SDK SNPE typically uses 50 to 100 images to estimate the maximum and minimum values of the distribution [43].

The third technique is required for the on-device inference method corresponding to pruning. The pruning technique removes the connection between neurons in adjacent layers to skip the computation while maintaining the target accuracy level. Figure 23 briefly illustrates weight pruning, neuron pruning, and filter pruning. Weight pruning makes several elements in the weights matrix zero, so we require a particular HW architecture to support the weight-zero-skipping feature. Neuron pruning removes neurons and all connections to the neurons. The advantage of neuron pruning is that it does not require any HW modification to support it. Filter pruning removes several convolutional filters, so it does not need any HW changes for the processing.

As of 2021, the state-of-the-art filter pruning technique [44] achieved an accuracy drop of ImageNet less than 1% when pruning ResNet 50 with 41.8% connection removed. However, in many neural networks, accuracy drops cannot be ignored. Therefore, as far as the authors know, the implementing inference for the general pruning technique has not been fully employed yet in the on-device AI of smartphones.

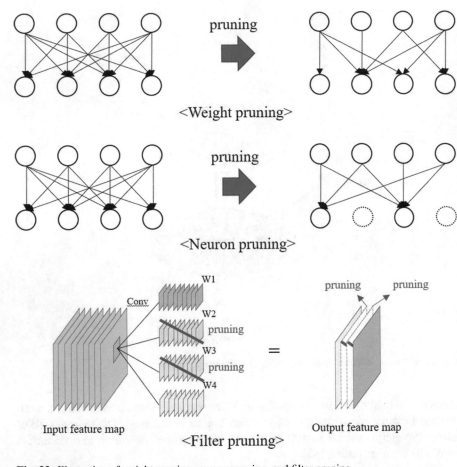

Fig. 23 Illustration of weight pruning, neuron pruning, and filter pruning

The last technique for on-device inference is data compression. The authors will introduce the run-length data compression technique employed Eyeriss ML accelerator [15], shown in Fig. 24. The core strategy of NPUs is to reduce DRAM BW. It can be achieved by enhancing data reuse or data compression. To reduce DRAM BW, the Eyeriss contains circuits for decompression and compression in NPU, which is positioned between DRAM and the internal memory of NPU. DRAM transmits compressed data to the NPU, and it is decompressed by decompress unit. Conversely, when data is transmitted from NPU to DRAM, the data is compressed by a compress unit.

A series of data is compressed in 64-bit format. The compressing rule is as follows. The first 5-bit contains the number of zeros, and the following 16-bit includes a non-zero number. This pattern is repeated three times. The last 2-bit holds '00', which gives information on termination of 64-bit compressed data. With this technique, Eyeriss reduced DRAM transactions of convolution layers of AlexNet from 1 to 5 by 16% to 46%.

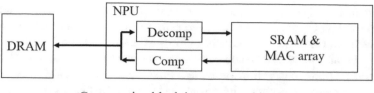

<Compression block in system architecture>

Uncompressed data: 0, 0, 23, 0, 0, 42, 0, 0, 0, 0, 11, ...

Compressed data (64b) :

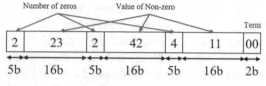

<Example of compression>

Fig. 24 Run-length data compression of Eyeriss ML accelerator. (Figure adapted from Chen et al. [15])

6   Discussion

This chapter introduced NPU trends, NPU architectures of smartphone vendors, and techniques for on-device inference. State-of-the-art NPUs allow on-device inference of machine learning models for classification, detection, segmentation, super-resolution, image enhancement, frame rate interpolation, etc. The readers may want to know the current limitations of NPUs. This discussion section will describe the HW constraint of NPUs for a smartphone and the authors' opinion of the future of NPUs.

As introduced in this chapter, commercial NPUs in a smartphone typically contain scalar-, vector-, and matrix-computing units. Scalar-computing units can accelerate activations such as Sigmoid, Switch, Tanh, etc. The vector-computing unit can accelerate dense operation, also known as the fully connected (FC) layer. Matrix-computing unit computes 3D and 4D convolutions efficiently. However, several matrix-computing units targeting the standard convolution do not efficiently compute dilated or depthwise convolution or cannot support those operations. Also, many new operations are being created in machine learning, and accelerators in NPU may not support the newly created operations.

A possible solution to support the specific operations is architecture modification by adding some HW features. For example, adding some HWs can allow the matrix-computing units to compute dilated or depthwise convolution efficiently. However, it cannot cope with new operations. Another possible solution is to include general-purpose computing units. However, the general-purpose architecture requires expensive hardware cost and cause inefficient processing, although it can process new operations. The solutions mentioned above need a hardware modification. It typically takes more than 1 year in the industry. Also, it requires an increase in chip area and power consumption. In the AP chip, the area and power for the NPU are strictly limited. To sum up, NPU cannot accelerate all operations with limited resources. Therefore, chip vendors should consider the trade-off.

What should AP vendors consider for next-generation NPU? The authors' personal opinions will be described. First, increasing SRAM size would improve performance rather than increasing MAC units in the future. Memory-intensive Unet-based models are increasingly employed, where the intermediate feature maps of the early layers have to be loaded from DRAM for the computation of the subsequent layers, resulting in DRAM access of typically tens of megabytes or more. Also, the smartphone's camera resolution has increased and will increase, demanding size increase the feature maps of neural networks. Second, NPUs accelerating 'transformer-based model' rather than CNN-based models may become more popular. Since 2020, not only have the transformer-based vision NN models achieved state-of-the-art (SOTA) in classification on ImageNet [45, 46], but also, in 2022, transformer-based NN model called EfficientFormer has improved accuracy by 5% compared to MobileNet V2 under similar computational speed [47].

7 Conclusion

This chapter introduces an overview of the history of the latest development of on-device AI as of 2021. The architecture of SIMD processing units for accelerating inference, CPU, GPU, DSP, and NPU, is described. Also, techniques for efficient inference under the power constraint of a smartphone, which are re-architect network, quantization, pruning, and data compression, are explained. At last, we discussed the limitations and the future of NPUs.

Acknowledgments This work was supported by two Institute of Information & Communications Technology Planning & Evaluation (IITP) grants funded by the Korea government (MSIT) ("No.2020-0-00056, To create AI systems that act appropriately and effectively in novel situations that occur in open worlds").

References

1. LeCun, Y., Bottou, L., Bengio, Y., Haffner, P.: Gradient-based learning applied to document recognition. Proc. IEEE. **86**(11), 2278–2324 (1998)
2. Krizhevsky, A., Sutskever, I., Hinton, G.E.: Imagenet classification with deep convolutional neural networks. Adv. Neural Inf. Proces. Syst. **25**, 1097–1105 (2012)
3. Simonyan, K., Zisserman, A.: Very deep convolutional networks for large-scale image recognition. arXiv preprint arXiv:1409.1556 (2014)
4. Galaxy Unpacked January 2021: Official Replay | Samsung, YouTube. https://youtube.be/TD_BZN0bn_U. Accessed 5 Jan 2022
5. Exynos 990 Mobile Processor: Official Introduction, YouTube. https://www.youtube.com/watch?v=13RgDxD83vl. Accessed 5 Jan 2022

6. Samsung Exynos 2100 official introduction webpage, Samsung. https://www.samsung.com/semiconductor/minisite/exynos/products/mobileprocessor/exynos-2100/. Accessed 5 Jan 2022
7. Qualcomm's Snapdragon 820 processor product brief document, Qualcomm. https://www.qualcomm.com/media/documents/files/snapdragon-820-processor-product-brief.pdf. Accessed 5 Jan 2022
8. Huawei Kirin 970 AI processor presentation 10/16/17, YouTube. https://www.youtube.com/watch?v=5v2D_QddnFc. Accessed 5 Jan 2022
9. Qualcomm's introduction material of Snapdragon 865 5G, Qualcomm. https://www.qualcomm.com/media/documents/files/2019-snapdragon-865-5g-ai-deep-dive-ziad-asghar-jeff-gehlhaar.pdf. Accessed 5 Jan 2022
10. Qualcomm Redefines Premium with the Flagship Snapdragon 888 5G Mobile Platform, Qualcomm. https://www.qualcomm.com/news/releases/2020/12/02/qualcomm-redefines-premium-flagship-snapdragon-888-5g-mobile-platform. Accessed 5 Jan 2022
11. Antony, A., Sarika, S.: A review on IoT operating systems. Int. J. Comput. Appl. **975**, 8887
12. Jang, M., Kim, K., Kim, K.: The performance analysis of ARM NEON technology for mobile platforms. In: Proceedings of the 2011 ACM Symposium on Research in Applied Computation, pp. 104–106 (2011)
13. The A14 Bionic | Apple Special Event 2020, YouTube. https://www.youtube.com/watch?v=5toYNtqbiwg
14. Han, S., Pool, J., Tran, J., Dally, W.J.: Learning both weights and connections for efficient neural networks. arXiv preprint arXiv:1506.02626 (2015)
15. Chen, Y.-H., Krishna, T., Emer, J.S., Sze, V.: Eyeriss: An energy-efficient reconfigurable accelerator for deep convolutional neural networks. IEEE J. Solid State Circuits. **52**(1), 127–138 (2016)
16. Horowitz, M.: 1.1 computing's energy problem (and what we can do about it). In: 2014 IEEE International Solid-State Circuits Conference Digest of Technical Papers (ISSCC), pp. 10–14. IEEE (2014)
17. Overton, M.L.: Numerical computing with IEEE floating point arithmetic. Society for Industrial and Applied Mathematics (2001)
18. Gholami, A., Kim, S., Dong, Z., Yao, Z., Mahoney, M.W., Keutzer, K.: A survey of quantization methods for efficient neural network inference. arXiv preprint arXiv:2103.13630 (2021)
19. Tools in Snapdragon Neural Processing Engine SDK Reference Guide, Qualcomm. https://developer.qualcomm.com/sites/default/files/docs/snpe/tools.html. Accessed 5 Jan 2022
20. September Event 2018 – Apple, YouTube. https://www.youtube.com/watch?v=wFTmQ27S7OQ. Accessed 5 Jan 2022
21. MediaTek Dimensity 800 introduction, Mediatek. https://www.mediatek.com/products/smartphones/dimensity-800. Accessed 5 Jan 2022
22. Huawei Kirin 990 5G launch, YouTube. https://www.youtube.com/watch?v=9zkxuxpObKI. Accessed 5 Jan 2022
23. Song, J., Cho, Y., Park, J.-S., Jang, J.-W., Lee, S., Song, J.-H., Lee, J.-G., Kang, I.: 7.1 an 11.5 tops/w 1024-mac butterfly structure dual-core sparsity-aware neural processing unit in 8nm flagship mobile soc. In: 2019 IEEE International Solid-State Circuits Conference-(ISSCC), pp. 130–132. IEEE (2019)
24. Liao, H., Tu, J., Xia, J., Zhou, X.: Davinci: A scalable architecture for neural network computing. In: 2019 IEEE Hot Chips 31 Symposium (HCS), pp. 1–44. IEEE Computer Society (2019)
25. Lin, C.-H., Cheng, C.-C., Tsai, Y.-M., Hung, S.-J., Kuo, Y.-T., Wang, P.H., Tsung, P.-K., et al.: 7.1 a 3.4-to-13.3 tops/w 3.6 tops dual-core deep-learning accelerator for versatile AI applications in 7nm 5g smartphone SOC. In: IEEE International Solid-State Circuits Conference-(ISSCC) 2020, pp. 134–136. IEEE (2020)
26. Introducing Neon for Armv8-A, Arm. https://developer.arm.com/architectures/instruction-sets/simd-isas/neon/neon-programmers-guide-for-armv8-a/introducing-neon-for-armv8-a/single-page. Accessed 5 Jan 2022 [27]

27. Codrescu, L.: Architecture of the Hexagon™ 680 DSP for mobile imaging and computer vision. In: 2015 IEEE Hot Chips 27 Symposium (HCS), pp. 1–26. IEEE (2015)
28. Arm Mali-T760 GPU, Arm. https://developer.arm.com/ip-products/graphics-and-multimedia/mali-gpus/mali-t760-gpu. Accessed 5 Jan 2022
29. Howard, A.G., Zhu, M., Chen, B., Kalenichenko, D., Wang, W., Weyand, T., Andreetto, M., Adam, H.: Mobilenets: Efficient convolutional neural networks for mobile vision applications. arXiv preprint arXiv:1704.04861 (2017) [31]
30. Sandler, M., Howard, A., Zhu, M., Zhmoginov, A., Chen, L.-C.: Mobilenetv2: Inverted residuals and linear bottlenecks. In: Proceedings of the IEEE conference on computer vision and pattern recognition, pp. 4510–4520 (2018)
31. Howard, A., Sandler, M., Chu, G., Chen, L.-C., Chen, B., Tan, M., Wang, W., et al.: Searching for mobilenetv3. In: Proceedings of the IEEE/CVF International Conference on Computer Vision, pp. 1314–1324 (2019)
32. Jolivet, R., Lewis, T.J., Gerstner, W.: Generalized integrate-and-fire models of neuronal activity approximate spike trains of a detailed model to a high degree of accuracy. J. Neurophysiol. **92**(2), 959–976 (2004)
33. Peng, P., Mingyu, Y., Weisheng, X.: Running 8-bit dynamic fixed-point convolutional neural network on low-cost arm platforms. In: 2017 Chinese Automation Congress (CAC), pp. 4564–4568. IEEE (2017)
34. Courbariaux, M., Bengio, Y., David, J.-P.: Binaryconnect: Training deep neural networks with binary weights during propagations. In: Advances in neural information processing systems, pp. 3123–3131 (2015)
35. Rastegari, M., Ordonez, V., Redmon, J., Farhadi, A.: Xnor-net: Imagenet classification using binary convolutional neural networks. In: European Conference on Computer Vision, pp. 525–542. Springer, Cham (2016) [36]
36. Li, F., Zhang, B., Liu, B.: Ternary weight networks. arXiv preprint arXiv:1605.04711 (2016)
37. Hubara, I., Courbariaux, M., Soudry, D., El-Yaniv, R., Bengio, Y.: Quantized neural networks: Training neural networks with low precision weights and activations. J. Mach. Learn. Res. **18**(1), 6869–6898 (2017)
38. Zhou, S., Wu, Y., Ni, Z., Zhou, X., Wen, H., Zou, Y.: Dorefa-net: Training low bitwidth convolutional neural networks with low bitwidth gradients. arXiv preprint arXiv:1606.06160 (2016) [39]
39. Zhou, A., Yao, A., Guo, Y., Xu, L., Chen, Y.: Incremental network quantization: Towards lossless CNNs with low-precision weights. arXiv preprint arXiv:1702.03044 (2017)
40. Szymon Migacz: 8-bit inference with TensorRT. NVIDIA. https://on-demand.gputechconf.com/gtc/2017/presentation/s7310-8-bit-inference-with-tensorrt.pdf
41. Park, H., Kim, D., Kim, S.: TMA: Tera-MACs/W neural hardware inference accelerator with a multiplier-less massive parallel processor. Int. J. Circuit Theory Appl. **49**(5), 1399–1409 (2021)
42. Model optimization in TensorFlow Lite guide: Google. https://www.tensorflow.org/lite/performance/model_optimization. Accessed 5 Jan 2022.
43. Quantizing a Model in Snapdragon Neural Processing Engine SDK Reference Guide, Qualcomm. https://developer.qualcomm.com/sites/default/files/docs/snpe/model_conversion.html. Accessed 5 Jan 2022.
44. He, Y., Dong, X., Kang, G., Yanwei, F., Yan, C., Yang, Y.: Asymptotic soft filter pruning for deep convolutional neural networks. IEEE Trans. Cybernet. **50**(8), 3594–3604 (2019)
45. Wu, B., Xu, C., Dai, X., Wan, A., Zhang, P., Yan, Z., Tomizuka, M., Gonzalez, J., Keutzer, K., Vajda, P.: Visual transformers: Token-based image representation and processing for computer vision. arXiv preprint arXiv:2006.03677 (2020)
46. Dosovitskiy, A., Beyer, L., Kolesnikov, A., Weissenborn, D., Zhai, X., Unterthiner, T., Dehghani, M., et al.: An image is worth 16x16 words: Transformers for image recognition at scale. arXiv preprint arXiv:2010.11929 (2020)
47. Li, Y., Yuan, G., Wen, Y., Hu, E., Evangelidis, G., Tulyakov, S., Wang, Y., Ren, J.: Efficient former: vision transformers at MobileNet speed. arXiv preprint arXiv:2206.01191 (2022)

Software Overview for On-Device AI and ML Benchmark in Smartphones

Hyunbin Park and Shiho Kim

1 Introduction

Hardware architectures to accelerate machine learning (ML) and various lightweight techniques of ML models for on-device inference are known to ML- and NPU-related researchers in academia, and various related papers have been published. Application programming interface (API), software development kit (SDK), and ML benchmark are considered essential factors in the smartphone industry, but these are not well known in academia. ML benchmark Apps or websites rank the AI performance of smartphone models. However, the performance of NPU shown in the benchmarks varies widely depending on which API and SDK are used.

Figure 1 shows the hierarchy of software (SW) and hardware (HW) for running on-device AI in smartphones. Developers can make a pre-trained model with various ML frameworks, TensorFlow, Pytorch, Caffe, Open Neural Network Exchange (ONNX), etc. AP chipset vendors have their own SDK/API for on-device AI, such as Qualcomm Snapdragon Neural Processing Engine (SNPE), Samsung ENN, Apple Core ML, MediaTek Neuro Pilot, Intel OpenVINO, and Huawei HiAI. AP vendor's ML SDK and API convert the pre-trained models into binary files for running HWs such as CPU, GPU, DSP, and NPU. The SDKs provide model conversion, analyzing, compiling, and initializing/executing HW.

All vendors of the Android realm also have tried to support Google's Android Neural Network Application Programming Interface (NNAPI) for a unified solution. However, NNAPI does not provide operations that are recently created. The operations are provided after the NNAPI update, which is bundled with the Android OS update. AP vendors prefer to support the latest operations in their SDK rather

H. Park · S. Kim (✉)
School of Integrated Technology, Yonsei University, Incheon, South Korea
e-mail: bin9000@yonsei.ac.kr; shiho@yonsei.ac.kr

© The Author(s), under exclusive license to Springer Nature Switzerland AG 2023
A. Mishra et al. (eds.), *Artificial Intelligence and Hardware Accelerators*,
https://doi.org/10.1007/978-3-031-22170-5_5

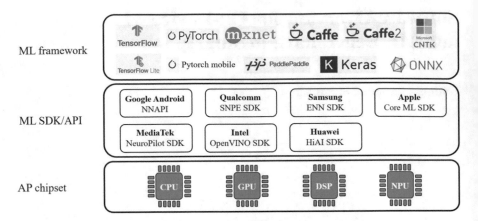

ML framework

ML SDK/API

AP chipset

Fig. 1 Hierarchical view of framework and development platform for on-device AI in smartphones

than NNAPI due to a delay of several months. Sometimes, AP vendors do not provide the latest operations to NNAPI due to human resource limitations and the cost of support. Therefore, several smartphones would not support an on-device inference of the latest ML models using NNAPI.

Developers should know this story to evaluate or apply their ML model on-device with the best performance of NPU and other ML-accelerating HWs in a smartphone. ML benchmarks for smartphones have recently been working closely with vendors to obtain each vendor's smartphone's maximum performance and achieve objective comparisons. This chapter will introduce three representative ML benchmarks for a smartphone, AI benchmark, MLperf, and UL Procyon. In addition, this chapter will introduce an overview of on-device AI SW and ML benchmark in smartphones for Google's Android NNAPI, TensorFlow Lite, and one of the representative vendors, SDK Qualcomm's SNPE. Section 2 will introduce the structure of Google's Android NNAPI. Section 3 will describe various functions of Qualcomm's SNPE SDK. Section 4 will introduce the representative ML benchmarks for a smartphone. Section 5 will discuss what we should consider evaluating NPUs, and summarize this chapter.

2   Google's Android NNAPI

This section introduces Google's Android NNAPI and how to perform on-device inference using Google's NNAPI on Android OS–based smartphones. NNAPI is the abbreviation of neural network (NN) and application programming interface (API), and it is the interface between the pre-trained model file and deep-learning-accelerating-HWs. Users and developers of Android OS–based smartphones can execute ML models using NNAPI on their devices.

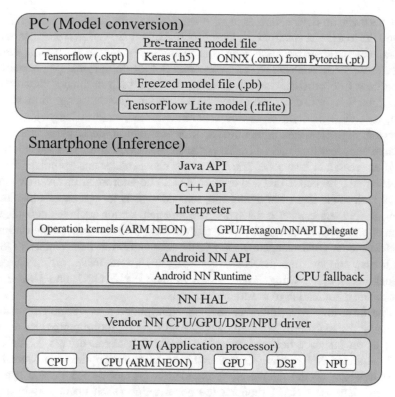

Fig. 2 The hierarchy from application to HW layer of NNAPI in Android-based smartphones

Figure 2 shows an example of the hierarchy from application to HW using NNAPI for on-device AI. The trained neural network model of TensorFlow Lite is an extension format of ".tflite". Pre-trained models from Keras, TensorFlow, Pytorch, or ONNX can be converted to TensorFlow Lite model representations.

The conversion to the tflite file format can be performed on a Linux PC platform. Before converting to the tflite file, the pre-trained model must be frozen to make the pb file format. Freezing means creating a binary file with a predetermined format containing graphs, weights, and other parameters. However, several pre-trained model files are not supported for conversion to the pb file. In this case, the files such as the pt. file created by Pytorch can be converted to the ONNX file first. The ONNX format is an open source format for ML models [1], which is compatible with most frameworks, so developers widely use it. TensorFlow Lite library supports the conversion from pb to tflite file format. However, not all pre-trained models can be converted to tflite files. Conversion to tflite of some latest operations in the pre-trained models is not supported. In this case, developers can replace the operations with compatible operations. Readers can find Python scripts for freezing and converting to a tflite file as an open source on the web.

To perform on-device inference, the tflite file is required to transfer from a desktop PC to a target smartphone. Developers may utilize various functions such as camera access, image processing, and hardware assignment to infer using the Java library provided in Android OS on their smartphones. In the smartphone platform, TensorFlow Lite provides various user applications such as image classification, object detection, pose estimation, speech recognition, gesture recognition, segmentation, text classification, on-device recommendation, natural language question answering, digit classifier, style transfer, super-resolution, audio classification, and reinforcement learning tasks [2]. To perform on-device inference, an interpreter should be created in Java code. The interpreter encapsulates and optimizes the tflite model with the limited kernel and operation set. The interpreter receives two inputs in Java code. One is a tflite model file, and another is an option of tflite. In this option, developers can set a delegate. The delegate option includes default, GPU delegate, and NNAPI delegate. The default setting provides processing optimized for ARM Neon. GPU delegate allows GPU acceleration in both Android OS and iOS. It also supports the precision of FP32, FP16, and INT8. NNAPI delegate provides acceleration using HWs, which include NPU, GPU, and DSP. It also supports the quantization of pre-trained models.

To perform on-device inference using a smartphone app, developers write the code using the high-level ML framework. Android NN runtime provides a shared library as an interface between an application and backend drivers, as shown in Fig. 3. Android NN runtime assigns ML model to appropriate hardware based on the delegate option. However, it is not always performed on the specified hardware. Suppose NNAPI supports the acceleration of all operations in an ML model with accelerating HWs such as NPU, DSP, or GPU. In that case, all operations can be accelerated using the HW accelerators mentioned. However, NNAPI does not support the acceleration of several operations with the accelerating HWs; the unsupported operations are processed by the CPU, which is called a CPU fallback. Figure 4 illustrates an extreme case of the CPU fallback. Overhead of data movement caused by the CPU fallback degrades performances of inference per second and power consumption significantly. In some cases, CPU processing could be faster than NPU acceleration with the CPU fallback. Therefore, developers should carefully look into any operations that cause the CPU fallback in their ML model.

Although NNAPI supports more and more newly created operations to be accelerated by vendors' accelerating HWs, operations not supported by NNAPI continue to be created. In this case, developers can choose one of two options. One is to change the unsupported operations to the supported operations in NNAPI. Another is to use the vendor's SDK, which supports the operations since the vendor's SDK typically supports more of the latest operations than NNAPI. To speed up the inference of the ML model, both NNAPI and HW vendors should optimize operations and provide acceleration of newly created operations in their hardware accelerators via update.

Android NN HAL (Hardware Abstraction Layer) defines an abstraction of the various devices in the vendor's smartphone such as CPU, GPU, DSP, or NPU, as

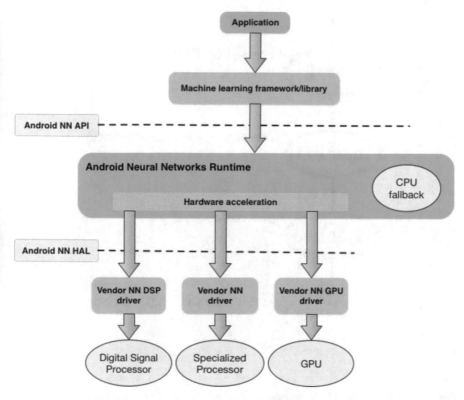

Fig. 3 Android NN runtime. (Figure is imported from Ref. [3])

shown in Fig. 5. The Android framework queries, at the initialization, the driver for its capabilities, including information of the vendor's hardware and computational performance provided by NNAPI benchmark App. Next, in the compilation step, the Android framework asks for information on operations of the neural network model that can execute in hardware. After that, it asks to prepare the execution of the model subset. The preparation includes generating code in the driver or creating re-formatted weights. In the following process, the execution of inference of the ML model is performed. The execution process starts inference in the assigned hardware, and the driver returns a callback of execution. The callback contains information about whether the execution is successful or an error occurred.

3 Qualcomm's SNPE SDK

This section introduces Qualcomm's SNPE SDK, a representative smartphone SDK. The introduction of SNPE SDK is based on the SNPE reference guide with SNPE version 1.56.2 written in October 2021 [5]. Users need a smartphone equipped with

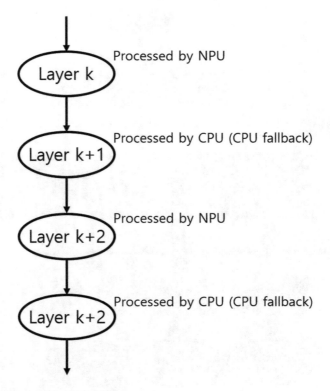

Fig. 4 An extreme case of CPU fallback

Qualcomm's AP to use this SDK. This SDK provides convenience for users such as ML model quantization, performance measurement, and on-device inference. Also, it provides a test environment of various operational modes of frequency. All SDK operations are performed by typing a predetermined script on the terminal via Android Debug Bridge (adb).

The first thing to do for on-device inference using the SNPE SDK, after the environment setting for SDK is finished, is to prepare a pre-trained model. SNPE SDK supports six formats of the pre-trained model: caffemodel files of Caffe, pb files of TensorFlow and Caffe2, ONNX files, pt. files of Pytorch, and tflite files of TensorFlow Lite. Pre-trained model files containing several unsupported libraries such as Keras can be transformed into ONNX files using some open source codes. These files of six formats are converted pb file to dlc file; users should type the script "snpe-tensorflow-to-dlc", after entering "adb shell" in the terminal window, as shown in Fig. 6.

After converting the dlc file, users should quantize the dlc file using the command "snpe-dlc-quantize", as shown in Fig. 7. For the quantization, some input images are required. With these inputs, the quantization algorithm can set the scaling factor of layers of the ML model, estimating the range of activation of the layers.

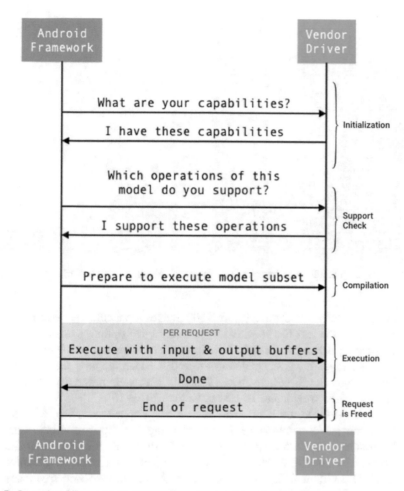

Fig. 5 Operational flow of Android NN HAL. (Figure imported from Ref. [4])

```
snpe-tensorflow-to-dlc --input_network $SNPE_ROOT/models/inception_v3/tensorflow/inception_v3_2016_08_28_frozen.pb \
                       --input_dim input "1,299,299,3" --out_node "InceptionV3/Predictions/Reshape_1" \
                       --output_path inception_v3.dlc
```

Fig. 6 An example Script for converting pb to the dlc file format. (Figure is imported from Ref. [5])

```
snpe-dlc-quantize --input_dlc inception_v3.dlc --input_list image_file_list.txt
                  --output_dlc inception_v3_quantized.dlc
```

Fig. 7 Script for quantization of dlc file. (Figure is imported from Ref. [5])

158   H. Park and S. Kim

	CPU_ub_float_timing(2 runs)				CPU_ub_tf8_timing(2 runs)				GPU_ub_float_timing(2 runs)			
	avg (us)	max (us)	min (us)	runtime	avg (us)	max (us)	min (us)	runtime	avg (us)	max (us)	min (us)	runtime
Load	168	178	159	CPU	535	541	529	CPU	216	218	214	CPU
Deserialize	43783	44037	43529	CPU	43517	43525	43510	CPU	43539	43757	43321	CPU
Create	284169	284756	283583	CPU	284869	285136	284602	CPU	1668233	1672069	1664398	CPU
Init	334161	334473	333849	CPU	333740	334046	333434	CPU	1722558	1726831	1718286	CPU
De-init	92654	94240	91088	CPU	94061	96144	91979	CPU	20458	20844	20072	CPU
Create Network(s)	232383	233174	231592	CPU	232643	232896	232391	CPU	170395	170549	170237	CPU
RPC Init Time	0	0	0	CPU	0	0	0	CPU	0	0	0	CPU
Snpe Accelerator Init Time	0	0	0	CPU	0	0	0	CPU	0	0	0	CPU
Accelerator Init Time	0	0	0	CPU	0	0	0	CPU	0	0	0	CPU
Total Inference Time	465089	473922	456257	CPU	458832	457514	456150	CPU	22445	22454	22437	CPU|GPU
Forward Propagate	463871	472715	455028	CPU	452557	453357	451757	CPU	22166	22175	22157	CPU|GPU
RPC Execute	0	0	0	CPU	0	0	0	CPU	0	0	0	CPU|GPU
Snpe Accelerator	0	0	0	CPU	0	0	0	CPU	0	0	0	CPU|GPU
Accelerator	0	0	0	CPU	0	0	0	CPU	0	0	0	CPU|GPU
Misc Accelerator	0	0	0	CPU	0	0	0	CPU	0	0	0	CPU|GPU
layer_000 [Name:data Type:data]	1	1	1	CPU	1	1	1	CPU	89	89	89	GPU
layer_001 [Name:conv0 Type:convolutional]	3605	3706	3501	CPU	3318	3353	3283	CPU	334	335	333	GPU
layer_002 [Name:conv0/relu Type:neuron]	375	397	353	CPU	372	374	371	CPU	131	132	130	GPU
layer_003 [Name:conv1/dw Type:convolutional]	71856	72866	70866	CPU	70747	71256	70238	CPU	249	251	247	GPU
layer_004 [Name:conv1/dw/relu Type:neuron]	418	446	391	CPU	428	440	416	CPU	0	0	0	GPU
layer_005 [Name:conv1 Type:convolutional]	8588	8774	8403	CPU	6205	6278	6132	CPU	676	677	676	GPU
layer_006 [Name:conv1/relu Type:neuron]	756	773	740	CPU	763	767	759	CPU	0	0	0	GPU
layer_007 [Name:conv2/dw Type:convolutional]	35439	36237	34641	CPU	34005	34033	33978	CPU	304	307	301	GPU
layer_008 [Name:conv2/dw/relu Type:neuron]	153	156	151	CPU	160	162	159	CPU	0	0	0	GPU
layer_009 [Name:conv2 Type:convolutional]	3960	3970	3951	CPU	3993	4003	3983	CPU	584	585	584	GPU
layer_010 [Name:conv2/relu Type:neuron]	374	403	346	CPU	348	350	347	CPU	0	0	0	GPU
layer_011 [Name:conv3/dw Type:convolutional]	76525	77420	75630	CPU	75138	75364	74912	CPU	269	272	267	GPU

Fig. 8 Qualcomm SNPE's benchmark results. (Figure is imported from Ref. [5])

According to Qualcomm's guide [5], 50–100 images are recommended for accurate quantization. In the text file specified by the input_list option, paths of the image files are written, where the format of the image files is binary, which are converted from original input images. Qualcomm's SNPE SDK supports the quantization into 8-bit weight and 8-bit activation (W8A8), 8-bit weight and 16-bit activation (W8A16), or 16-bit weight and 16-bit activation (A16W16).

Users can get a dlc file if no error message is popped up. Now, with the dlc file, users can perform on-device inference using the "snpe-net-run" command [5]. The required arguments for "snpe-net-run" are the dlc file and a text file that specify a path to input files. Optionally, hardware that performs on-device inference such as GPU, DSP, or CPU can be specified. Also, the hardware frequency can be adjusted with several operation modes such as system_settings, power_saver, balanced, default, high_performance, sustained_high_performance, and burst.

Qualcomm SNPE provides a metric of inference throughput in a specified HW using the "throughput-net-run" command [5]. It executes multiple instances during a time interval that users set using the "–duration" option. For the time interval, it executes the inferences consecutively and measures the number of executions.

Qualcomm's SNPE provides a Python script for profiling, which is snpe_bench.py. The Python script creates a csv file containing the measured time of each layer of an executed ML model, as shown in Fig. 8. Users can analyze which layer causes a bottleneck in the execution and replace the operation of the layer with another operation. It also provides time for loading an ML model binary file and initializing HW. In addition, it shows a performance comparison of supported HWs such as CPU, GPU, and DSP. Users can choose the optimal HW with this profiling.

Qualcomm SNPE can show the structure of the ML model contained in a dlc file using the "snpe-dlc-viewer" command, as shown in Fig. 9. Users can check if their ML model is well converted, and therefore it allows them to debug easily. When they

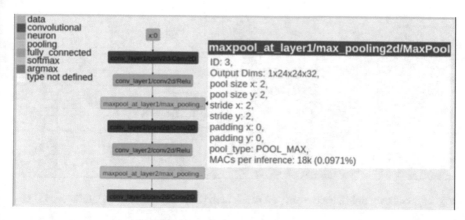

Fig. 9 Performance analysis flow of Qualcomm's SNPE benchmarking tools. (Figure is imported from Ref. [6])

point out a layer with their mouse, information about the layer such as input/output dimension, padding, stride, filter size, and Multiply-ACcumulate (MAC) counts. In order to improve readability, operations of the same type are displayed in the same color.

Qualcomm SNPE's update cycle is several weeks to several months, and the update cycle is much shorter than TensorFlow Lite's update cycle. The update fixes some bugs or supports operations that were not supported before and supports new operations. Users can check what is updated in the release notes.

4 ML Benchmarks for On-Device Inference in Smartphone

In order to evaluate the AI performance of smartphones that equips with different vendors' NPUs, ML benchmarks perform various AI tasks on-device on the smartphone based on the same criteria. ML benchmarks execute predetermined AI tasks consecutively on a smartphone, showing the scores of several performances. It ranks the scores of various smartphone models, and the ranking is able to provide information about which device shows the best AI performance.

As of 2021, less than 5 years have passed since the first release of commercial NPU for a smartphone. HW architecture, SW, and SDK are all different, and TensorFlow Lite, the representative unified SDK, still does not bring out the best performance of each vendor's AI accelerating HWs. Also, the precision of the HWs is not the same. Some NPUs only support 8-bit precision [7], and some NPUs support 16-bit precision [8, 9]. Therefore, it is not easy for ML benchmarks to compare AI performances fairly. Therefore, users should look closely into the policy of the ML benchmarks and the features of NPU of each vendor if they want to compare the performance of NPU fairly.

Benchmark	API/SDK	Neural Network models				Precision	Input resolution
		Classification	Detection	Image enhancement/segmentation	Language/Text		
AI benchmark	NNAPI or Vendor SDK	MobileNet V2				INT8, FP16	224x224
		Inception V3				INT8, FP16	346x346
		MobileNet V3 Large-M				INT8, FP16	512x512
					CRNN/Bi-LSTM	FP16, FP32	64x200
				PyNET-Mini		INT8, FP16	96x96
				VGG19		INT8, FP16	256x256
				SRGAN		INT8, FP16	512x512
				U-Net		INT8, FP16	384x384
				DeepLab V3+		INT8, FP16	513x513
				DPED-ResNet		INT8, FP16	128x192
					Static Rnn/LSTM	FP16	32x500
MIperf	NNAPI or Vendor SDK	MobileNetEdgeTPU				Chosen by Vendors	224x224
			MobileNet V2-SSD				300x300
				DeepLab V3+			512x512
					Mobile-BERT		
UL Procyon	NNAPI	MobileNet V3				INT, FP	
		Inception V4					
			MobileNet V3-SSDLite				
				Deeplab V3			

Fig. 10 Summary of ML benchmarks, AI benchmark [10], MLperf [11], and UL Procyon [12]

This section explains how policies of ML benchmarks and characteristics of NPUs affect scores to provide a guide for users and developers to more objectively evaluate the AI performances of smartphones. This section introduces representative ML benchmarks for on-device inference in smartphones: AI benchmark, MLperf, and UL Procyon. This section explains how policies and NPU characteristics of various ML benchmarks affect scores. Therefore, it provides a guide for users and developers to more objectively compare the AI performance of smartphones.

Figure 10 shows a summary of representative ML benchmarks, AI benchmark [10], MLperf [11], and UL Procyon [12]. AI benchmark v4 perform 14 benchmark sections, which are subdivided into 46 AI tests according to precision and executing HW. The tests cover various AI areas practically utilized in smartphones such as classification, face recognition, OCR, deblurring, super resolution, bokeh rendering, semantic segmentation, image enhancement, and text completion. The tested ML models of 14 sections are as follows: ModbileNet-V2, Inception V3, MobileNet V3, MobileNet V2, Convolution Recurrent Neural Networks (CRNN), PyNET, Visual Geometry Group 19 (VGG19), Super Resolution Generative Adversarial Network (SRGAN), U-Net, DeepLab-V3+, DSLR (Digital Single-Lens Reflex) Photo Enhancement Dataset (DPED) ResNet, Long Short-Term Memory (LSTM), and Super-Resolution Convolutional Neural Networks (SRCNN). Sections 1, 4, and 11 test the performance of parallel execution of the ML models. The parallel execution is a known scenario of real-time inference for camera application of smartphone. Section 11 performs a memory test. It executes inferences of SRCNN model of various input resolutions from 200×200 pixels to 2000×2000 pixels [13], and calculates a score based on which input resolution throws Out-Of-Memory-Error. Each test consumes time between 20 and 30 seconds according to test section, in the version 3.0.2 (Figs. 11 and 12).

After finishing every test, inference time, initialization time, and a score converted from each test section are displayed. The score of each section consists of 5 scores as follows: FP16/32 score @ CPU, INT8 score, FP16 score, L1 error of INT8, and L1 error of FP16. The L1 error is computed from the difference between inference and ground truth output and used to represent penalizing the scores. Some APs with embedded NPU only support integer format of data processing. The NPU

Fig. 11 Sample images of AI Benchmark App. (Figure is adapted from Google Play store)

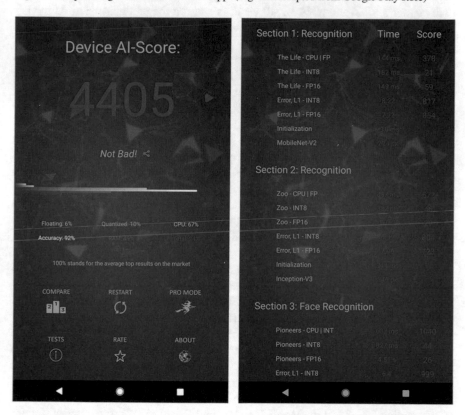

Fig. 12 An example of ML benchmark results. (Figure is imported from Ref. [13])

shows the best performance for the integer-based ML model but cannot support the FP-based ML model. AI benchmark App executes FP32 test on GPU or CPU. Therefore, smartphones integrating INT8-based-NPU can receive lower scores than those integrating FP16-based-NPU, resulting in considerable damage to the total score.

The AI benchmark's early versions, v1 to v3, only supported NNAPI. As the authors explained in Sect. 2, as of 2021, most vendors' accelerating HWs still cannot perform on-device AI tasks in optimal performance on NNAPI. For this reason, the scores of several AP vendors did not reflect the AP's optimal AI performance. Therefore, AI benchmark v4 begins to permit the vendor's SDK.

MLperf Mobile inference benchmark is an open-source benchmark developed by academia and industry [11]. As the authors said in the earlier section, NNAPI's long update period and various HW features of each vendor have made it difficult for APs to run with the optimal performance based on NNAPI. Therefore, benchmarks based on NNAPI have not shown the optimal AI performance of vendors' AP. In response to the needs of the industry for a fair comparison between APs, MLperf allows both vendor's SDK and NNAPI. AP vendors can run the predetermined AI models flexibly to maximize their performance while obeying the MLperf organization's rules.

MLperf Mobile v1.0 performs benchmark tests in 5 criteria: image classification single stream with MobileNetEdgeTPU model, image classification offline with MobileNetEdgeTPU model, object detection with MobileDetSSD model, image segmentation with Deeplab V3+ model, and natural language processing with MobilneBert model [14].

Figure 13 shows a screenshot of the MLperf App showing scores of the test results. Vendors can choose any precision of the AI model as long as they achieve the predetermined accuracy. Therefore, an INT8-based HW may have an advantage in scores over FP-based HW. Each test is performed for 60 seconds. Therefore, vendors should allow running inference with sustained maximum performance rather than peak performance, considering thermal throttling. Therefore, not only is the AP's performance important, but also the heat dissipation capability of their smartphones itself is also concerned.

Figure 14 shows the possible combination of ML models, precisions, SDKs, and HWs of the MLperf benchmark [15]. Vendors can choose the optimal combination for achieving the best benchmark score. For example, Qualcomm can choose the combination as follows: MobileNetV2 – INT8 – SNPE – DSP.

UL Benchmark was established in 1997 and has performed benchmarks in various areas such as smartphones, storage, virtual reality (VR), and so on [16]. UL launched Procyon in 2020, a benchmark service only for paid professional users. Figure 15 shows Screenshots of the UL Procyon AI inference benchmark App [16]. It executes inferences of four ML models consecutively using NNAPI: MobileNet V3, Inception V4, SSDLite MobileNet V3, and DeepLab V3. It produces a score based on latency and accuracy and shows scores of float- and integer-ML models. Since it does not support the vendor's SDK, smartphones with low optimization of NNAPI may show low scores.

Fig. 13 A screenshot of MLperf Mobile inference benchmark App. (Figure is imported from Ref. [15])

5 Summary and Discussion

Each of the vendors' NPUs has its own strengths and weaknesses. Some NPUs may be efficient in INT8 performance, and others may be strong in FP16 performance. Also, the support of various precisions in NPU can be a factor that lowers the performance of a specific precision, causing inefficiency in HW. In addition, NPUs with a well-designed zero-skipping feature may show outstanding performance when executing ML models containing the ReLU activation. Also, some NPUs do not support specific operations, so they will inefficiently combine them into several other operations to process within NPUs, or cause the CPU fallback. Even if the NPU is well designed, it may show low performance depending on the optimization of SDK and API.

ML benchmarks attempt to objectively evaluate the performance of NPUs with these individual characteristics, but it is not an easy task. The distinct characteristics

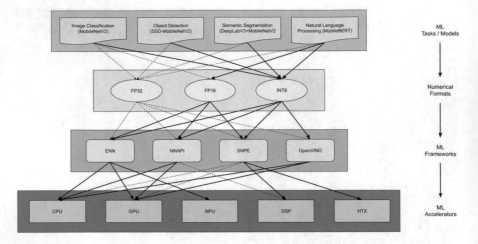

Fig. 14 Possible combination of ML models, precisions, SDKs, and HWs of the MLperf benchmark test. (Figure is imported from Ref. [15])

Fig. 15 Screenshots of UL Procyon AI inference benchmark App. (Figure is imported from Ref. [12])

of NPU allow scores and rankings to change according to the policy of ML benchmarks. MLperf gives each vendor a degree of freedom to achieve maximum performance. However, some AP vendors still do not participate in the audition of the MLperf benchmark, and the scores depend on vendors' efforts to this benchmark than NPU's performance. Therefore, for users and developers to make a fair evaluation of NPUs, it is necessary to identify the strengths and weaknesses of each vendor's NPU and the evaluation policy of ML benchmarks.

In order to compare and understand the AI performance of each vendor's AP, in addition to identifying the NPU HW structure, SDK/API must be identified. The evaluation policy of ML benchmarks must be identified for objective performance comparison. To deliver the background information related to ML benchmark tests, this chapter introduced the structure of NNAPI and Qualcomm SNPE, one of the representative vendor SDKs. In addition, three representative ML benchmarks for smartphone on-device conference evaluation were introduced.

Acknowledgments This work was supported by two Institute of Information & Communications Technology Planning & Evaluation (IITP) grants funded by the Korea government (MSIT) ("No.2020-0-00056, To create AI systems that act appropriately and effectively in novel situations that occur in open worlds").

References

1. Open Neural Network Exchange: The Linux Foundation. https://onnx.ai/. Accessed 5 Jan 2022
2. TensorFlow Lite example apps: Google. https://www.tensorflow.org/lite/examples. Accessed 5 Jan 2022
3. Android NDK guide: Google. https://developer.android.com/ndk/guides/neuralnetworks. Accessed 5 Jan 2022
4. Neural Networks API driver: Google. https://source.android.com/devices/neural-networks. Accessed 5 Jan 2022
5. Snapdragon Neural Processing Engine SDK reference guide: Qualcomm. https://developer.qualcomm.com/sites/default/files/docs/snpe/. Accessed 5 Jan 2022
6. Qualcomm SNPE's performance analysis using benchmarking tools: Qualcomm. https://developer.qualcomm.com/software/qualcomm-neural-processing-sdk/learning-resources/vision-based-ai-use-cases/performance-analysis-using-benchmarking-tools. Accessed 5 Jan 2022
7. Song, J., Cho, Y., Park, J.-S., Jang, J.-W., Lee, S., Song, J.-H., Lee, J.-G., Kang, I.: 7.1 an 11.5 tops/w 1024-mac butterfly structure dual-core sparsity-aware neural processing unit in 8nm flagship mobile soc. In: 2019 IEEE International Solid-State Circuits Conference-(ISSCC), pp. 130–132. IEEE (2019)
8. MediaTek Dimensity 800 introduction: MediaTek. https://www.mediatek.com/products/smartphones/dimensity-800. Accessed 5 Jan 2022
9. Huawei Kirin 990 5G launch event: YouTube. https://www.youtube.com/watch?v=9zkxuxpObKI. Accessed 5 Jan 2022
10. Ignatov, A., Timofte, R., Chou, W., Wang, K., Wu, M., Hartley, T., Van Gool, L.: Ai benchmark: Running deep neural networks on android smartphones. In: Proceedings of the European Conference on Computer Vision (ECCV) Workshops (2018)
11. Reddi, V.J., Cheng, C., Kanter, D., Mattson, P., Schmuelling, G., Wu, C.-J., Anderson, B., et al.: Mlperf inference benchmark. In: 2020 ACM/IEEE 47th Annual International Symposium on Computer Architecture (ISCA), pp. 446–459. IEEE (2020)
12. UL Procyon AI Inference Benchmark: UL Procyon. https://benchmarks.ul.com/procyon/ai-inference-benchmark. Accessed 5 Jan 2022
13. Ignatov, A., Timofte, R., Kulik, A., Yang, S., Wang, K., Baum, F., Wu, M., Xu, L., Van Gool, L.: Ai benchmark: All about deep learning on smartphones in 2019. In: 2019 IEEE/CVF International Conference on Computer Vision Workshop (ICCVW), pp. 3617–3635. IEEE (2019)
14. MLperf v1.0 Results (May 19, 2019): MLPerf. https://mlcommons.org/en/inference-mobile-10/. Accessed 5 Jan 2022
15. Reddi, V.J., Kanter, D., Mattson, P., Duke, J., Nguyen, T., Chukka, R., Shiring, K., et al.: MLPerf mobile inference benchmark. arXiv preprint arXiv:2012.02328v4 (2022)
16. UL Benchmarks: UL. https://benchmarks.ul.com/. Accessed 5 Jan 2022

Hardware Accelerators in Embedded Systems

Jinhyuk Kim and Shiho Kim

1 Introduction

1.1 Introduction to Embedded Systems

The empirical law for embedded system design is that adding hardware increases a power requirement. However, with hardware accelerators, adding hardware can affect performance without severely increasing power requirements. Analyzing algorithms of programmable logic and implementing appropriate accelerators allow designers to increase design performance while reducing power consumption in embedded computing systems. Various test results show that the accelerator extends the trade-off option from up to $200\times$ performance improvement for the same performance to 90% performance reduction for the same performance [1].

It is somewhat unfair that programmable logic has maintained its reputation throughout its early history as a method of designing logic that consumes a lot of power. Typically, for the Complementary Metal Oxide Semiconductor (CMOS) process technology, the power consumption of an integrated circuit is approximately proportional to the area of the chip, and designs implemented using programmable logic tend to be larger than those implemented using wired logic. However, these two factors are apparent but misleading.

More important than the area-dependent performance dependency is the frequency-dependent performance dependency of the integrated circuits. Because CMOS circuits consume most of the current when the logic gate changes states, the circuit's frequency has a much more significant impact on power consumption than a simple chip size. The higher the frequency, the greater the power requirement.

J. Kim · S. Kim (✉)
School of Integrated Technology, Yonsei University, Incheon, South Korea
e-mail: jinhyuk.kim@yonsei.ac.kr; shiho@yonsei.ac.kr

© The Author(s), under exclusive license to Springer Nature Switzerland AG 2023
A. Mishra et al. (eds.), *Artificial Intelligence and Hardware Accelerators*,
https://doi.org/10.1007/978-3-031-22170-5_6

This frequency dependency of power consumption allows designers to reduce chip power consumption by adding circuitry when hardware additions significantly reduce clock speeds.

For years, embedded processors have relied on custom hardware features to accelerate routine algorithms such as graphics or signal processing and perform more tasks per clock cycle. This approach increases system performance without reducing the system clock or dynamic power consumption. If one can apply hardware to accelerate software algorithms and reduce clock speeds, we can save power while maintaining system performance.

However, not all functions are equally applicable to frequency trading circuits. Sequential processes that require one step to be completed before the next begins often do not see the benefit of additional circuitry. On the other hand, if the hardware can perform multiple steps at the same time, the parallel operation capability can be performed faster. The result is a higher performance for a given clock speed but lower clock speeds for a given performance class. Therefore, adding hardware to the chip design can reduce power requirements while maintaining performance.

Commonly used hardware accelerators range from fully customizable application-specific integrated circuits (ASICs) designed for specific functions, such as floating point units, to more flexible graphics processing units (GPUs) and highly programmable field programmable gate arrays (FPGAs). Since these devices require different programming models and have different system-level interfaces, it is natural that there will be different trade-offs between versatility, performance, and power consumption.

1.2   What Are the Components of the Embedded System?

Embedded systems are configured differently for different purposes but typically consist of three components: hardware, software, and real-time operating system (RTOS) [2]. The *embedded system's hardware* is integrated with memory, DSP, and CPU on a single chip. In general, hardware based on a microprocessor or microcontroller is used. *Software and firmware* for embedded systems typically require an elementary level of performance. Software required for IoT systems or industrial microcontrollers requires little memory. *RTOS* is required to manage and supervise the operation of the software during the execution of the embedded system. Not all embedded systems are included; this feature is rarely needed, especially for small systems.

Figure 1 shows an example of how an embedded system is constructed. First, the embedded system receives a user's analog or digital input. Input devices include sensors, keypads, touch screens, push-button switches, etc. After that, it is processed according to the input inside the system. Depending on the purpose of the system, it may be a simple operation or complex learning may be required. The final result is converted into a digital format and outputted through a device or system, such as a monitor or actuators. Throughout this process, it is possible to interact by being connected to other subsystems.

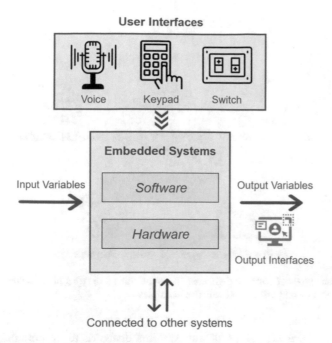

Fig. 1 Brief description of typical embedded systems. Embedded systems might have inputs and corresponding outputs with various user interfaces and can connect to other systems or subsystems. (Figure adapted from Ref. [3])

Embedded systems have different characteristics depending on their purpose, but typically they have the following characteristics:

- In most cases, it consists of hardware, software, and firmware.
- It can be embedded in parts of the system to serve some purpose.
- It is a microprocessor or microcontroller-based board with computing power as a standalone.
- They can be utilized with real-time sensing, computing, or IoT communication functions.
- Complexity and functionality vary, affecting the system's hardware, software, and firmware types.
- There may be some constraints to the smooth operation of the entire system. For example, it needs to perform some tasks within a specific time frame.

1.3 Types of Embedded Systems

Embedded systems can be categorized into several categories depending on their functionality. *Network embedded systems* are connected to a network and operate for communication with other systems, including a housing security system. *Mobile*

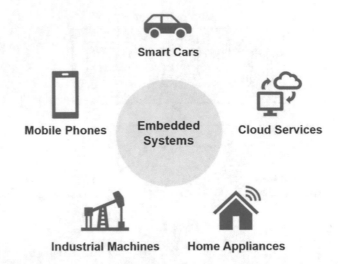

Fig. 2 Various types of embedded systems. There are far more types of embedded systems, but here we introduced only a few representative examples

embedded systems are a type of tiny systems designed to be portable, including smartphones and digital cameras. *Real-time embedded systems* perform an error-free operation, such as a traffic control system, over a specific interval or time. *Standalone embedded systems* operate independently of the entire system. It works to perform a particular task, but unlike other embedded systems, it does not belong to the main system. Examples of *standalone embedded systems* are calculators and cameras mounted inside smartphones.

As shown in Fig. 2, various types of embedded systems depend on the base technology and the board's purpose.

- *Autonomous vehicles*: Modern vehicles are equipped with several embedded systems for various purposes. There are also embedded systems for performing essential functions of the vehicle, such as cruise control, navigation, and airbags, but they are also equipped with embedded systems for additional purposes such as air circulation, temperature control, and entertainment.
- *Mobile phones*: Mobile includes various embedded systems such as communication, GUI software, OS, camera, microphone, input/output module, etc.
- *Cloud services* can reduce individual IT resources by allowing significant functions such as controllers, data processing, and memory on cloud computing servers. Individuals can use various types of cloud functions on PCs by maintaining minimal control and communication functions.
- *Industrial machines*: They may include an embedded system or may have the characteristics of an embedded system. It aims to implement functions to perform specific purposes, such as surveillance cameras and delivery robots, within the machine. Ultimately, it aims to design so that onboard training is possible, but in reality, it is implemented based on information.

2   Hardware Accelerating in Embedded Systems

This chapter will discuss challenges faced by hardware accelerators for embedded systems. We will review the various commercial hardware accelerators for embedded systems by discussing commercialized accelerators' specifications, performance, and power consumption.

2.1   Current Issues in Embedded Systems

Embedded systems are typically designed to perform only selected functions at low cost; some require high-speed processing, while others do not care about speed. Many components of embedded systems often perform poorly. Compared with the hardware of general-purpose computer systems, the overall structure of the embedded system is intentionally simplified to reduce the unit cost. Since many embedded systems are mass-produced in millions, reducing production costs is inevitably a significant concern. Some embedded systems do not require considerable processing power and resources and thus can be designed with relatively slow processors and less memory to reduce cost. In other words, there are many constraints to consider when designing an embedded system—the following are the considerations for designing embedded systems [3, 4].

- *Power consumption*: It is a critical issue in most embedded systems. The power is closely related to the operating voltage and the clock frequency, representing the operation's strength. Therefore, finding the optimal operating point according to the system is a way to minimize power consumption. Under a stable power supply, low-power execution is generally one of the factors that secure life reliability. Power control may be implemented to maximize efficiency for given power consumption. Power gating refers to reducing leakage power in a circuit not used in the current standby state [5]. Clock gating reduces dynamic power by removing clock signals when a circuit is not in use [6]. These methods can reduce the switching power consumption of the circuit.
- *Energy efficiency*: The energy consumption of embedded systems affects battery use. That is, by saving energy, the battery life may be increased, and the length of the battery charging interval may be extended. Since energy represents the total amount of work done, energy efficiency refers to the ability to perform the necessary work with as little energy as possible. If complex neural computing is dependent on cloud computing, it consumes less computation energy but a lot of energy in data transmission over the network. Conversely, performing complex computations on its own in the system reduces the energy consumed by networking but increases the energy consumed by the computation. In other words, the optimization method of the embedded system varies depending on the conditions required by the entire system.

- *Connectivity*: Because traditional embedded systems cannot perform complex neural network computing, they rely on the performance of remote computing clusters using methods such as cloud computing. Therefore, constant and stable network connectivity must be secured. The loss of network connectivity causes significant problems in the system and further provides a low level of user experience for ordinary users. Network-dependent services include autonomous vehicles, unmanned aircraft, drones, and satellites. Therefore, securing network connectivity is indispensable to providing a smooth user experience.
- *Reliability*: Reliability becomes a factor in evaluating embedded systems in various aspects. Unlike modular systems, embedded systems are difficult to reprogram in the field. Even if a defect is found, it is not very easy to upgrade software/hardware, and adjusting it can affect other systems by itself. Thus, embedded systems must be able to accurately and continuously perform the desired performance of the entire system, and for this purpose, embedded CPUs with high reliability and low power consumption, such as Advanced RISC (Reduced Instruction Set Computer) Machines (ARM) cores, are used. Reliability is also closely related to other factors mentioned above (e.g., connectivity). To increase reliability by relying less on network connectivity and external systems, some embedded systems process some or all of the neural network computations on local processors.

2.2   Commercial Options for Hardware Accelerators

Various hardware solutions have been proposed to accelerate neural computing in embedded system domains, ranging from hardware accelerators of various methods such as GPUs, NPUs, and VPUs to systems of various types such as mobile and HPC [4, 7]. Table 1 summarizes embedded systems' commercial devices used for neural network acceleration. This section will cover the rest of the three boards (Rockchip RK3399Pro [10], Intel Movidius Myriad X [11], and NVIDIA Jetson AGX Xavier [12]) that will be featured in Sect. 3.

- *Raspberry Pi 4*: The Raspberry Pi 4 Model B [8] is an affordable single-board computing device with a small form factor. It uses a Broadcom processor with a quad-core Cortex-A72 64-bit ARM CPU, Video core VI GPU, 2-8GB LPDDR4–3200 SDRAM, and various on-board I/Os including wireless LAN, Bluetooth, Gigabit Ethernet network, HDMI, USB-C, etc. The GPU in the Raspberry Pi 4 does not support the General Purpose Graphics Processing Unit (GPGPU) or specialized accelerators in the SoC.
- *Arduino Portenta H7*: The Arduino Portenta H7 [9] provides an affordable embedded hardware-acceleration solution. It contains a dual-core ARM Cortex-M7 CPU and a 32-bit M4 MCU with Chrome ART GPU. The Arduino Portenta H7 does not have GPGPU support or dedicated accelerators in the SoC for neural network computations. It provides practical processing solutions for low-end embedded or IoT devices.

Table 1 Configuration of several types of commercial embedded systems

Commercial device	CPU	Accelerator	Max power
Raspberry Pi 4Model B [8]	4 ARM Cortex-A72 cores	None	10 W
Arduino Portenta H7 [9]	2 ARM Cortex-M7 cores 1 Cortex-M4 MCU core	None	N/A
Rockchip RK3399Pro [10]	2 ARM Cortex-A72 cores 4 ARM Cortex-A53 cores	Mali-T860MP4 GPU	5 W
Intel Movidius Myriad X [11]	2× LEON4 cores	20+ VPUs with Neural Compute Engine	N/A
NVIDIA Jetson AGX Xavier [12]	8 ARM v8.2 cores	512-core Volta GPU 64 Tensor Cores	30 W
Google Edge TPU [13]	4 ARM Cortex-A53 cores 1 Cortex-M4F core	TPU coprocessor	2 W
Xilinx PYNQ-Z1 FPGA [14]	2 ARM Cortex-A9 cores	Programmable fabric	5 W
Graphcore IPU GC200 [15]	None	1216 IPU-cores	N/A
Qualcomm Cloud AI 100 [16]	None	16 AI cores	75 W

- *Google Edge TPU*: An edge tensor processing unit (TPU) [13] is a neural network accelerator for small, low-power platforms. The Edge TPU coprocessor is equipped with quad-core Cortex-A53 and Cortex-M4F ARM CPU, GC7000 Lite graphics card, 1GB LPDDR4 memory, and various I/O peripherals such as USB-C, Gigabit Ethernet, HDMI, etc. It supports TensorFlow Light and is the only Google-made TPU product currently available to consumers as a general commercial product, not for data centers. Details of the accelerator used in the TPU have not yet been released.
- *Xilinx PYNQ-Z2 FPGA*: Xilinx PYNQ-Z1 FPGA [14] is a programmable SoC for embedded systems. It integrates reconfigurable fabric and dual-core Cortex-A9 ARM CPU with 512 MB DDR3 memory. PYNQ is a Zynq series board that supports programming in Python languages. PYNQ-Z1 FPGA board supports 13,300 logic slices and 630 KB fast block RAM.
- *Graphcore IPU GC200*: With 59.4 billion transistors and the latest TSMC 7 nm process, the Colossus MK2GC200 IPU [15] is an advanced processor. Each MK2 IPU has 1472 processor cores running approximately 9000 independent parallel program threads. Each IPU has 900 MB of processor memory and 250 TeraFLOPS AI calculations in FP16.16 and FP16.SR (probably rounded). The GC200 is designed to support FP32 computing.

- *Qualcomm Cloud AI 100*: The Qualcomm Cloud AI 100 chip [16], which comprises 16 AI cores, achieves up to 400TOPs INT8 inference. The on-chip memory subsystem is supported by four 64-bit LPDDR4X memory controllers (LPDDR4X-4200) running at 2100 MHz, each driving one 4 × 16-bit channel for total system bandwidth of 134 GB/s. To overcome the limitations of relatively small bandwidth, Qualcomm designed to keep on-chip memory traffic as much as possible, using 144 MB of on-chip SRAM cache.

3   Recent Trends of Hardware Accelerators in Embedded Systems

3.1   General Purpose Graphics Processing Unit

General Purpose Graphics Processing Units are techniques used to calculate GPUs typically only calculated for computer graphics and traditionally for applications held by CPUs [17]. What makes this possible is connecting programmable layers and fixed-level operations to the graphics pipeline, which allows software developers to use stream processing for non-graphics data.

The GPU is designed to accelerate 3D computer graphics. Therefore, the GPU can only be used for general operations in very limited cases. The GPU can handle only independent vertices and fragments, but it can handle many in parallel. The GPU is beneficial when a programmer wants to process a large number of vertices or fragments in the same way. Typically, the GPU is effective in solving problems using stream processing, and recently, thanks to the development of technology, the types of problems that can be handled are increasing. In other words, the dataset of an ideal GPGPU application should be large, have high parallelism, and have minimal dependence between data elements. However, despite the efforts of graphic card manufacturers and researchers in related fields, GPU utilization areas and methods are still limited compared to CPU. One of the critical issues of GPGPU is the implementation of ultra-low power. Intel's Myriad X is one of the hardware accelerators that reflects this trend (Fig. 3).

Movidius Myriad X is a 4TOPS-performance Vision Processing Unit (VPU) for artificial intelligence (AI), an SoC-type product that incorporates its own Neural Compute Engine for deep learning acceleration at the edge. The neural computing engine integrated into Myriad X SoC is designed to respond in real-time based on the device's perception and understanding of the surrounding environment and exhibits 1TOPS deep neural network inference performance and 4TOPS comprehensive performance. The main features of this chip are as follows [18]:

16 Vector Processors Optimized for Computer Vision Workloads: Myriad X's Vector Processor is 128-bit VLIW programmable and has the flexibility to run multiple image and vision application pipelines simultaneously.

Fig. 3 Specifications of Intel Movidius Myriad X. (Figure imported from Ref. [11])

16 MIPI channels: Myriad X has a configurable MIPI channel. In line with VPU, it can process image signals of 700 million pixels per second from eight HD cameras.

Vision-Specific Hardware Accelerator: Myriad X consists of more than 20 hardware accelerators with the ability to perform optical flow or depth problems independently.

Centralized On-Chip Memory: Myriad X has 2.5 MB of memory that can support 450GB/s of internal bandwidth to minimize off-chip data transmission and achieve extremely low power and low latency.

3.2   NPU with CPUs and GPUs

Neuromorphic chips and neural processing units (NPUs), also known as intelligent processing units (IPUs), are particular network application package processors that use a data-driven parallel computing architecture to process extensive multimedia data, especially video and images. Traditional CPU/GPU works similarly, but neural network–optimized NPUs can deliver better performance [19]. Dedicated NPU devices perform more and more neural network-like operations. The NPU is also an integrated circuit, but network processing is more complex and flexible than the single functionality of the ASIC. Typically, NPUs can be used in software or hardware, depending on the type of network computing custom programming, to achieve a specific purpose in the network.

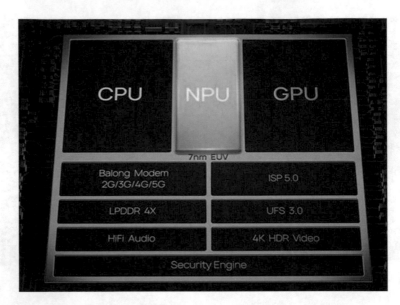

Fig. 4 Huawei Kirin 980 chip with 7 nm SoC. (Figure imported from Ref. [20])

The highlight of the NPU is the ability to run multiple parallel threads. The NPU moves in different layers, with several specific hardware-level optimizations to provide some caching schemes that are easily accessed by several different processing cores. These high-performance cores are more superficial than regular processors because they do not require multiple types of work. This optimization set makes the NPU more efficient, which is why ASICs invest heavily in R&D. One of the advantages of NPUs is that they spend most of their time in low-precision algorithms, new dataflow architectures, or in-memory computing power. Unlike GPUs, NPUs care more about throughput than latency.

NPU is an optimized processor for computation for AI. The operation for AI is a large matrix operation, and the existing CPU is not optimized for this matrix operation, so even the latest chipset did not perform well. However, the NPU is optimized for such AI operations so that they can be performed in real-time without delay. In other words, high-end AI technologies that require complex and high computation can be implemented at low power.

Deep learning inference should be made on end equipment with NPU. However, the deep learning model is more efficient for learning from large servers equipped with GPUs. In other words, GPU is required to speed up model learning, NPU is required to infer from terminal equipment at low power, and CPU is required to operate as a standalone. Therefore, as shown in Fig. 4, commercialized hardware accelerators take the form of SoC with a GPU and NPU, along with ARM core CPUs (Fig. 5).

Released in Q1 2019, the RK3399Pro has been improved by adding an NPU to the existing RK3399 board. As shown in Fig. 6, it consists of a 64-bit CPU

Fig. 5 Rockchip RK3399Pro board. (Figure imported from Ref. [10])

with two Cortex-A72, four Cortex-A534, and an ARM Mali-T864 GPU. This CPU architecture is called the big.LITTLE architecture and is typically installed to improve power efficiency by converting four low-power, low-performance cores into two high-performance cores when high-performance information is needed. The RK3399 Pro supports USB Type-C, audio/video output through the display port, H265/VP9 10bit 60fps HDMI 2.0 decoding, PCI-e, and 8-channel microphone input. It is widely compatible and can operate on Android, Linux, Chrome OS, and Windows 10.

Geekbench 4 is a complete benchmark platform that includes testing to compare the board's performance in a number of individual items, including data compression, image processing, 3D object simulation, and memory testing. In an Android 64-bit environment, comparing the average performance of 250 benchmark software, the RK3399Pro showed a 3% performance improvement on a single core and a 2% on a multi-core (Fig. 7).

Geekbench 5 is a benchmark platform for comparing computing power, making it possible to test different workloads consisting of real-world applications or tasks and compare them to each other across multiple systems. In an Android environment, the RK3399Pro increased performance by 7% on a single core and 4% on a multi-core (Table 2).

The Jetson AGX Xavier GPU is NVIDIA's highly efficient mobile integrated graphics solution released in October 2018. This device is based on a 12 nm process based on the GV10B GPU and supports DirectX 12. There are 32 texture mapping units, 16 ROPs, and 512 shading units. It also includes 64 tensor cores to accelerate

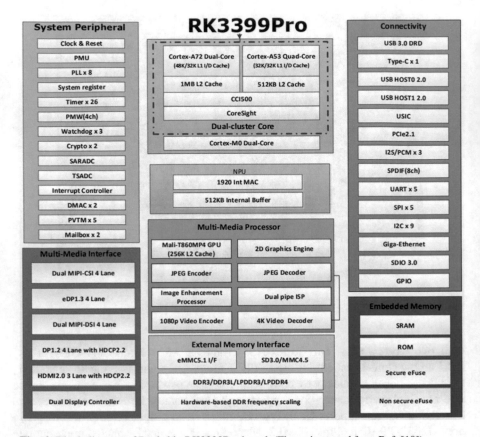

Fig. 6 Block diagram of Rockchip RK3399Pro board. (Figure imported from Ref. [10])

various types of machine learning. The GPU operates at 854 MHz and can scale up to 1377 MHz [22].

Jetson AGX Xavier is a small form factor SOM (System-on-Module) that provides supercomputing performance to edge devices. This single board can integrate digital and analog functions and places general applications in the embedded system area. It runs state-of-the-art neural networks in parallel and provides the performance needed to process data transmitted from multiple high-resolution sensors, meeting the requirements of popular AI frameworks.

According to benchmark results provided by NVIDIA [23], the Xavier series with vision accelerators performed overwhelmingly on image processing benchmarks such as Inception-V4, VGG-19, Super Resolution, U-Net, and ResNet-50. It also showed excellent performance for various object detection in real-time high-frame environments over 60 FPS, including OpenPose, Tiny YOLO V3, and SSD MobileNet-V1. SSSD ResNet34, a benchmark for high-resolution image processing of 1200×1200, shows that the Xavier series and higher specifications are essential. Since BERT, a natural language processing model, is only available on CPUs via

Fig. 7 NVIDIA Jetson AGX Xavier. (Figure imported from Ref. [12])

Table 2 Geekbench performance comparison for Rockchip RK3399 series [21]

Benchmark	Rockchip	
	RK3399	RK3399Pro
Geekbench 4 single core (Android 64-bit)	1144	1175
Geekbench 4 multi-core (Android 64-bit)	2776	2819
Geekbench 5 single core (Android)	269	251
Geekbench 5 multi-core (Android)	615	642
GFLOPS performance (SGEMM)	23.9 GFLOPS	24.56 GFLOPS

Volta, Nano and TX2 models have no benchmark results. When processing natural languages with sequence length 128, it can be seen that AGX Xavier performed better than twice in both BASE and LARGE benchmarks. Jetson Nano is suitable for learning simple AI models, and Jetson TX2 can be used for low-resolution machine vision learning. Jetson Xavier NX is sufficient for intermediate resolution machine vision and simple natural language processing, but high-resolution machine vision and general natural language processing require more than Jetson AGX Xavier (Fig. 8).

4 Conclusion

This chapter describes the definitions, characteristics, and types of embedded systems. This chapter also examined embedded systems issues and reviewed several commercial boards. One of the trends in embedded systems is the GPGPU-like

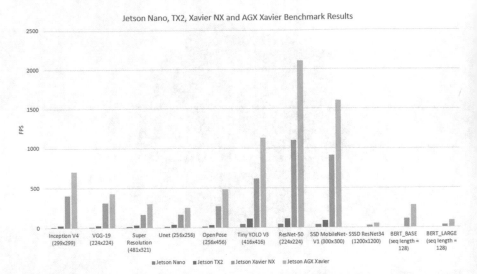

Fig. 8 Comparison of benchmark results across Jetson Nano, TX2, Xavier NX, and AGX Xavier. This figure shows inference rates per second for 11 AI inference categories: Inception V4, VGG-19, Super Resolution, U-Net, OpenPose, Tiny YOLO V3, ResNet-50, SSD MobileNet-V1, SSSD ResNet34, BERT_BASE, and BERT_LARGE [23]

Movidius Myriad X board. GPGPU aims to enable on-device AI applications to run with ultra-low power consumption rather than overwhelming performance. Another trend is the integration of ARM core CPUs and GPUs into NPUs. Some developers are looking for lightweight, affordable boards like the Rockchip RK3399Pro board, while others are looking for ultra-high performance like the Jetson AGX Xavier. Most embedded boards now focus on interference-centric performance to accommodate surveillance cameras, robots, and consumer electronics. On-board training is a remaining challenge for embedded systems to move forward.

Acknowledgments This work was supported by the Institute of Information & communications Technology Planning & Evaluation (IITP) grant funded by the Korean government (MSIT) (No.2017-0-00244, HMD Facial Expression Recognition Sensor and Cyber-Interaction Interface Technology).

References

1. Intel: Adding Hardware Accelerators to Reduce Power in Embedded Systems. Available online: https://www.intel.com/content/dam/support/cn/zh/programmable/support-resources/bulk-container/pdfs/literature/wp/wp-01112-hw-reduce-power.pdf. Accessed on 17 July 2022
2. TechTarget: Embedded system. Available online: https://www.techtarget.com/iotagenda/definition/embedded-system. Accessed on 17 July 2022

3. Codrey Electronics: What is Embedded System and How it Works? Available online: https://www.codrey.com/embedded-systems/embedded-systems-introduction/. Accessed on 17 July 2022
4. Song, W.J.: Hardware accelerator systems for embedded systems. In: Advances in Computers, vol. 122, pp. 23–49. Elsevier (2021)
5. Jiang, H., Marek-Sadowska, M., Nassif, S.R.: Benefits and costs of power-gating technique. In: 2005 International Conference on Computer Design. IEEE (2005)
6. Kathuria, J., Ayoubkhan, M., Noor, A.: A review of clock gating techniques. MIT Int. J. Electron. Commun. Eng. 1(2), 106–114 (2011)
7. Park, H., Kim, S.: Hardware accelerator systems for artificial intelligence and machine learning. In: Advances in Computers, vol. 122, pp. 51–95. Elsevier (2021)
8. Raspberry Pi: Raspberry Pi Documentation. Available online: https://www.raspberrypi.com/documentation/. Accessed on 17 July 2022
9. Arduino: Meet Portenta H7, designed for high performance. Available online: https://www.arduino.cc/pro/hardware/product/portenta-h7. Accessed on 17 July 2022
10. Rockchip: RK3399Pro. Available online: https://www.rock-chips.com/a/en/products/RK33_Series/2018/0130/874.html. Accessed on 17 July 2022
11. Hruska, J.: New Movidius Myriad X VPU Packs a Custom Neural Compute Engine. Available online: https://www.extremetech.com/computing/254772-new-movidiusmyriad-x-vpu-packs-custom-neural-compute-engine. Accessed on 17 July 2022
12. NVIDIA: Jetson AGX Xavier series module. Available online: https://www.nvidia.com/ko-kr/autonomous-machines/embedded-systems/jetson-agx-xavier/. Accessed on 17 July 2022
13. Google Cloud: Edge TPU. Available online: https://cloud.google.com/edge-tpu/. Accessed on 17 July 2022
14. Xilinx: Python Productivity for Zynq (Pynq). Available online: https://pynq.readthedocs.io/en/v2.5.1/. Accessed on 17 July 2022
15. Graphcore: Intelligence processing unit. Available online: https://www.graphcore.ai/products/ipu/. Accessed on 17 July 2022
16. Qualcomm: Cloud AI 100. Available online: https://www.qualcomm.com/products/technology/processors/cloud-artificial-intelligence/cloud-ai-100/. Accessed on 17 July 2022
17. TechTarget: GPGPU (general purpose graphics processing unit). Available online: https://www.techtarget.com/whatis/definition/GPGPU-general-purpose-graphics-processing-unit. Accessed on 17 July 2022.
18. Intel: Intel Unveils Neural Compute Engine in Movidius Myriad X VPU to Unleash AI at the Edge. Available online: https://newsroom.intel.com/news/intel-unveils-neural-compute-engine-movidius-myriad-x-vpu-unleash-ai-edge/#gs.6joc5z. Accessed on 17 July 2022
19. Utmel Electronic: Neural Processing Unit (NPU) Explained. Available online: https://www.utmel.com/blog/categories/integrated%20circuit/neural-processing-unit-npu-explained. Accessed on 17 July 2022
20. Hisilicon: Kirin 980. Available online: https://www.hisilicon.com/en/products/Kirin/Kirin-flagship-chips/Kirin-980. Accessed on 17 July 2022
21. GadgetVersus: Rockchip RK3399 vs. Rockchip RK3399Pro. Available online: https://gadgetversus.com/processor/rockchip-rk3399-vs-rockchip-rk3399pro/. Accessed on 17 July 2022
22. Techpowerup: NVIDIA Jetson AGX Xavier GPU. Available online: https://www.techpowerup.com/gpu-specs/jetson-agx-xavier-gpu.c3232/. Accessed on 17 July 2022
23. NVIDIA: Jetson Benchmarks. Available online: https://developer.nvidia.com/embedded/jetson-benchmarks/. Accessed on 17 July 2022

Application-Specific and Reconfigurable AI Accelerator

Hao Zhang, Deivalakshmi Subbian, G. Lakshminarayanan, and Seok-Bum Ko

1 Introduction

Artificial intelligence (AI) nowadays becomes ubiquitous in many fields of applications. Deep learning, as one of the most popular AI technologies, can provide state-of-the-art performance for many applications. However, their superior performance comes with high hardware cost. Deep learning models are both computation- and memory-intensive, which makes their implementation slow and have high energy consumption. To reduce the hardware cost, researchers tend to design application-specific hardware accelerators using custom design platforms [1]. In those designs, optimizations specific to the deep learning processing can be applied, and thus the performance and energy efficiency of deep learning processing can be improved.

The field-programmable gate array (FPGA) and the application-specific integrated circuit (ASIC) are two of the most popularly used custom hardware design platforms. The design of FPGA or ASIC accelerator usually needs to be performed with hardware description language (HDL), such as Verilog HDL, SystemVerilog, or Very High-Speed Integrated Circuit HDL (VHDL). In addition, the design is performed directly on the logic resources, and thus the designers need to have a

H. Zhang
Faculty of Information Science and Engineering, Ocean University of China, Qingdao, China
e-mail: hao.zhang@ouc.edu.cn

D. Subbian · G. Lakshminarayanan
Department of Electronics and Communication Engineering, National Institute of Technology, Tiruchirappalli, India
e-mail: deiva@nitt.edu; laksh@nitt.edu

S.-B. Ko (✉)
Department of Electrical and Computer Engineering, University of Saskatchewan, Saskatoon, SK, Canada
e-mail: seokbum.ko@usask.ca

© The Author(s), under exclusive license to Springer Nature Switzerland AG 2023
A. Mishra et al. (eds.), *Artificial Intelligence and Hardware Accelerators*,
https://doi.org/10.1007/978-3-031-22170-5_7

good knowledge of those hardware devices. In recent years, the high-level synthesis (HLS) design flow for both FPGA and ASIC is proposed. With this HLS flow, the development of FPGA and ASIC can be performed with high-level programming language, such as C/C++. Therefore, with a general understanding of hardware resource and architecture, software developers can easily deploy their models on FPGA and ASIC by HLS-based design flow. In this chapter, the basic structure of FPGA and ASIC will be discussed. The tools used to design on FPGA and ASIC and their corresponding design flows (both HDL-based and HLS-based flows) will also be presented.

There are already many FPGA-based or ASIC-based AI accelerators available in the literature. Similar to the classic computer architecture, the general AI accelerator is composed of the computation module, the memory module, and the control module. For those designs available in the literature, optimizations are performed for all these three major modules. In this chapter, the basic AI accelerator architecture will be discussed. In addition, optimizations proposed for computation, memory, and control modules will be presented as well. Due to the structure differences between FPGA and ASIC, the AI accelerators designed on these two platforms are different in terms of design objectives and hardware metrics. The comparison between FPGA and ASIC accelerators will be performed in this chapter.

In addition to those basic hardware optimizations, the software and hardware co-design method is also popularly applied in the literature. In the software and hardware co-design, hardware characteristics are considered during the software development. In hardware implementation, performance and energy efficiency can be improved by having fewer memory data transfer, having fewer computations, and doing computations with smaller bit-width numeric format. During model development, the model can be compressed by removing unnecessary operations, and the bit width of numeric format can also be reduced due to the error tolerance feature of AI computation. We will cover both software and hardware co-design methods, model compression and reduced-precision computation, in this chapter.

Besides manually designing machine learning models, recent automated machine learning (AutoML) technique generates AI models and applies them to specific applications automatically. It has the ability to provide more efficient models with better performance than the manually designed models. Neural architecture search (NAS) is one of the most important techniques in AutoML in which the generation of AI models is automatically performed. Model accuracy is usually set as the optimization goal. Within a certain search space, models are searched by applying certain searching strategies to try to meet the accuracy requirement. In recent years, hardware metrics are also considered in NAS searching process in order to generate a hardware-efficient AI model. This is also the software and hardware co-design field, and the software developers can contribute a lot in such hardware-aware NAS methodology.

The rest of this chapter is organized as follows: Sect. 2 presents the FPGA basics, its design tools and design flow, and some AI accelerators designed on FPGA platforms. The ASIC counterparts are presented in Sect. 3. In Sect. 4, general AI accelerator architectures and the optimizations for each component are presented.

ASIC- and FPGA-based AI accelerators are also compared in this section. In Sects. 5 and 6, the software and hardware co-design methods and the NAS-based method for AI accelerators are discussed, respectively. After reading this chapter, the readers are expected to have a good understanding of the FPGA and ASIC design platforms and the methodologies and optimizations that can be performed to design efficient AI accelerators.

2   FPGA Platform for AI Acceleration

Field-programmable gate array (FPGA) is a custom hardware design platform. It is a type of integrated circuit (IC) fabricated and provided by FPGA vendors and can be later programmed for different applications and algorithms. Different from microprocessor, where the reprogramming for specific application is achieved by software program, the programming for FPGA is to reconfigure its hardware architecture to fit into the specific application. The reconfigurability is an advantage of FPGA that makes them less expensive to adapt to new applications without fabricating a new chip.

The traditional FPGA design flow is similar to the ASIC or other IC design, where the designers need to use HDL to design and implement their expected circuits. The new HLS-based process enables the use of C/C++ programming language to design FPGA. In both design methodologies, the understanding of FPGA architecture and resource is necessary to design an efficient architecture for the target application.

2.1   FPGA Architecture

The basic structure of an FPGA is composed of four components: the look-up table (LUT), the flip-flop (FF), the input/output (IO) pads, and the wire. LUT is the basic element in FPGA to perform logic operations. FF is the register element to store LUT outputs and to enable sequential operations in FPGA. IO pads are physical ports to receive inputs and send out processing results. Wires are used to connect LUTs, FFs, and IO pads. The configuration or reconfiguration of FPGA is basically to program the LUTs and the FFs to realize logic functions and modify the way they are connected with each other using wires. FPGAs from different vendors or belonging to different product series may have different functions or various amounts of available resources; however, their basic structures are similar.

Next we will introduce the structures of the FPGA components. Xilinx and Intel are the two major FPGA vendors nowadays. We will use Xilinx Zynq UltraScale+ (xczu9eg-ffvb1156-2-e, use XCZU9EG hereinafter) and Intel Arria-10 (10AS066N3F40E2SG, use A10SX660 hereinafter) as examples when introducing the logic components in FPGA. These are the FPGA chips embedded in the

Xilinx Zynq UltraScale+ MPSoC ZCU102 Evaluation Kit and Intel Arria-10 SoC Development Kit, respectively, which are popularly used in AI accelerators design.

2.1.1 LUTs

The mapping from HDL code to LUT resources is usually performed by the design tools automatically. However, having the knowledge of LUT architecture will be helpful in designing efficient implementation.

The logic resources in XCZU9EG device are organized into configurable logic blocks (CLBs) where each CLB contains one slice and each slice has eight 6-input LUTs and sixteen FFs. Those LUTs can be used as one 6-input LUT with one output or two 5-input LUTs with separate outputs. When used as two 5-input LUTs, the two LUTs share the same inputs.

The LUTs in XCZU9EG device are able to perform any logic function of 5 or 6 Boolean variables. The LUT is actually implementing a truth table of the relationships between the input Boolean variables and the outputs. Therefore, the LUT can be thought of a collection of memory cells, and during implementation, one of the cells is selected to generate the correct output, where the selection signals are the input variables. In modern FPGA devices, part of the LUTs can also be configured as distributed memory blocks, and in this case, the inputs of the LUTs can be treated as the memory address.

In A10SX660 device, the resources are organized into logic array block (LAB). Each LAB is composed of adaptive logic modules (ALMs), and each ALM contains two adaptive LUTs (ALUTs) and four registers. The two ALUTs have up to 8 inputs, and thus each ALUT can have up to 4 independent inputs. Two ALUTs in the same ALMs can support up to six input functions. Within these limits, the ALUTs can support many functions with various amount of inputs.

Although the organization of LUTs in Xilinx and Intel devices is different, their basic structures are quite similar. Use a simple 4-bit adder as an example. For each bit position, three inputs (two operand bits a_i and b_i and the input carry bit c_i) are used to generate two outputs that are the sum bit s_i and the output carry c_{i+1}:

$$s_i = a_i \oplus b_i \oplus c_i$$
$$c_{i+1} = (a_i \cdot b_i) + (a_i \oplus b_i) \cdot c_i, \tag{1}$$

where $\oplus$ represents the exclusive-OR operation, $\cdot$ represents the logic-AND operation, and $+$ here represents the logic-OR operation. As there are only three inputs for each bit position, one LUT is enough to generate both sum and output carry in both XCZU9EG and A10S660 devices. Therefore, implementing a 4-bit adder will require four LUTs. The diagram of this 4-bit adder is shown in Fig. 1. The internal logic for each bit position, as shown in Eq. (1), is implemented as a truth table inside each LUT.

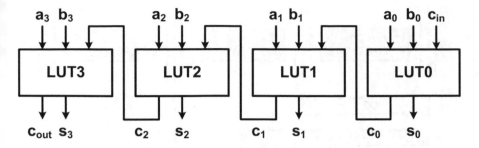

Fig. 1 Simple 4-bit adder using LUTs

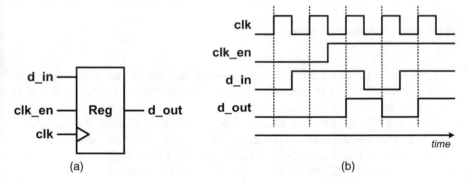

Fig. 2 FF structure and example waveform during operation. (**a**) FF structure. (**b**) Operation waveform

2.1.2 Flip-Flops

The FF is the storage unit within the FPGA logic. The diagram of a typical FF is shown in Fig. 2a. The input data d_in is stored in the FF and only passed to the output d_out when an edge of the clock signal clk arrives. The FFs in modern FPGA also have clock enable port clk_en where only when the clock enable signal is set, the input can be passed to the output at the clock edge. This clock enable port allows the data signal to be latched for more than one clock cycle. A sample waveform of those signals during operation is shown in Fig. 2b.

In addition to data temporary storage, the FF can also be used to realize pipeline design. Pipeline design is to divide a whole combinational circuit into multiple smaller circuits and use registers to pass data in-between small circuits, as shown in Fig. 3. The benefit of pipeline design is that different small circuits can be used by different operations at the same time, and thus the operation throughput can be improved. As shown in the bottom part of Fig. 3, at time t_1, operation 1, $OP1$, is using combinational circuit 1. In normal combination circuit, the second operation, $OP2$, can only start after t_3 when $OP1$ goes through all three combinational circuits. In pipelined design, as the three circuits are isolated by registers, at t_2, when $OP1$ goes to the second circuit, $OP2$ can start using circuit 1. And similarly,

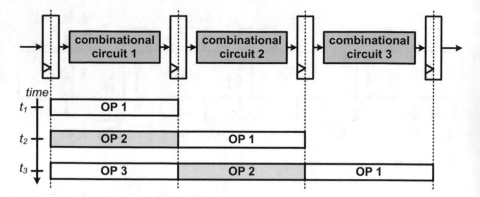

Fig. 3 Example of pipeline operation

at t_3, the third operation, $OP3$, can start. Therefore, after the initial latency, a new result can be generated at every clock cycle. The technique of pipeline design is helpful to increase the system throughput.

LUTs and FFs are the most abundant logic resources in FPGA. In XCZU9EG device, there are about 270K LUTs and 540K FFs. In A10SX660, there are 250K ALMs that are equivalent to 500K ALUTs and 1M FFs. Except some arithmetic operations that are performed using DSPs, all remaining arithmetic operations that cannot be adapted into DSPs and all other logic operations are performed by LUTs and FFs.

With these basic elements, almost all logic functions for any applications can be realized on FPGA. However, most of the applications contain a large amount of arithmetic operations and have data or parameters to be read from or written to memory. Although the LUT-based implementation can realize arithmetic operations, its speed performance is slow, which cannot meet the requirements of certain applications. In addition, the data transfer between the logic and external memory consumes a lot of power that will make the application on FPGA very power-consuming. To solve these problems, modern FPGA devices introduce more elements other than the four basic elements: including embedded block memory, DSP blocks, phase-locked loops (PLLs), transceivers, and external memory controller.

The PLLs are used to enable different clock rates for different FPGA fabrics that are useful for clock domain crossing (CDC) designs. The transceivers are designed for communication applications where FPGAs are used to receive, process, and then send data with high speed. The external memory controller can ease the use of external memory. Many popularly used data communication protocols are embedded so that the designers do not need to implement memory controller by themselves. External memory, such as double data rate (DDR) memory or modern 3D high-bandwidth memory (HBM), is essential in AI accelerator because AI model usually has a large amount of data and parameters that need to be stored in a large-

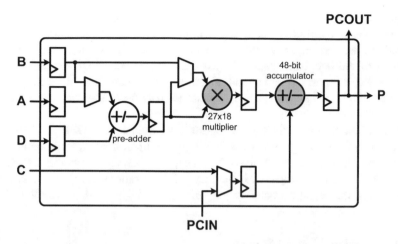

Fig. 4 Xilinx FPGA DSP architecture

capacity memory. In addition to the external memory controller, the embedded block memory and DSP blocks are also important in AI accelerators.

2.1.3   DSP Blocks

The DSP blocks are designed for performing high-speed arithmetic operations. In XCZU9EG device, the DSP block is called DSP48E2, and its main architecture for computation is shown in Fig. 4. It has four data input ports A, B, C, and D and one data output port P. $PCIN$ and $PCOUT$ are DSP internal signals used to cascade multiple DSPs for larger bit-width computation, and they are not accessible from external signals. The DSP48E2 has a 27×18 multiplier and a 48-bit accumulator, and it can be used to accomplish operations in the form of

$$P = A \times B + C. \tag{2}$$

In addition, A or B can be pre-added with D and then perform multiplication operation. The 27×18 multiplier supports 27×18 signed fixed-point multiplication or 26×17 unsigned fixed-point multiplication. The product is sent to the accumulator and is then accumulated in 48-bit fixed-point format. The DSP blocks also support standalone multiplication or addition. However, for AI accelerator, the multiply-accumulate-like operation is more common. When the AI model is quantized to 16 bit, 8 bit, or even 4 bit, the computation can be efficiently supported inside the DSP blocks. When using reduced-precision operations, special INT8 and INT4 operations are supported, which can significantly improve the computation throughput. The INT8 and INT4 operation will be discussed in detail later.

The DSP blocks in A10SX660 are designed to support different computations. For fixed-point multiplication, it supports one 27×27 or two parallel 18×19 operations. Moreover, 18×19 multiplier adder mode is supported where the products of the two 18×19 multiplications can be added together inside a DSP block. It also supports 18×18 multiplication added to another 36-bit input. There is another mode that supports 18×19 systolic array operations and can be used in designing filters.

In addition to those fixed-point operational modes, the DSP in A10SX660 also has support for single-precision floating-point operations, where floating-point multiplication, addition, multiply-accumulate, or other vector operations (such as dot product) can be performed. If floating-point operations are required in the AI applications, they can be efficiently supported by DSPs in the A10SX660 device.

2.1.4 Embedded Block Memory

Modern FPGA has embedded memory elements. In Xilinx UltraScale+ FPGA devices, block random-access memory (BRAM) is embedded. BRAM is a dual-port RAM module fabricated into the FPGA chip as the on-chip storage. For XCZU9EG device, it has 912 BRAMs, each can be configured as either one 36Kb RAM or two independent 18Kb RAMs, and thus the total amount of on-chip memory is about 32Mb. When working as 36Kb RAM, each BRAM can be configured to $32K \times 1$, $16K \times 2, 8K \times 4, 4K \times 9, 2K \times 18$, or $1K \times 36$, where the first number represents the depth of the BRAM and the second number is the interface bandwidth (in bit) of the BRAM. The maximum effective data bit width is always 2^n. When the interface bandwidths are 9, 18, or 36 bits, the extra bits are used for parity check. Similarly, in 18Kb RAM mode, it can be configured to $16K \times 1, 8K \times 2, 4K \times 4, 2K \times 9$, or $1K \times 18$.

When using BRAMs, they can be configured to single-port mode, simple dual-port mode, or true dual-port mode. In single-port mode, there is only one data port that will be used for both data read and write operations. Therefore, the data read and write operations cannot happen at the same time. For two dual-port modes, there are two data ports enabled, and thus data read and write can happen at the same time. For simple dual-port mode, one port is designated to data read port and the other one is fixed to data write port, whereas in true dual-port mode, each of the data port can read or write data and the two ports are working independently. In dual-port modes, when read and write on the same address happen, users need to define which operation comes first. However, this usually cannot happen during AI processing if the datapath is correctly configured.

The block memory in A10SX660 device is called M20K, which is 20Kb in capacity for each. In A10SX660, a total of 2131 M20Ks are available. They can be configured to $512 \times 32(40)$, $1K \times 16(20)$, $2K \times 8(10)$, $4K \times 4(5)$, $8K \times 2$, or $16K \times 1$, where the number in parentheses is the bandwidth when parity check is enabled. Similar as the BRAMs in Xilinx FPGA, the M20K can be configured to single-port mode, simple dual-port mode, or true dual-port mode.

In addition to those block RAMs, part of the LUTs can also be configured as memory, which are termed SLICEM and MLAB in Xilinx and Intel FPGAs, respectively. These are the distributed RAMs available in FPGA and can be used when the capacity of block RAM is not enough or the design needs some faster memory. Shift register, which is a chain of registers, is another type of storage elements available in FPGA. It can provide the opportunity for data reuse that is quite useful for AI applications where both data and weights are reused during, for example, convolution computation.

2.1.5   Overall Architecture

Combining all the resources, the general architecture of an FPGA is shown in Fig. 5. The logic resources are organized into multiple columns. The main component is the CLB in Xilinx FPGA, or LAB in Intel FPGA. Block RAM and DSP are distributed in some columns. Those I/O pads are located around the border of the chip. High-speed transceivers are also near the border. The external memory controller is usually placed inside I/O pads. In addition, the clock tree and PLL are distributed inside the chip. These resources provide FPGA the flexibility to realize any algorithms.

2.1.6   AI-Optimized FPGA

AI is one of the typical applications for FPGAs. As deep learning is more frequently used in many applications, FPGA vendors also focus on optimizing AI operations in FPGA: both novel computing methods in old generation FPGA and AI-optimized new generation FPGA devices are available.

INT8 Computation

The 8-bit integer operation in Xilinx FPGA [2] is one of the optimized computing methods for AI applications. In deep learning inference, 8-bit integer operations can provide enough accuracy for most of the models. In addition, in deep learning computation, single-input data is reused by multiple weights for different outputs. As DSP48E2 in Xilinx has wide enough input bit width, it is possible to perform two independent 8-bit multiplications in a single DSP when sharing one common operand.

When performing two independent 8-bit multiplications, $a \times c$ and $b \times c$, the inputs to the DSP block are shown in Fig. 6. Basically, a is left shifted by 18 bit and added with b using the pre-adder. The addition here just combines a and b into a single operand. And then the combined operand is multiplied with c using the multiplier. In the output port P, two products for $a \times c$ and $b \times c$ can be correctly separated.

The datapath of this operation is shown in Fig. 7. During the computation, a and b are assumed to be weight values in a deep learning model, and c is assumed to be

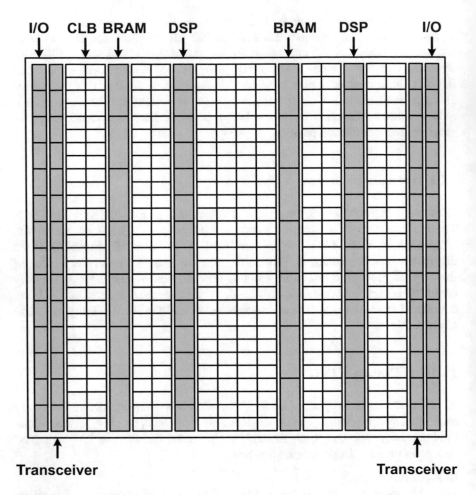

Fig. 5 FPGA overall architecture

activation of a model. As the ReLU activation function is popularly used in modern deep learning models, it is common to consider a and b as signed numbers, while c as an unsigned number. Therefore, when extending a, b, and c to larger bit width, a and b need to be sign extended and c can be extended with zeros. a is left shifted and added with the extended b to form the 27-bit operand. c is extended with zeros to form the other 18-bit operand.

After multiplication, as all operands are 8 bit, the lower product is at most 16 bit in bit width. As a is shifted by 18 bit, the lower product will not be affected by the higher multiplication. In DSP48E2, the bit width of the multiplier output is 45 bit, and thus the second product also has enough bit positions. When both a and b are positive numbers, the two products can be separated easily. When only a is negative, the higher order product is still correct since the DSP is able to perform

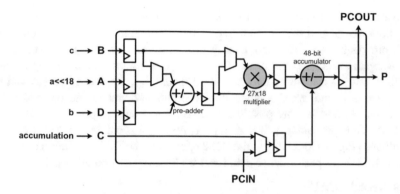

Fig. 6 Inputs to the DSP when performing INT8 operations

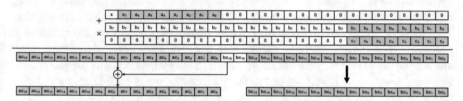

Fig. 7 Datapath of the INT8 operations

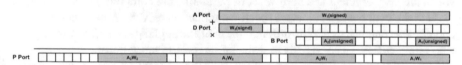

Fig. 8 Datapath of the INT4 operations

signed operations. The lower order product is not affected. When b is negative, as a sequence of one bits will be added with a in the pre-adder, a simple separation may not give the correct higher order product. However, the correction for the higher order product is still simple: adding the sign bit of the lower order product with the higher order product. By using this INT8 method, the peak throughput can be improved by 1.78 times compared to the normal DSP usage [2].

INT4 Computation

When even lower precision computation can provide acceptable accuracy, INT4 operations [3] can be used to achieve higher speed performance. The design methodology is similar to the INT8 operation that is to use the pre-adder to pack multiple operands into a single operand and arrange the data position in the combined operand to ensure the parallel products can be correctly separated at the final output port.

The datapath of the INT4 operation is shown in Fig. 8. Here two sets of signed weights, W_1 and W_2, are used to multiply with two sets of unsigned activations, A_1 and A_2, for four different outputs. When using 4-bit operands, the product is at most 8 bit in bit width. To ensure all products can be separated correctly, two activations are placed with a 7-bit gap in between. In addition, W_2 is left shifted by 22 bit. As a result, two neighboring products will have a 3-bit separation. As both weights are signed numbers, they are sign extended to fill the higher order positions. During operation, two weights, W_1 and W_2, are sent to port A and port D, respectively, and are combined into a single operand in the pre-adder. Two activations, A_1 and A_2, are positioned correctly and sent to port B. At the output port P, the four products, A_1W_1, A_2W_1, A_1W_2, A_2W_2, can be separated correctly according to Fig. 8. Compared to the INT8 operation, the INT4 operation can further achieve 1.52 times performance gain for deep learning computation [3].

Xilinx Versal ACAP

The Xilinx Versal adaptive compute acceleration platform (ACAP) is the next generation FPGA optimized for AI applications. It is a heterogeneous computing architecture, containing scalar engines, adaptable engines, and intelligent engines. The scalar engines are composed of ARM dual-core CPUs that are efficient at processing complex algorithms with diverse decision trees. The adaptable engines are similar to previous generations of FPGA devices that contain programmable logic and memory cells and can be precisely customized to a particular computation function. The intelligent engines are an array of very long instruction word (VLIW) and single instruction multiple data (SIMD) processing engines and memories, which are very efficient at a narrow set of parallelizable computing functions, such as AI and signal processing. For a specific application, developer can choose to use one of these engines or deploy different parts of the application to different engines and tightly couple those engines to realize more performance boost.

The optimizations for AI applications in Xilinx Versal ACAP devices mainly come from two architecture designs: the AI engine, and the DSP engine inside adaptable engine. Each AI engine contains vector processors, scalar processors, dedicated program and data memory, dedicated AXI data channels, and support for direct memory access (DMA) and locks. It supports SIMD and performs VLIW operations, providing high degree of parallelism for various kinds of operations. They are optimized for real-time DSP and AI computation. Compared to the design using programmable logic, the AI engine can achieve a 3 times to 8 times smaller area and 50% reduction in power consumption. The AI engine can be used with the adaptive engine together for AI applications, where AI engine can provide high-performance computation module, while the adaptive engine can be used to realize programmable memory hierarchy that can be reconfigured to fit the memory hierarchy requirements of different AI models.

The DSP engine in ACAP also provides many new computing features for AI computation. Compared to the previous generation DSP48E2, the new DSP58 increases the size of the multiplier to 27×24 and the size of the accumulator to 58 bit. INT8 vector dot-product mode is enabled to support low-precision dot-product operations widely used in AI computations. Hardened floating-point operations are included where the multiplier can perform the multiplication of single-precision operands or half-precision operands and the accumulation is performed in single-precision format. All these provide more flexibility and computing power for AI applications.

Intel Stratix-10 NX

Intel Stratix-10 NX is another FPGA device optimized for AI computations. Stratix-10 NX device belongs to the Intel Stratix-10 family. Besides the normal features of Stratix-10 devices, the NX series has three specialized designs that make it suitable for AI acceleration: the high-performance AI tensor block, the near-compute memory resources, and the high-bandwidth networking.

The AI tensor block is a type of AI-optimized arithmetic block. Each block contains three dot-product units. Each dot product has 10 multipliers and accumulators. This organization supports high-performance matrix–matrix or vector–matrix multiplications for high-performance AI computations. The AI tensor block supports INT4 and INT8 multiplications. In addition, 16-bit and 12-bit block floating-point multiplications are also supported. Block floating point refers to a floating-point format that has a shared exponent, and thus the cost of block floating-point multiplier is much smaller than a normal floating-point multiplier, while the format can provide much larger dynamic range than a fixed-point format. The accumulation in the AI tensor block is performed with a higher precision format, which is either INT32 or single-precision format.

The Stratix-10 NX FPGA also has abundant embedded memory for model persistence where the parameters of the model are all saved in the embedded memory at runtime, and thus the costly external memory data transfer can be avoided. For the external memory, the Stratix-10 NX FPGA adopts high-bandwidth memory that can provide a much higher bandwidth than conventional DDR memory. In addition, the high-performance networking components enable fast and flexible interconnection among multiple devices for better scalability.

Intel Agilex F-Series SoC

The Agilex series FPGA is Intel's next generation FPGA device. The F-Series FPGA is optimized for power and performance, which is suitable for AI acceleration. Although, at the time of writing this chapter, Intel has not released detailed user guide for Agilex FPGA yet, some features of the core fabric and DSP blocks are available in Intel website. For the core fabric, Agilex adopts the second generation HyperFlex core and Hyper-registers that are optimized for performance and power consumption. For the DSP block, it still supports both fixed-point operations and floating-point operations as its previous generation FPGA does; however, more numeric precision support and further flexibility are achieved. For floating-point format, it adds the supports for half-precision and BFloat-16 formats. For fixed-point format, it supports the multiplications of sizes from 9×9 to 54×54, while the amount of 9×9 multipliers is much more than previous generation FPGAs. All these improvements enable a fast and flexible computation for AI applications.

2.2   FPGA Tools and Design Flow

FPGA design tools are provided by the FPGA vendors to manage the design process on their FPGA devices. For Xilinx FPGAs, the design can be performed with Xilinx Vivado or Xilinx Vitis. For design in Intel FPGAs devices, the Intel Quartus Prime can be used. These tools are designed for relatively new generations of FPGA devices. If you are using old generations of FPGA devices, you may need to use Xilinx ISE and Altera Quartus II, respectively, for Xilinx and Intel (Altera) FPGAs. Licenses are required to unlock all functions and device supports for these tools. However, both vendors provide a free lite version of their design tools that has limited device support and function support, but they are good enough for study or evaluation purpose.

There are two different design flows available for FPGA design: the RTL-based flow and the HLS-based flow. The RTL-based flow is the conventional hardware design flow using HDL to directly design on the hardware platform. The HLS-based flow is more friendly to the software developers where the design is performed in high-level programming languages, such as C/C++, and the design process does not require too much hardware-specific knowledge. When designing using the RTL-based flow, you can use Xilinx Vivado or Intel Quartus Prime, respectively, for Xilinx and Intel FPGA devices. For HLS-based design, Xilinx Vitis can be used for Xilinx devices. For Intel FPGAs, the high-level synthesis compiler embedded in Quartus Prime can be used for HLS-based design. For Intel FPGAs, in addition to the HLS-based flow, the OpenCL-based design flow is also available.

We will now discuss the general process of developing in RTL-based flow, which contains the following steps:

- Create a project.
- Develop HDL code.
- Perform functional simulation.
- Apply design constraints.
- Perform logic synthesis.
- Perform place and route.
- Generate bitstream and program device.

All FPGA design tools provide graphical user interface (GUI). When creating a project in FPGA tools, you will need to specify the project location where you want to save all your design files, specify the target FPGA device, and import design source files and constraint files if you already have.

The next step is to develop your HDL code. Code editors are available in FPGA tools with syntax highlight and basic formatting for developing HDL code. FPGA tools also have abundant intellectual property (IP) resources to reduce the workload of your design. IP cores are designed and verified by FPGA vendors and provide supports for many popularly used functions, such as block memory and floating-point computation. When you need to use IP cores, you can use the IP tools available in the FPGA design tools to build and configure your IP. An instantiation model is then generated that can be integrated into your HDL code.

After the HDL development complete, functional simulation is required to verify your HDL code. Xilinx Vivado has embedded simulator so that you can perform simulation in Vivado. For Intel Quartus Prime, you will need to specify an external simulator. A free version of Mentor Graphics ModelSim is installed with Quartus Prime and can be used to perform simulation. If some functions are not correct, you should debug your HDL code with the help of the information from simulation process, such as the waveform of all signals.

After all HDL codes are verified, design constraints are needed to be provided for further logic synthesis and place and route. Design constraints usually come from your design specification, including timing, area, and power consumption. During synthesis, the FPGA tools will try to optimize your design to meet your timing, area, and power consumption requirements. In addition to those hardware metrics, in FPGA designs, you will also need to specify pin assignments that is to specify which I/O pins of the FPGA should be associated with the I/O ports of your design. The constraints can be edited in a text file using several lines of Tcl commands. Or the FPGA tools provide GUI for configuring constraints.

The synthesis process is to map your HDL code to those basic elements available in FPGA. In your HDL code, you could express your logic or functions in any abstraction levels. During synthesis process, the synthesis tools will determine how they can be realized using FPGA basic elements, such as LUTs, DSP, and BRAMs. The synthesis process will generate a netlist that only contains instantiations of those FPGA basic elements. Optimizations of the design are also partially performed during the synthesis process. Those optimizations are guided by the design constraints provided. If you have tight area constraints, then the tools will try to merge LUTs or other resources to reduce the area. Or if you have tight timing constraints, the tools will use more logic to increase the parallelism to reduce the latency of your design. As a result, the synthesized design may have a different architecture from your original design, and thus a post-synthesis simulation is required to verify the designated functions still work properly.

The next step is to perform place and route (PNR), which is called implementation in Xilinx Vivado and is called fitter in Intel Quartus Prime. It is a combination of two processes. The placement process is to find a proper location for those utilized FPGA elements specified in the synthesis netlist. Then, the routing process is to interconnect those elements using wires and find an optimal way to arrange the trace of those wires. In this process, optimizations are also performed based on the design constraints. As shown in Fig. 5, there are many possible locations for each of the element, and there are many ways to interconnect them. If they are sparsely distributed in different locations in the chip, the wires to interconnect them become long, and thus the latency will be large. So, during PNR process, the tools will try to meet your design constraints by finding proper locations and wiring methods for your design elements. As optimizations are performed during PNR, a post-PNR simulation is required. After the PNR process, reports will be generated that contain timing information, resource consumption, and power consumption of your design.

You may use these information to evaluate your design. If the design constraints are not met, you may change your constraints and perform synthesis and PNR again to find the constraints that your design can meet. If those metrics are not satisfied by your design specifications, you may need to tune your design accordingly.

Finally, you can generate the bitstream and download it to your FPGA board to finish the programming of your FPGA. A physical test is required to verify whether all functions work properly in actual devices. This completes the whole process of the RTL-based design flow.

The HLS-based design flow is more friendly to software developers. The HLS flow starts with developing algorithms using C/C++ languages. Although the development is performed in C/C++, the coding style is a bit different from the conventional C/C++ program. Some hardware-specific libraries need to be imported and used to realize hardware-specific requirements, for example, the library defining customized numeric format. In addition, loop unrolling for iterative operations is required to make use of the hardware parallel processing. Moreover, some design constraints are needed, which can be used to realize specific hardware architecture, for example, to realize a pipelined architecture or to design customized memory hierarchy. The design can be performed in a pure software programming way. However, without hardware consideration and design constraint, the generated hardware architecture using default configurations will not be efficient. Therefore, hardware-specific consideration, such as loop unrolling, and design constraints are recommended during the code development.

With HLS, the simulation of the design can also be performed with another C/C++ program. The designed architecture can be treated as a subfunction, and the testing program can be treated as the main function for the program. Therefore, this process is also friendly to software developers.

The HLS compiler will then be used to generate a hardware architecture for the algorithm with the consideration of design constraints. The estimated timing, area, and power consumption can be reported. The HLS compiler generates a whole set of Vivado project, including the HDL code generated from C/C++ designs and design constraints. They can be easily ported to Vivado for further synthesis, PNR, generating bitstream, and downloading to the FPGA board. The design in HLS can also be packed into IP cores and used by other Vivado projects.

2.3 AI Accelerators on FPGA Platform

FPGA devices have been widely used in AI acceleration because of its low cost and its reconfigurability. Many FPGA-based AI accelerator architectures have been proposed in recent years [4–6].

As FPGA is reconfigurable, many of the FPGA-based architectures are designed and optimized for a specific deep neural network model in an application. In addition, as fixed-point multiplication can be implemented in DSP blocks, which is much more efficient than floating-point operations, many designs quantize the deep learning models to use either 8-bit or 16-bit fixed-point numbers and implement an accelerator using fixed-point format. For some application scenarios where accuracy degradation is not that sensitive, binary or ternary networks are used. In hardware accelerators, as parameters are in binary of ternary, no hardware multiplier is required, which makes the design very efficient in terms of energy consumption.

FPGAs are used in some cloud servers as well. Project Brainwave [7] is the deep learning platform proposed by Microsoft for real-time AI inference in the cloud and on the edge. It optimizes the computations of batch=1 operations that are the typical inference workloads and thus reduces the processing time for inference task. The Amazon AWS F1 instance [8] is another cloud service using FPGA devices.

Due to the highly parallel computing architecture, FPGA-based AI accelerators can achieve much better performance than the CPU implementations. Moreover, due to the specialized memory hierarchy and computing unit designs, the energy efficiency of FPGA-based AI accelerator is much better than GPU implementations. Therefore, FPGA is a good candidate platform for AI accelerator design.

In addition, the evolution of AI algorithms is much faster than the design cycle of hardware processor. It usually takes 3–5 years for the IC industry to design and fabricate a hardware processor. However, the new AI algorithms come out nearly every month. To fit the high evolution frequency of AI algorithms, the reconfigurability of FPGA devices becomes critical. With FPGA devices, developers can quickly reconfigure the architecture to fit new algorithms instead of designing and fabricating a new chip. As a result, FPGA-based AI accelerators are becoming more and more popular.

3 ASIC Platform for AI Acceleration

Application-specific integrated circuit (ASIC) is another custom hardware design platform. Unlike FPGA design that is to reconfigure the already fabricated IC chip, ASIC design is to design the architecture of the chip and then fabricate the chip. The basic elements of ASIC design are CMOS logic gates provided by a semiconductor technology library instead of LUTs in FPGA. FPGA design is to use pre-built blocks to realize the target functions. ASIC design is to use the fundamental design elements to achieve fully customized designs, which has less overhead compared to FPGA designs. Therefore, ASIC-based architecture can achieve much better performance and energy efficiency than FPGA-based design.

3.1 ASIC Tools and Design Flow

As the basic elements of ASIC are logic gates, we will not discuss them in detail and directly start the discussion of ASIC tools and design flow.

Fig. 9 ASIC design flow

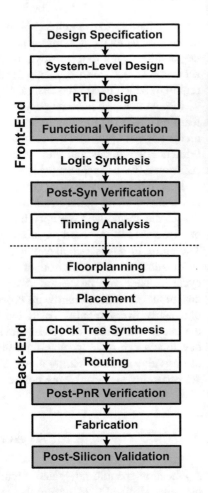

The whole ASIC design flow is shown in Fig. 9. The whole process is divided into two phases: the front-end design and the back-end design. The front end mainly focuses on the logic and architecture designs, and the back end performs physical designs. Those important verification phases are highlighted with gray color in Fig. 9. The front-end design process is similar to the first few steps in the FPGA design process. The design specification is basically to define the functions that to be realized and the target performance and efficiency. The system-level design is to divide the functions into multiple design modules and to define the interaction among different modules. Then, each module is designed using HDL. In ASIC, IP cores created by tool vendors are also available to be used in HDL design process. One of them is the Synopsys DesignWare library that contains many basic logic functions such as arithmetic operations and memory functions.

Functional verification is performed through simulation after HDL modules are developed. There are many tools available to perform ASIC simulation, including the Mentor Graphics ModelSim, Synopsys VCS, and Icarus Verilog, and so on.

This simulation process is similar to that in FPGA design since it is only to verify the functionality of the design without any details in physical implementations.

Similar to the FPGA design flow, the next step is to perform logic synthesis for the verified RTL design. The synthesis here in ASIC is to map the design to logic gates provided by the semiconductor technology library. In FPGA design, you need to specify the available logic resources by choosing the correct device name. In ASIC design, you should provide a semiconductor library during synthesis, and that library will provide logic resources for the synthesis tools to realize your RTL design. Some industrial level library providers include STMicroelectronics (STM) and Taiwan Semiconductor Manufacturing Company (TSMC). There are also some free options for academic research, including the FreePDK 45nm provided by Oklahoma State University and NanGate 15nm provided by NanGate. After synthesis, a netlist is generated, which only contains the instantiations of the logic gates available in the target semiconductor library. Hardware metrics, such as timing, area, and power consumption, of the design can be obtained through synthesis reports. Similar to the FPGA design flow, another stage of verification is required for the synthesized netlist.

Timing analysis of the circuit is required after synthesis. Timing analysis is to check timing violations for all possible paths to validate the timing performance of the design. In ASIC design flow, static timing analysis (STA) method is usually used in this process. In STA method, a circuit design is broken down to a set of timing paths. The delay along each path is calculated, and then any violation is checked to verify the timing performance. Synopsys PrimeTime STA tool can be used in this STA process. The STA process concludes the front-end design phase.

In the back-end phase, the first step is the floorplanning that is to arrange the tentative positions of the main functional blocks and macros in the chip. Using Fig. 5 as an example, the floorplanning is to decide where to put the I/O, CLB, BRAM, DSP, and others. The next step is placement that is to arrange detailed logic gates into each block. In Fig. 5, the placement is to put the detailed design of, for example, DSPs into the DSP column determined during the floorplanning stage.

The next step is to interconnect these placed logic resources using wires. This includes the connections between clock inputs of the design to the clock source. As the clock inputs may be placed anywhere in the chip, the length of the clock path will be different, which leads to imbalanced clock delay to each of the clock ports. To solve this problem, a certain amount of buffers or inverters are inserted along clock paths in order to make the clock delay balanced. This process is the clock tree synthesis process. Then, the routing process is to interconnect other logic resources. The tool will look for optimal interconnections to reduce the overall delay of the design. The tool that can help in the back-end design phases includes Cadence SoC Encounter and Synopsys IC Compiler, both of which provide GUI for floorplanning and PNR.

The physical chip is fabricated after the post-PNR verification is completed. After that, the final post-silicon validation of the physical chip is performed before the chip can be deployed in the practical application.

3.2 Comparison Between FPGA and ASIC Design

Until now we have discussed the design flows and tools for both FPGA and ASIC platform. In this section, we are going to compare the two platforms.

The main difference between these two platforms is the logic resource available for the designer. In FPGA platform, the available design resources are pre-built functional blocks, such as LUT, BRAM, DSP, and I/O, and so on, while in ASIC platform, the basic resources are the logic gates provided by the semiconductor technology library. Recall Fig. 1 that is the diagram of a simple 4-bit adder in FPGA. As the basic element in FPGA is LUT, the adder can be realized by cascading 4 LUTs.

In ASIC platform, as the basic element is logic gate, the architecture to realize the logic functions of Eq. (1) needs to be designed based on logic gates instead of LUTs as in Fig. 1. The diagram of the 4-bit adder is shown in Fig. 10. This adder needs 4 full adder (FA) cells, and each FA cell contains two XOR gates, two AND gates, and one OR gate.

Comparing the design in Figs. 1 and 10, the ASIC design in Fig. 10 uses just a right amount of logic gates to realize the logic functions, while in Fig. 1, the complexity of one LUT is much higher than an FA cell. Therefore, the overhead of using FPGA is higher than that of ASIC. Although the example 4-bit adder is a small design, other large-scale designs follow the same trend. As a result, the speed performance and energy efficiency of an ASIC design are usually better than those of an FPGA design for the same application.

In terms of design flow, the cost of FPGA design is much cheaper than that of ASIC design. FPGA provides pre-built function blocks, and the design on FPGA is just to reconfigure those functional blocks and use wires to interconnect them. All these design processes can be performed automatically by the design tools with some user constraints. However, for ASIC, the design process is complex and the design cycle is long. Each design process needs to be performed manually. Moreover, as everything is designed from scratch, full verification is needed for each of the design phase. In addition, the fabrication process is quite long. It may take about 3–5 years to design an ASIC chip before deploying it in designated applications.

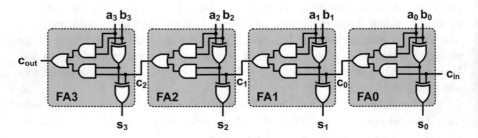

Fig. 10 Simple 4-bit adder in ASIC using logic gates

For an ASIC chip, its architecture cannot be modified after fabrication. When the designated algorithm is replaced by a new one, another full design and fabrication processes need to be performed to produce a new ASIC chip for the new algorithm. The process is costly and takes long time. However, for FPGA, as it is reconfigurable, when a new algorithm comes, the designer can easily reconfigure the FPGA and deploy it in practical applications. The time-to-market for FPGA is much shorter.

In summary, ASIC designs can provide better speed performance and energy efficiency, while its design cost is high, its design cycle is long, and its architecture cannot be tuned for new algorithms after fabrication. For FPGA, it is reconfigurable, and thus it can be easily tuned to fit for new algorithms and its design cost is low. Although its performance and power efficiency cannot reach the level of ASIC designs, they are still much better than CPUs and GPUs and are good enough to be used for designing domain-specific processors for applications.

3.3 AI Accelerators on ASIC Platform

As AI-based method is ubiquitous in many fields of applications, designing domain-specific ASIC chip for high-performance and low-power AI processing becomes reasonable. Many companies and academic institutions have developed ASIC chips for AI computation [1], including the TPU [9] proposed by Google, the DianNao series AI chips [10, 11] proposed by the Chinese Academy of Sciences, the Cambricon series AI chips and instruction-set architectures [12–14] proposed by Cambricon Tech., the Ascend series AI chips [15] proposed by Huawei, and the Eyeriss chip [16, 17] proposed by MIT, and so on.

ASIC chips cannot be modified after fabrication. As a result, many ASIC-based AI chips are designed to be flexible to accommodate many different algorithms. For one method, the base hardware architecture is designed to provide many different computation modes to fit many different AI models. In another method, the base architecture is designed with general matrix multiply (GEMM) architecture or systolic array architecture, and it relies on the specialized compiler to efficiently map different AI models onto the base hardware architecture.

As discussed before, ASIC chip can provide better speed performance than FPGA. In addition, the memory hierarchy can be manually designed to fit the application requirements, and thus the ASIC can provide higher memory bandwidth for computation. With these advantages, many ASIC-based AI chips have the ability to perform AI training. Due to the high efficiency and the ability to perform both AI training and inference, ASIC-based AI processors are widely deployed in cloud servers and datacenters.

4 Architecture and Datapath of AI Accelerator

After having the general knowledge of FPGA and ASIC platforms, in this section, we are going to discuss the architecture and datapath of the AI accelerators. A general architecture of an AI accelerator is presented. Then, optimizations of each design module are briefly discussed. FPGA-based and ASIC-based AI accelerators will be compared in this section as well.

4.1 AI Accelerator Architecture and Datapath

The general architecture of an AI accelerator is presented in Fig. 11. It can be divided into three major components: the computation module, the memory module, and the control module. The computation module is composed of a GEMM engine, an accumulator, and a vector engine. The memory module contains two buffers for input data and algorithm parameters and output results, respectively. A large-capacity external memory, such as DDR memory, is usually used as the main memory. The control module mainly includes an instruction decoder and the controller for each of the functional module.

All the computations of the AI model are performed in the computation module. The GEMM is typically used for the computations of convolution layer and fully connected layer that are the most computation-intensive operations in AI and deep learning models. The GEMM is a large array of processing elements (PEs), for example, 256 × 256 PEs in TPU v1 [9]. Each PE contains a multiply-accumulate (MAC) unit to compute the products of the input features and weights and then to

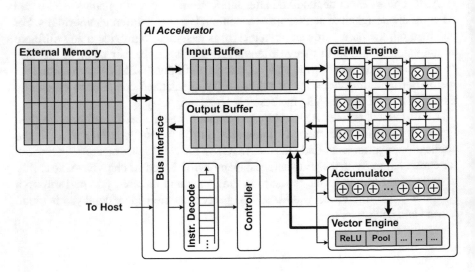

Fig. 11 Architecture of a general AI accelerator

accumulate with products from other PEs. Some registers are also usually included in PE to provide data to the MAC unit or to transfer data among different PEs.

The accumulator unit contains parallel adders to add the results in the current round of computation to the partial results calculated in previous computation round. If bias is enabled in the AI model, the accumulator can also be used to add bias to the computation result. If full accumulation is not finished yet, the results are sent back to the output buffer for further accumulation. If accumulation is done and activation functions or pooling operations are required, the results are sent to the vector engine for further processing.

The vector engine contains parallel functional blocks to process vector inputs. Typical operations that are handled by the vector engine include the rectified linear unit (ReLU) activation function and maximum pooling operation. The results of this engine are sent back to the output buffer.

The memory module contains a 3-level hierarchy: the external memory, the input/output buffer, and registers. Modern AI models usually contain millions of parameters, which exceed the capacity of the on-chip memory, and thus an external DDR memory is usually required. The DDR memory has large enough capacity to save all model parameters and required input and output data. Some recent designs use the high-bandwidth memory (HBM) to take the place of the DDR memory to achieve higher memory interface bandwidth and in-memory processing. The next level is the data buffer built using on-chip memory modules. The input buffer is small in capacity and can only cache the data and parameters required for the current rounds of computation. At runtime, part of the data and parameters are read from external memory to the input buffer. Part of those data are further sent to the registers in the GEMM engine to perform the current computation. In one layer of the deep learning model, to generate one output, a large-sized dot-product operation is required, which usually needs multiple computation cycles. The partial results after each cycle are saved in the output buffer for further processing. When a complete output is generated, it is sent to the output buffer and then sent back to the external memory. Registers are close to the MAC units and are responsible to provide direct data to the MAC units. When data or parameter sharing is enabled among different computations, the registers can pass their data to neighboring registers instead of reading them again from the buffer.

The control module is the coordinator of all the modules. It receives the instructions of a program sent from the host PC, saves them into a FIFO, and then one by one decodes those instructions. The instructions are decoded into configuration parameters and are distributed to the corresponding controllers of the functional blocks. The controller then controls the operation of the corresponding module according to the configuration parameters. By using different configuration parameters, the hardware accelerator can be configured to process different AI models with different layers and different sizes of layers.

During operation, the host PC sends input and parameters of the model to the external memory and sends program instructions to the instruction decoder. Then, the accelerator starts processing the AI model. The first step is to read part of the data and parameters from external memory into the input buffer. Those data are then transferred to the GEMM engine to participate the main computations. The partial results are added with the bias in the accumulator and sent back to the output buffer. In the next round of computation, another part of data or parameters are read from input buffer to the GEMM engine. After computation, the previous partial results are read from output buffer to the accumulator to be added with the GEMM outputs. These processes are repeated until a complete result is generated. It is then sent to the vector engine for performing activation function and/or performing pooling operations. After vector operations, the complete results are sent back to the output buffer, and when the external memory data bus is free, the results are transferred back to the external memory. After the operation of the whole AI model is completed, host PC can read results from external memory for further processing.

4.2 Optimizations for AI Accelerators

AI models, especially the deep learning models, are both computation-intensive and memory-intensive, which lead to large latency and high power consumption. In AI accelerators, specific optimizations are required to improve the speed performance and energy efficiency. In this section, the optimizations for computation, memory, and control modules for the AI accelerator shown in Fig. 11 are going to be discussed.

4.2.1 Computation Module

Arithmetic units are the main component of the computation module, and thus the optimizations are focused on the design of arithmetic units. Reduced-precision computation and special convolution algorithms are proposed in the literature. Reduced-precision computation belongs to the software–hardware co-design category and will be discussed in the next section. Here the special convolution algorithms will be discussed.

The convolution operation and dot-product operation in fully connected layers can be converted into matrix multiplication operation and are performed with GEMM architecture. This conversion is useful for hardware devices that have native matrix multiplication support, such as GPU. For custom hardware design, further optimization can be performed to reduce the amount of computation. Multiplication is much more expensive than the addition, and thus the main focus is put on the multiplication.

Winograd algorithm is one of the optimizations for convolution operations. Suppose we are doing convolution for 4×4 images using 3×3 filters to generate 2×2 outputs. When using normal convolution operations, for each of the output, 9 multiplications are required, and thus a total of 36 multiplications are needed for the whole output.

When using Winograd algorithm, both the filter weights and the inputs need to be transformed first. And then, an element-wise multiplication is performed. And finally, the results are transformed back to obtain the convolution result. Suppose the input is represented by d, the filter is represented by g, the transform matrices for input and filter are represented by B and G, respectively, and the transform matrix for result is represented by A. In our case, d is a 4×4 matrix, and g is a 3×3 matrix. According to the Winograd algorithm, B is a 4×4 matrix, G is a 4×3 matrix, and A is a 4×2 matrix.

$$B = \begin{bmatrix} 1 & 0 & 0 & 0 \\ 0 & 1 & -1 & 1 \\ -1 & 1 & 1 & 0 \\ 0 & 0 & 0 & -1 \end{bmatrix} \quad G = \begin{bmatrix} 1 & 0 & 0 \\ \frac{1}{2} & \frac{1}{2} & \frac{1}{2} \\ \frac{1}{2} & -\frac{1}{2} & \frac{1}{2} \\ 0 & 0 & 1 \end{bmatrix} \quad A = \begin{bmatrix} 1 & 0 \\ 1 & 1 \\ 1 & -1 \\ 0 & -1 \end{bmatrix}.$$

Then, the convolution result Y can be obtained by

$$Y = A^T ((GgG^T) \odot (B^T dB))A, \tag{3}$$

where $\odot$ represents element-wise multiplication. Therefore, only 16 multiplications can accomplish this convolution with extra matrix multiplications. However, as the entries of the transform matrix are either 0, 1, or 0.5, the matrix multiplication can be performed with only shift and add operations, which is very efficient in hardware. In addition, after training, the filter weight values will not be changed, and thus the transformation for weights can be performed offline, which further reduces the computation cost at runtime.

For different filter sizes and output sizes, the transformation matrix of the Winograd algorithm is different. However, they could achieve a fewer number of multiplications for convolution operations. As small filter size, such as 3×3 filters are widely used in modern AI models, Winograd can be applied to efficiently reduce the computation cost.

4.2.2 Memory Module

AI models contain millions of parameters. At runtime, they need to be read from memory to the computation module. In addition, each layer of the AI model will generate a large amount of data at runtime. As a result, AI model is memory-intensive. There are many memory accesses happen during AI computation, including input reading, parameter reading, and (partial) output reading and writing. The external memory and on-chip memory access consumes much higher energy compared to the computation, and thus the optimization of memory module is to find a method to reduce the access to the memory.

Fortunately, there are many data and parameter reuse opportunities in AI computation. Convolution layer is the most computation-intensive operations in AI models. Based on the convolution algorithm, there are many opportunities to reuse data and parameters:

- Convolution operation is performed by sliding the filters throughout the input feature maps, and thus the same set of filters is reused by the whole input feature maps. As a result, one set of filters can be read from external memory only once and reused by multiple input feature maps.
- When computing different channels of the output feature maps, the same set of input feature maps is used with different sets of filters. As a result, one set of input feature maps can be read from external memory once and reused by multiple sets of filters for different output channels.
- Considering batch operations, many different input feature maps are processed at the same time with the same set of filters. As a result, one set of filters can be reused by multiple input feature maps in batch operation.

By using registers, the data or parameters to be reused can be read from memory and stored in the registers near the computation units for multiple rounds of computations.

Dataflow of the computation can be carefully designed to achieve higher degree of data reuse and thus to reduce energy consumption. Some basic dataflow includes weight stationary (WS), output stationary (OS), input stationary (IS), and row stationary (RS) [1]. We will now use the architecture shown in Fig. 11 as the base architecture to describe these dataflow methods.

The WS dataflow is to reuse the weights in the registers as much as possible to reduce the energy consumed by reading weights from memory. Weight values, W_0 to W_8, are read from memory and kept in the corresponding registers for a while. Input is read from memory every cycle, while the partial result is streaming through PEs to participate accumulation in each PE. The final partial sum is then sent to output buffer, and it will be further accumulated or written back to external memory. This WS dataflow is used by TPU [9].

In the OS dataflow, the partial result of the same output is kept in the local register to minimize the memory access for the partial result. In this implementation, each PE reads different inputs from input buffer, while all PEs use the same weight at each time. The inputs are streaming through the PEs. In this implementation, the output buffer may not need to cache partial sums. Inputs and weights for the same output can be kept reading from the input buffer until the full sum is completed. The DianNao series AI processors [10, 11] use this OS dataflow.

The IS dataflow is to keep the inputs in the local registers for a while to reduce the energy consumed by reading inputs from the memory. In this dataflow, weight values are read from input buffer and multiplied with the inputs kept in each PE. The partial sum is streaming through PEs to finish the full accumulation. The SCNN accelerator [18], which is an accelerator for sparse convolutional neural networks, uses this IS dataflow.

Each of the WS, OS, and IS dataflow is designed to reduce the memory access for weights, partial result, and inputs, respectively. The RS dataflow, on the other hand, is able to maximize the reuse of all three types of data in the registers.

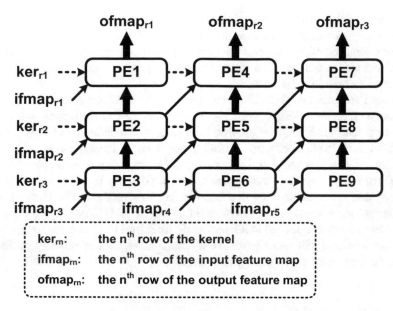

Fig. 12 Example of convolution computation using row stationary dataflow

A 2D convolution example using the RS dataflow is shown in Fig. 12. In this RS dataflow, each column of PEs processes one row of the 2D convolution. Each row of the filter is kept inside the corresponding PE. Each row of the input is streaming into the PEs. Inside each PE, the 1D row convolution is performed. The row of input is then diagonally passed to the next column of PEs. The partial sum from one PE is sent vertically to other PEs in the same column to perform accumulation. For three rows of output, three columns of PEs are required. Moreover, as the filter has three rows, three PEs are needed for each column. By using this datapath, each row of weight is kept inside the PE registers and reused by multiple inputs. The row of weight is also shared among PEs in the same row. The input data are also shared among different Pes, which reduces the number of memory access. In addition, partial sums are streaming through a column of PEs, and thus the memory access for partial sum is also avoided. Therefore, the data reuse and accumulation are maximized inside the PE array. This RS dataflow is used by Eyeriss series AI chips [16, 17].

Besides the dataflow optimization, recent 3-dimensional memory, such as the high-bandwidth memory (HBM) and hybrid memory cube (HMC), can be used in AI accelerator to take the place of the conventional DDR memory to improve energy efficiency.

The 3D memory is composed of a vertical stack of memory dies communicating using through silicon vias (TSVs). It provides a logic layer at the bottom of the memory stack to realize memory controller and other simple logic functions. As a result, some simple operations on data in AI computation can be realized by this logic layer instead of transferring these data to the main processor for computation. In addition, the memory interface and the main processor are connected through silicon interposer, which can provide much higher bandwidth than the conventional DDR memory interface. Therefore, the speed performance is expected to be improved by using 3D memory.

In the literature, the TETRIS accelerator [19] is an AI accelerator design using 3D memory. In this design, HMC is used to increase performance and energy efficiency. The 3D HMC can provide high enough bandwidth, and thus some of the data can be read directly from HMC to computation units bypassing the global data buffer. This can make full use of the buffer capacity to maximize the data reuse. For others with low reuse chance, they are directly transferred between the memory and the computation unit without going through the buffer. In addition, in this design, the accumulation is performed inside the HMC logic layer. Therefore, the partial sum does not need to be read again from the HMC to the main processor to be accumulated with other partial sum, which saves a lot of memory traffic and thus effectively reduces the energy consumption.

4.2.3 Control Module

The optimization for the control module mainly focuses on the design of AI-specific instruction sets. In [12], an instruction-set architecture designed specifically for AI computation is proposed. When designing AI-specific instruction sets, the characteristics of AI computation are analyzed, and some optimizations over conventional instruction sets are required. As AI models are organized into layers and the computation pattern for each layer is highly uniformed, data-level parallelism will be helpful in improve the performance. In addition, convolution operation is the main computation in AI models, and thus customized instructions for matrix operations or vector operations will be suitable. Scratchpad memory access for variable-length data is also needed in AI computation.

By properly designing the instruction sets, the efficiency and the flexibility of AI computation can be improved. On one hand, with proper instruction sets, such as matrix operations or vector operations, the density of the code for AI computation can be increased. On the other hand, when implementing different AI models, different configurations can be generated for corresponding instructions, and thus different AI models could be handled efficiently using the same architecture.

4.3 Comparison Between FPGA- and ASIC-Based AI Accelerators

The basic architecture of AI accelerators on FPGA and ASIC platforms is the same. On both design platforms, a highly parallel computation module is used for the main computations of the AI models. A 3-level memory hierarchy is applied, which includes a large-capacity external memory, a high-speed on-chip memory, and register files. Model parameters are saved in the external memory and are transferred to the on-chip memory and then the computation units. The input data are also cached in the external memory before being processed. It is then transferred to the on-chip memory and then the registers of the computation units. Results or partial results are saved in the on-chip memory as well. When their computations are completed, they are sent back to the external memory and will then be used by host PCs. When mapping the designed architecture onto the design platforms, different logic resources to realize the architecture will be considered, and thus the differences between FPGA and ASIC accelerators still mainly come from the differences between FPGA and ASIC platforms.

The speed performance and energy efficiency of ASIC-based AI accelerators are usually better than those of FPGA-based accelerators. ASIC platform provides higher level of freedom for the design. The developers can use just a right amount of logic resources for the computation. Memory capacity and pattern can also be customized to make an optimal match with the computation module. As a result, the whole design is highly optimized and can provide high speed performance and high energy efficiency. Compared with GPU devices, both the performance and energy efficiency can be significantly improved.

On FPGA, the architecture is constrained by available resources as you cannot have more resources than the provided amount. In addition, the amount of each functional block is fixed, and it is not easy to make fully use of all resources at the same time. For example, certain FPGA device may have a large number of DSP blocks where a potentially large-computation module with high speed performance is expected. However, the on-chip memory capacity may be limited, and thus the memory cannot provide enough data bandwidth to the large-computation module. In this case, the scale of the computation module has to be reduced to match the memory capacity and bandwidth. As a result, the logic resources on FPGA chip may not be fully utilized, which creates some resource and energy overhead, and the performance and efficiency of FPGA-based accelerator are affected. However, a carefully designed FPGA-based accelerator can still provide good enough performance for AI applications. Compared to GPU devices, the FPGA-based accelerator can achieve comparable performance, but the energy efficiency is significantly improved.

ASIC-based accelerators can provide high computation performance, but its architecture cannot be changed after fabrication. Although some ASIC-based architecture can be designed to be reconfigurable by providing different operational modes, however, such merged design will bring extra overhead, which may affect the performance. As a result, the ASIC-based accelerator is more suitable for applications where the algorithms are already optimal and big changes on algorithms are not expected.

FPGA-based accelerators, on the other hand, are suitable for applications that the algorithms are still fast revolutionized due to its reconfigurability. In addition, the design cycle is shorter, and the cost of the development is lower for FPGA devices.

5 Software and Hardware Co-design for AI Accelerator

In the last section, we discussed the architecture of a general AI accelerator, and some optimizations for each main module are introduced from hardware design perspectives. In addition to those hardware-only optimizations, software and hardware co-optimization strategy can be applied for further efficiency improvement. In this section, we will introduce the two popular software and hardware co-design methods for AI accelerators: the model compression and the reduced-precision computation. The methodologies of these two methods are discussed first, and the development tools to help in these design methods are then introduced.

5.1 AI Model Compression

Modern AI models contain a large amount of parameters. During implementation, they need to be transferred to and computed by the computation module. This large amount of computations will lead to a large latency for processing the AI model. In addition, transferring a large amount of data between the memory and the computing unit consumes a lot of energy. Therefore, it is desirable to reduce the amount of parameters while keeping the same level of performance. Model compression is one of the techniques to reduce the amount of parameters for AI model. It makes use of the sparsity characteristics of AI models to remove some unnecessary parameters of neurons.

The method to compress deep learning models is first proposed in [20]. By performing statistical analysis on the parameters of a deep neural network model, the authors found that many of the parameters are very small numbers and cannot contribute too much for the computation results. By setting a pre-defined threshold value, all parameter values that are smaller than the threshold value are removed, and the corresponding neurons may also be removed if all the weights associated to it are set to zeros. This process is called network pruning. According to the results presented in [20], the pruning process can reduce the network size by 9 to 13 times. On one hand, this can reduce the amount of computation to increase the speed performance of the computation. On the other hand, the memory capacity requirement is reduced, and data to be transferred during computation are reduced. This can help in the reduction of energy consumption.

To further reduce the network model size, quantization of the network parameters can be performed. Quantization is to map all the parameters into a smaller value range and use a fewer amount of numbers to represent all the parameters. A set of quantized values can be set according to the statistical analysis of the parameters. Then, each of the parameters is then replaced with one of the quantized value that is the most close to its original value.

In the literature, two quantizations are widely used: symmetric quantization and asymmetric quantization. In symmetric quantization, positive range and negative range are the same. Those original parameters are mapped into this symmetric range. This method is simple in computation but may not be efficient in some cases where the values are biased, for example, many more positive values than negative values. In this case, the negative range cannot be fully utilized. The other method is the asymmetric method that is to set independent range for positive and negative values. In this case, the range can be more precisely defined according to the data statistical characteristics, and thus smaller bit-width format can be used to represent the quantized data.

After quantization, only those quantized values need to be stored in the memory. As those values are shared, small-sized memory is enough to store all the parameters. In addition, with shared parameter values, the input data corresponding to the same weight value can be added first, and a single multiplication can be performed after accumulation. As the cost of hardware multiplier is much higher than that of the adder, the weight sharing operation can significantly reduce the computation cost. Moreover, when 3D memory is used, the addition operation of this weight sharing operation can be performed inside the memory, and only the accumulated data need to be transferred through the memory interface to the computation unit and are then multiplied with the shared weight value in the computation unit. This helps in reducing the amount of memory data transfer and can thus reduce the energy consumption of the computation.

Combining pruning and quantization can reduce the memory requirement by 27 to 31 times [20]. Due to its effectiveness, it is widely applied in many deep learning frameworks. Ristretto is the quantization tools proposed for the Caffe framework. TensorFlow has embedded quantization function. A recent tool called QKeras is another quantization option for TensorFlow framework. For PyTorch, Intel has developed Distiller for model quantization and reduced-precision computation.

The memory space can be further reduced by using some special encoding method. In [20], Huffman coding is used for the parameters after pruning and quantization. The Huffman algorithm uses a longer code for those infrequent values while using a shorter code for values that frequently appear. Therefore, the overall code size for all values is reduced, which can save memory space.

The pruning, quantization, and Huffman coding are performed for the AI models from the software development side. In hardware side, the EIE processor [21] is one of the neural network inference processors for compressed neural networks. After performing compression methods, the neural network model becomes sparse, and the multiplication with zero value does not need to be performed. In EIE, on one hand, only the non-zero values are stored and used at runtime. On the other hand, only the index of the first non-zero value and the relative position of the next non-zero value are stored. During inference, these information will be used to search for the next non-zero values for computation, and all other computations are skipped. For input data, similar detection methods are performed. With these methods, the energy efficiency of EIE processors is significantly improved.

5.2   Reduced-Precision Computation

Computation using reduced-precision numeric format is another software and hardware co-design method. This is applicable to AI computation due to the characteristics of data distribution of AI models.

An example of AI model data distribution is shown in Fig. 13, where the weight and data distribution in convolution layer 2 of the AlexNet are presented. As shown in Fig. 13, although the amount of parameters and data is large, they all distribute in a relatively small range. As a result, a format with small bit width can be enough to represent all these numbers. With small bit-width format, on one hand, the cost of the arithmetic unit can be reduced. On the other hand, the total amount of memory data transfer can also be reduced. Both of them help in improving speed performance and energy efficiency.

By default, on CPU or GPU, AI models are trained with single-precision floating-point numeric format. This format can provide high precision and large dynamic range to represent parameters and data in AI models. This single-precision format can provide good accuracy performance for AI models; however, the complexity of single-precision arithmetic units is high, and it consumes a high energy to transfer single-precision data between memory and computing unit.

The training process and inference process of the AI model have different requirements on numeric formats. For AI inference, as shown in Fig. 13, the required dynamic range is small, and thus, fixed-point formats can provide a good accuracy performance. In the literature, 8-bit or 16-bit fixed-point formats are popularly used in deep learning inference. For some AI models, even 4-bit fixed-point format can be used to trade off for a better speed performance and energy efficiency. In some extreme cases, when accuracy degradation can be tolerated, binary neural network or ternary neural network are used where the model parameters are represented by only 2 (± 1) or 3 (0 and ± 1) numbers, respectively. The benefit of using binary or ternary network is that the multiplication of ± 1 can be achieved by XNOR gate,

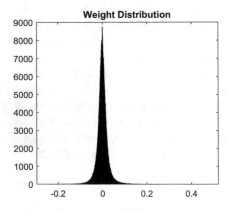
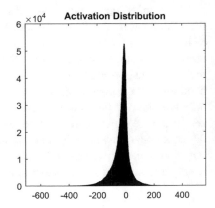

Fig. 13 Weight and data distribution of AlexNet convolution layer 2

and thus the hardware multiplier can be completely eliminated to achieve significant performance boost.

For the training process, floating-point format is still needed. The main reason to use floating-point format is to correctly represent the gradient values during the training process. The gradient is used to tune the weight parameters of an AI model, and it will become smaller as the training process goes. At the last few epochs of the training, the gradient will be a very tiny number. Therefore, a numeric format with large dynamic range is needed to correctly represent the gradient values for the training process.

Although floating-point format is needed, the single-precision format is still too large. The floating-point format is composed of three components: the sign, the exponent, and the mantissa. The exponent mainly controls the dynamic range of the format, and the mantissa mainly controls the representation precision of the format. The work [22] has evaluated the importance of dynamic range and representation precision for AI accuracy. The results show that dynamic range is more important than representation range for AI model. As a result, the bit-width requirement of the exponent should be met, while the bit width of the mantissa can be reduced. The recent BFloat16 format proposed by Google is designed according to the findings in [22]. The BFloat16 format is generated from single-precision format by directly truncating the mantissa from 23 bit to only 7 bit, while maintaining the 8-bit exponent unchanged. Another format, DLFloat [23] proposed by IBM, uses 6-bit exponent and 9-bit mantissa.

In addition to conventional fixed-point and floating-point formats, some specially designed numeric formats are also suitable for AI computing. The posit format [24] proposed in 2017 is one of them. The posit format is similar to the floating-point format, which has a sign bit, an exponent component, and a mantissa component. In addition, the posit format uses another component called regime to extend the exponent for larger dynamic range. The regime is a series of 0s (or 1s) ended by the opposite bit. The number of consecutive 0s (or 1s) is directly related to the absolute value of the regime.

In floating-point format, the bit width of each component is static for a specific precision. However, in posit format, except the sign bit, other components share the remaining bit positions, and the bit width of each of the component is dynamic. The regime component always exists, while the exponent and the mantissa will appear only if there exist bit positions that are not occupied by the regime. In this way, when the value of the number is small, the regime will occupy fewer bit positions, and thus the mantissa can have more bit positions to represent the number with higher precision. This is suitable when the number density is high and needs high representation precision to distinguish them. When the value is large, the regime needs to occupy more bit positions, and thus the mantissa will have fewer bit positions. In this case, the number can still be represented, but the precision is reduced, which is suitable for sparsely distributed numbers. Recall the data distribution of AI model, shown in Fig. 13, and posit format is especially suitable for representing data with normal distribution. When using floating point, as the bit width of the components is static, to meet the requirements of both dynamic range and representation precision, a format with large exponent field and large mantissa field is needed, which makes the total bit width very large. In posit, as the bit width of regime and mantissa can be exchanged, all the numbers can still be correctly represented, while the total bit width is kept low. Some implementations in the literature have proved the effectiveness of posit in AI computation. By using 8-bit posit, the performance of AI models can reach almost the same level as the ones with 32-bit single-precision format. In inference, to achieve the same level of accuracy, posit can use even smaller bit width than fixed-point format.

6 Hardware-Aware Neural Architecture Search

With the development of AI algorithms, the model becomes deeper, and its operations become much more complex. The development of an AI model becomes complicated and thus error-prone. In this situation, the automated machine learning (AutoML) was proposed to automatically generate an AI model for the designated applications. The neural architecture search (NAS) technique is one of the AutoML design methods. With NAS, the neural network architecture can be automatically searched by setting search space and search rule. The neural network generated by NAS method can usually have better performance than manually designed neural network.

6.1 Neural Architecture Search Overview

The NAS method usually has three main components: the search space, the search strategy, and the evaluation method. The search space defines the basic structures that can be used to form a complex neural network. In some NAS methods, the network architecture is fixed, and the searching process is only to optimize its hyperparameters. In other NAS methods, the network architecture will no longer be fixed, and both the type of layer operations and the way different layers are connected can be optimized. The latter set of approaches can be further divided into three categories: layer-wise search space, cell-based search space, and hierarchical search space. In layer-wise search space, the model is created from a stack of layers that are generated from a set of pre-defined operators. In the cell-based search space, a fixed architecture cell is first created, and the whole model is then generated by repeating the created cell. In this method, the search of a proper cell architecture is the main focus. Last, in the hierarchical search space, a cell is first defined, and then a larger scale block is created using the cell, and finally, the whole network is designed using those blocks. Compared to the cell-based search space, the hierarchical search space is more flexible and the size of the search space is also smaller.

The search strategy will determine how to generate a candidate neural network architecture and how to refine the architecture based on the evaluation results. The commonly used methods include the reinforcement learning, the evolutionary algorithm, gradient-based method, and random search. Reinforcement learning is the most commonly used search strategy in NAS. The controller of the reinforcement learning will sample an architecture from the search space. The accuracy of the architecture is then evaluated, and the controller is then rewarded or penalized accordingly. The evolutionary algorithm will first generate a population of models, and then mutations are applied to create some new model generations. The new generations are then evaluated, and the best ones are added to the population. The gradient-based method is proposed to reduce the search time. It will create a supernetwork that shares weights with all child models. In this method, only the supernetwork needs to be trained, and the parameters of child networks are all obtained. Random search is the simplest method. It is usually combined with the Bayesian optimization method for hyperparameter optimization.

The evaluation method will determine how the candidate network is trained and evaluated. In conventional NAS method, each candidate network will be trained and its accuracy is evaluated.

Conventionally, the only target of the NAS process is the accuracy of the resulting neural network. To achieve a good accuracy for an application, a complex neural network architecture is usually generated. The generated neural network can meet the accuracy requirement; however, its architecture, especially the connections among basic structures, may be complex, which makes it difficult to be efficiently implemented in hardware processors. This will lead to increased latency and energy consumption. In recent years, the edge intelligence, which applies AI techniques in edge devices, becomes more and more popular. Different from the cloud datacenter, the edge devices do not have strong computing capability, and they have strict requirement on power consumption. Therefore, deploying complex NAS-generated neural network model at the edge will become challenging.

To generate hardware-friendly neural network models, while maintaining their accuracy, the hardware-aware NAS (HW-NAS) method is proposed. In HW-NAS method, some hardware implementation metrics in addition to the model accuracy are also used as the optimization target. As a result, the searching result is expected to be efficient in the designated hardware, while its accuracy can still meet the requirement of the application.

6.2   Hardware-Aware Neural Architecture Search

In HW-AWS, in addition to the accuracy, the hardware implementation metrics of the generated model are also considered during the searching process. There are multiple hardware platforms that are used for accelerating AI computation: (mobile) CPU, GPU, FPGA, and ASIC. For FPGA and ASIC, as their architectures are customized, there are also multiple hardware configurations available for a single

FPGA device or ASIC design. Based on these platforms and configurations, the HW-NAS can be divided into three main categories: single platform with fixed configuration, single platform with multiple configurations, and multiple platforms.

In HW-NAS for single platform with fixed configuration, the hardware platform and its configuration are fixed before the searching process. The target is then to find a neural network model for this hardware that can achieve good accuracy and low hardware implementation cost, in terms of speed performance and energy consumption. There are several NAS methods proposed in the literature [25–28] that can be classified into this category. In [25], the authors performed NAS for GPU, CPU, and mobile devices separately and found the optimal architectures for those platforms are also different. The NAS method targeting both accuracy and latency for mobile platform was proposed in [26]. The latency was measured in real mobile device implementation instead of estimating using the number of computations, such as floating-point operations per second (FLOPs). In [27], the search cost of the searching process for mobile devices was further reduced compared to [26]. A larger search space and better search algorithm were proposed in [28] to achieve better accuracy for different hardware platforms. A disadvantage of this type of HW-NAS is that when a new platform or a new configuration is used in the hardware, the whole searching process needs to be performed again for the new hardware.

When considering multiple configurations for custom design platforms, the HW-NAS becomes more complex. In the literature, multiple configurations are represented using pre-defined design templates, and the hardware metrics are evaluated using the combinations of these templates during the NAS process. In [29], the FPGA-based HW-NAS method was proposed for multiple FPGA configurations. The searching space considered the hardware configurations, such as data tiling and the amount of PEs. A performance abstraction model was provided to estimate the latency of the generated neural network architecture. Network pruning would be performed if the latency requirement was not met. Similar approach for ASIC was proposed in [30]. For ASIC, the design freedom is much larger than FPGA, and thus the searching space is much larger for ASIC NAS. In [30], multiple ASIC templates were built based on other ASIC accelerator designs to reduce the search space. Both neural network model with high accuracy and an ASIC architecture that can efficiently implement the generated model are produced during the search process.

When the neural network model is going to be deployed on different hardware devices, either among different devices of the same category or among heterogeneous devices, the HW-NAS for multiple platforms can be used [31, 32]. As the optimal model for different hardware platforms may be different, the complexity of HW-NAS is much higher when using multiple different HW-NAS strategies for different hardware. To reduce the level of complexity, in [31] and [32], a single neural network model is generated, respectively, for deployment in multiple different hardware platforms.

For HW-NAS, the search strategy is almost the same as the general NAS method, where reinforcement learning, evolutionary algorithm, gradient-based method, and random search methods are still applied. There are some extended search space and additional evaluation metrics required in HW-NAS, which will be discussed in the following sections.

6.2.1 Search Space

In HW-NAS, the architecture search spaces used in general NAS methods are still applicable. Some implementations restricted the layer operators or the cells to those that are efficient in designated hardware platforms [31, 32].

In addition to architecture search space, for searching for multiple configurations or multiple platforms, hardware search space is also needed to be defined. The hardware search space can be divided into two categories: hardware-configuration-based search space and template-based search space. The former one contains hardware design configurations, for example, the tiling parameter, loop unrolling parameter, pipeline stages, buffer size, memory interface bandwidth, and the amount of processing elements, and so on. For FPGA devices, all these parameters are constrained by the total available resources, such as DSP, memory, and LUTs. This search space is used in [29].

In the template-based search space, a set of pre-defined hardware architecture templates are used. For example, several ASIC accelerator designs exist in the literature can form a set of templates. In [30], the existing ASIC designs are used as templates to look for the optimal architecture for the generated neural network model.

In addition to these general hardware search spaces, searching for low-precision numeric format, or mixed-precision computation, and automatic pruning are also available in the literature. The numeric format search is to find the optimal low-precision format that can achieve efficient computation while maintaining the accuracy. The pruning method also targets for efficient processing by searching for the optimal compression method for the architecture.

6.2.2 Evaluation Metrics

In HW-NAS, in addition to the accuracy, some hardware metrics when implementing the neural network models are also needed to be evaluated to determine whether the model is efficient for the designated hardware. These hardware metrics can be measured by running a real execution on the target hardware or be estimated by using hardware models. Measuring in actual hardware is more accurate, but it is time-consuming and can slow down the search process. In addition, for multiple configuration search or multiple platform search, to perform real measurement, all target hardware designs are required, which will be costly. To solve this problem, many works investigated the methods to design accurate estimation models.

In HW-NAS, besides the accuracy, some hardware metrics are evaluated according to the target of the HW-NAS method. To optimize the latency, floating-point operation per second (FLOPs), model size, and latency are usually measured or evaluated. When a model size is small, the amount of computation, in terms of FLOPs, and thus the delay of execution will be small. However, sometimes, the latency of executing a model is not directly related to its model size. For models with the same amount of parameters or FLOPs, the one with more hardware-friendly architecture can have smaller execution latency. So it is more popular to directly measure the latency of the model. For some resource-limited hardware devices, the constraint on latency will be strict, and thus the trade-off between the accuracy and the latency needs to be balanced.

To optimize the energy consumption, energy consumption or memory footprint can be profiled. However, the estimation model for power consumption is much more complex than latency estimation model. Some hardware vendors provide energy estimation tools, such as the nvprof in CUDA toolkit provided by NVIDIA. An energy estimation tool for deep neural network is also provided by the MIT Eyeriss team [33]. If estimation model is not available, energy estimation by a real execution is necessary. Modeling the memory footprint can also reveal the energy consumption of the neural network. As discussed before, the energy consumed by the memory data transfer is the dominant factor of total energy consumption. By estimating the memory bandwidth and the amount of memory read and write, the rough level of energy consumption of the architecture can be obtained.

Area or logic resource usage can also be estimated if needed. For FPGA devices, a template architecture can be built, and the area can be estimated in terms of DSPs and BRAMs. For ASIC, a tool named MAESTRO [34] can be used to estimate the area and power consumption of the design. In some HW-NAS designs, the area is used to estimate the static power consumption that is closely related to the total area of the architecture.

7 Summary

In this chapter, we have reviewed the fundamentals, the design techniques, and the optimizations for AI accelerators on application-specific integrated circuit (ASIC) and field-programmable gate array (FPGA). We start from introducing the architecture of FPGA and discussing the logic resources available for AI accelerator design on FPGA. The HDL-based design flow and HLS-based design flow for FPGA and their corresponding design tools are presented. We also discuss the design flow and tools available for the ASIC platform.

The architecture and datapath of AI accelerator are then discussed. We first introduce the general architecture of the AI accelerator. It is composed of a highly parallel computing unit, a hierarchy of memory, and a controller. Then the optimizations for main modules, including the computation module, the memory module, and the control module, are presented. Optimizations for matrix multiplication and vector computations can be applied for the computation module. For the memory module, the optimizations focus on exploiting the reuse of data and parameters to reduce the memory transfer. For the control module, AI-specific instruction-set architecture is desired. Finally, the characteristics of FPGA-based and ASIC-based AI accelerators available in the literature are presented and compared.

In addition to the hardware optimization methods, the software and hardware co-optimization methods are also available in the literature. Two popularly used software and hardware co-design methods for AI accelerator, the AI model compression and the reduced-precision computation, are introduced. Then the hardware-aware neural architecture search method is presented. With these methods, the hardware characteristics are considered during AI model development, and thus hardware-friendly model with high accuracy can be expected.

References

1. Sze, V., Chen, Y.-H., Yang, T.-J., Emer, J.S.: Efficient processing of deep neural networks: A tutorial and survey. Proc. IEEE **105**(12), 2295–2329 (2017)
2. Fu, Y., Wu, E., Sirasao, A., Attia, S., Khan, K., Wittig, R.: Deep learning with INT8 optimization on Xilinx devices. Xilinx, White Paper WP486 (Apr 2017)
3. Xilinx: Convolutional neural network with INT4 optimization on Xilinx devices. Xilinx, White Paper WP521 (Jun 2020)
4. Shawahna, A., Sait, S.M., El-Maleh, A.: FPGA-based accelerators of deep learning networks for learning and classification: A review. IEEE Access **7**, 7823–7859 (2019)
5. Kim, S., Deka, G.C.: Hardware Accelerator Systems for Artificial Intelligence and Machine Learning, ser. Advances in Computers, vol. 122. Elsevier (2021)
6. Kim, J.-Y.: Chapter five - FPGA based neural network accelerators. In: Kim, S., Deka, G.C. (eds.) Hardware Accelerator Systems for Artificial Intelligence and Machine Learning, vol. 122, ser. Advances in Computers. Elsevier (2021), pp. 135–165
7. Chung, E., et al., Accelerating persistent neural networks at datacenter scale. In: HotChips 2017, Aug 2017, pp. 1–52
8. Pellerin, D.: FPGA accelerated computing using AWS F1 instances: Applications and development environment. In: HotChips 2017, Aug 2017, pp. 1–27
9. Jouppi, N.P., et al.: In-datacenter performance analysis of a tensor processing unit. In: Proceedings of the 44th Annual International Symposium on Computer Architecture, Jun 2017, pp. 1–12
10. Chen, T., Du, Z., Sun, N., Wang, J., Wu, C., Chen, Y., Temam, O.: DianNao: A small-footprint high-throughput accelerator for ubiquitous machine-learning. In: Proceedings of the 19th International Conference on Architectural Support for Programming Languages and Operating Systems, Feb 2014, pp. 269–284
11. Luo, T., Liu, S., Li, L., Wang, Y., Zhang, S., Chen, T., Xu, Z., Temam, O., Chen, Y.: DaDianNao A neural network supercomputer. IEEE Trans. Comput. **66**(1), 73–88 (2017)
12. Liu, S., Du, Z., Tao, J., Han, D., Luo, T., Xie, Y., Chen, Y., Chen, T.: Cambricon: An instruction set architecture for neural networks. In: 2016 ACM/IEEE 43rd Annual International Symposium on Computer Architecture (ISCA), Jun 2016, pp. 393–405
13. Zhang, S., Du, Z., Zhang, L., Lan, H., Liu, S., Li, L., Guo, Q., Chen, T., Chen, Y.: Cambricon-X: An accelerator for sparse neural networks. In: 2016 49th Annual IEEE/ACM International Symposium on Microarchitecture (MICRO), Oct 2016, pp. 1–12
14. Zhou, X., Du, Z., Guo, Q., Liu, S., Liu, C., Wang, C., Zhou, X., Li, L., Chen, T., Chen, Y.: Cambricon-S: Addressing irregularity in sparse neural networks through a cooperative software/hardware approach. In: 2018 51st Annual IEEE/ACM International Symposium on Microarchitecture (MICRO), Oct 2018, pp. 15–28

15. Liao, H., Tu, J., Xia, J., Zhou, X.: DaVinci: A scalable architecture for neural network computing. In: HotChips 2019, Aug 2019, pp. 1–44
16. Chen, Y.-H., Krishna, T., Emer, J.S., Sze, V.: Eyeriss: An energy-efficient reconfigurable accelerator for deep convolutional neural networks. IEEE J. Solid State Circ. **52**(1), 127–138 (2017)
17. Chen, Y.-H., Yang, T.-J., Emer, J., Sze, V.: Eyeriss v2: A flexible accelerator for emerging deep neural networks on mobile devices. IEEE J. Emerg. Sel. Top. Circ. Syst. **9**(2), 292–308 (2019)
18. Parashar, A., Rhu, M., Mukkara, A., Puglielli, A., Venkatesan, R., Khailany, B., Emer, J., Keckler, S.W., Dally, W.J.: SCNN: An accelerator for compressed-sparse convolutional neural networks. In: 2017 ACM/IEEE 44th Annual International Symposium on Computer Architecture (ISCA), Jun 2017, pp. 27–40
19. Gao, M., Pu, J., Yang, X., Horowitz, M., Kozyrakis, C.: TETRIS: Scalable and efficient neural network acceleration with 3D memory. SIGARCH Comput. Archit. News **45**(1), 751–764 (2017)
20. Han, S., Mao, H., Dally, W.J.: Deep compression: compressing deep neural network with pruning, trained quantization and Huffman coding. In: Bengio, Y., LeCun, Y. (eds.) 4th International Conference on Learning Representations, ICLR 2016, May 2016, pp. 1–14
21. Han, S., Liu, X., Mao, H., Pu, J., Pedram, A., Horowitz, M.A., Dally, W.J.: EIE: Efficient inference engine on compressed deep neural network. SIGARCH Comput. Archit. News **44**(3), 243–254 (2016)
22. Lai, L., Suda, N., Chandra, V.: Deep convolutional neural network inference with floating-point weights and fixed-point activations. CoRR **abs/1703.03073**, 1–10 (2017)
23. Agrawal, A., Mueller, S.M., Fleischer, B.M., Sun, X., Wang, N., Choi, J., Gopalakrishnan, K.: DLFloat: A 16-b floating point format designed for deep learning training and inference. In: 2019 IEEE 26th Symposium on Computer Arithmetic (ARITH), Jun 2019, pp. 92–95
24. Gustafson, J.L., Yonemoto, I.: Beating floating point at its own game: Posit arithmetic. Supercomput. Front. Innov. Int. J. **4**(2), 71–86 (2017)
25. Cai, H., Zhu, L., Han, S.: ProxylessNAS: direct neural architecture search on target task and hardware. In: International Conference on Learning Representations (ICLR), May 2019, pp. 1–13
26. Tan, M., Chen, B., Pang, R., Vasudevan, V., Sandler, M., Howard, A., Le, Q.V.: MnasNet: Platform-aware neural architecture search for mobile. In: 2019 IEEE/CVF Conference on Computer Vision and Pattern Recognition (CVPR), Jun 2019, pp. 2815–2823
27. Wu, B., Dai, X., Zhang, P., Wang, Y., Sun, F., Wu, Y., Tian, Y., Vajda, P., Jia, Y., Keutzer, K.: FBNet: Hardware-aware efficient convnet design via differentiable neural architecture search. In: 2019 IEEE/CVF Conference on Computer Vision and Pattern Recognition (CVPR), Jun 2019, pp. 10,726–10,734
28. Zhang, L.L., Yang, Y., Jiang, Y., Zhu, W., Liu, Y.: Fast hardware-aware neural architecture search. In: 2020 IEEE/CVF Conference on Computer Vision and Pattern Recognition Workshops (CVPRW), Jun 2020, pp. 2959–2967
29. Jiang, W., Zhang, X., Sha, E.H.-M., Yang, L., Zhuge, Q., Shi, Y., Hu, J.: Accuracy vs. Efficiency: achieving both through FPGA-implementation aware neural architecture search. In: 2019 56th ACM/IEEE Design Automation Conference (DAC), Jun 2019, pp. 1–6
30. Yang, L., Yan, Z., Li, M., Kwon, H., Lai, L., Krishna, T., Chandra, V., Jiang, W., Shi, Y.: Co-exploration of neural architectures and heterogeneous ASIC accelerator designs targeting multiple tasks. In: 2020 57th ACM/IEEE Design Automation Conference (DAC), Jul 2020, pp. 1–6
31. Jiang, Y., Wang, X., Zhu, W.: Hardware-aware transformable architecture search with efficient search space. In: 2020 IEEE International Conference on Multimedia and Expo (ICME), Jul 2020, pp. 1–6
32. Chu, G., Arikan, O., Bender, G., Wang, W., Brighton, A., Kindermans, P.-J., Liu, H., Akin, B., Gupta, S., Howard, A.: Discovering multi-hardware mobile models via architecture search. In: 2021 IEEE/CVF Conference on Computer Vision and Pattern Recognition Workshops (CVPRW), Jun 2021, pp. 3016–3025

33. Yang, T.-J., Chen, Y.-H., Sze, V.: Designing energy-efficient convolutional neural networks using energy-aware pruning. In: 2017 IEEE Conference on Computer Vision and Pattern Recognition (CVPR), Jul 2017, pp. 6071–6079
34. Kwon, H., Chatarasi, P., Sarkar, V., Krishna, T., Pellauer, M., Parashar, A.: MAESTRO: A data-centric approach to understand reuse, performance, and hardware cost of DNN mappings. IEEE Micro **40**(3), 20–29 (2020)

Neuromorphic Hardware Accelerators

Pamul Yadav, Ashutosh Mishra, and Shiho Kim

1 Introduction to Neuromorphic Computing

1.1 Neural-Inspired Computing

Since the early twentieth century, neuroscientists, psychologists, and artificial intelligence researchers have developed "neural-inspired" algorithms for solving complex tasks. These algorithms, popularly known as neural network algorithms, are based on a simplified approximation of the actual neural connections in the brain. A biological neuron is shown in Fig. 1 for ready reference of the reader. McCulloch and Pitts proposed one of the first computational models for neural networks in 1943 [1]. This model inspired the future development of artificial neural networks, such as Perceptron by Rosenblatt in 1958 [2]. Various advancements brought over by several researchers over approximately five decades led to compelling modern neural network algorithms. Many deep neural networks showed an ability to perform image recognition and achieved superhuman performance in the task [3]. With its popularity began a new subfield within artificial intelligence, known as deep learning. On the hardware side, the development of hardware accelerators such as faster CPUs and GPUs served as the backbone for accelerating the training of deep learning algorithms on very large datasets (> several million data points). Nonetheless, a crucial factor in the current computer architecture design is mainly responsible for dragging the performance down. We will discuss that architecture and its drawback in the next section.

P. Yadav · A. Mishra (✉) · S. Kim
School of Integrated Technology, Yonsei University, Incheon, South Korea
e-mail: pamul@yonsei.ac.kr; ashutoshmishra@yonsei.ac.kr; shiho@yonsei.ac.kr

© The Author(s), under exclusive license to Springer Nature Switzerland AG 2023
A. Mishra et al. (eds.), *Artificial Intelligence and Hardware Accelerators*,
https://doi.org/10.1007/978-3-031-22170-5_8

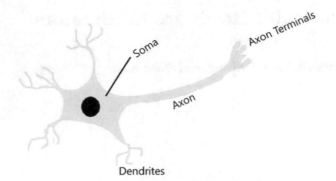

Fig. 1 A schematic diagram of a biological neuron

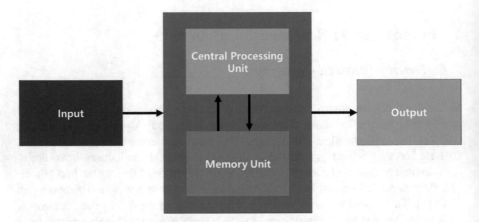

Fig. 2 The high-level design of von Neumann computer architecture

1.2   von Neumann Architecture

Early general-purpose computing machines such as ENIAC could perform complex but conceptually similar tasks. Rewriting the program on them was a tedious job that usually involved rewiring the components or physically redesigning the whole machine. To ease the programming process, John von Neumann introduced von Neumann Architecture (Fig. 2) in 1945 and subsequently proposed a computer based on the architecture called EDVAC, which laid the foundation for developing modern digital computers. The von Neumann architecture refers to any computing structure where the memory is placed separate from the processor, and the two communicate via a common bus.

von Neumann architecture consists of the following essential components and has been shown pictorially in Fig. 2:

(i) *Central Processing Unit (CPU)*: It is responsible for handling all of the processing tasks in a computer. It has various sub-parts:

- *The Arithmetic Logic Unit (ALU)*: It is responsible for arithmetic and logical operations.
- *Control Unit*: It is responsible for managing the movement of instructions between the CPU and memory and ensuring that the instructions are executed one at a time.
- *Registers*: They are referred to as the fast and temporary storage for instructions/data. They are broadly classified as accumulators, data registers, address registers, general-purpose registers, and special-purpose registers.

(ii) *Memory Unit*: It is the storage unit responsible for storing the instructions and data related to the program running on the processor. It is also called main memory or Random-Access Memory (RAM).

(iii) *Input/Output (I/O)*: It refers to the devices/methods used for inputting and outputting the data.

The reasons behind using von Neumann architecture even in today's modern computers are found in its fundamental advantages. Some of the primary advantages are:

- Its simple design allows for a more straightforward implementation of the overall computer architecture.
- Storage of instructions and data in the same memory allows for the cheaper implementation of the computer architecture.

Despite the advantages of the von Neumann architecture, it still drives its extensive applications in modern computer systems. There is a growing need to develop a different kind of architecture due to one of the fundamental drawbacks of von Neumann architecture, popularly known as the *von Neumann bottleneck*. This bottleneck dictates that the separation of memory and CPU forces the computer to limit its performance due to the data transfer overhead between the memory and CPU. In other words, no matter how fast a computer is made, it will always face a reduction in speed caused by the time consumed in transferring the instructions/data to and from memory. The utilization of parallelism has significantly enhanced computing performance by enabling several processors, often on the scale of thousands to millions, to be joined together to perform faster and more efficient computation. Nonetheless, the basic architecture remains the same as in a typical computer, carrying the von Neumann bottleneck even in robust parallel computing systems.

1.3 Need for Neuromorphic Computing

There is a tremendous desire for advanced computing systems over von Neuman architecture to develop next-generation processing systems that can perform faster computations compared to today's scale and consume less power than conventional computers. In contrast to neural-inspired computing, a different class of computing

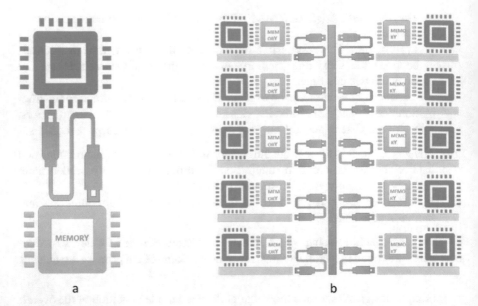

Fig. 3 An illustration of the comparison between (**a**) von Neumann architecture and (**b**) neuromorphic architecture

systems called neuromorphic computing takes direct inspiration from biological neural systems. They utilize the concepts of neuron, synapse, and spiking activity in the brain to mimic a biological neuron. Early works in this area were mostly carried out by neuroscientists who relied upon laboratory experiments and lacked any algorithmic framework for implementing an existing computing system. As with neural-inspired computing, the model proposed by McCulloch and Pitts also served as the starting point in neuromorphic computing; however, it was not until 1949 that Donald O. Hebbs proposed the Hebbian learning rule for modeling synaptic plasticity [4]. This rule was the foundation for one of the essential mechanisms in neuromorphic computing known as spike-timing-dependent plasticity (STDP) (discussed in Sect. 2.2). Figures 3a, b show the fundamental difference between von Neumann and neuromorphic architecture. In contrast to von Neumann's architecture, neuromorphic architectures can transmit information among the neurons using a single bus allowing for parallel communication among the neurons. Theoretically, such a design eliminates the von Neumann bottleneck by reducing the overhead time between processing and memory.

2 Building Blocks of Neuromorphic Computing

Neurons are the fundamental units in the brain responsible for storing and processing the information to perform various tasks enabling the diverse functionalities of a biological organism. Artificial neurons, a simpler approximation of the biological neurons (shown in Fig. 1), are the fundamental unit in developing all neural networks. However, the neuron in a neuromorphic algorithm is slightly different (discussed in Sect. 2.1) and is known as a spiking neuron. Synapses are the neuronal structures that allow one neuron to transmit the signal to another neuron. Transmission between the neurons and the synapse is unidirectional. Therefore, each spiking neuron can act as both a 'pre-synaptic' and a 'post-synaptic' neuron [5]. Spiking neurons and synapses together form the basis for building the neuromorphic systems at the algorithmic level. Section 2.1 discusses the mathematical model for implementing neuromorphic algorithms. At the hardware level, the processing units in the computing systems are implemented using Complementary Metal-Oxide-Semiconductor (CMOS) transistors. Despite the core contribution of CMOS transistors in digitizing the current world, they suffer from the limits of Moore's law [6], which dictates that "the number of transistors doubles every two years in the development of integrated circuit (IC)." Therefore, by virtue of their design, CMOS transistors are infeasible to develop miniaturized or small-scale powerful computers. To overcome this problem, researchers have proposed another fundamental circuit element called memristors for simulating the actual neuromorphic principle where the processing and memory storage is performed in the same unit. Doing so also eliminates the von Neumann bottleneck; thus, we get the best of both worlds. Section 2.2 discusses the basic principle and design of memristors.

2.1 Spiking Neural Networks

The spiking neural network (SNN) (shown in Fig. 4) is a kind of artificial neural network (ANN) with specific fundamental differences. SNNs store the information by encoding it in a spike train, allowing them to process spatiotemporal information that is potentially similar to the type of information processed by an animal

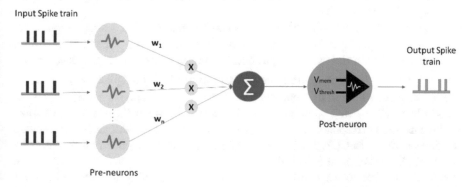

Fig. 4 An illustration of spiking neural network. (Figure adapted from Roy et al. [7])

brain. In contrast to an ANN neuron, SNN neurons are described using a set of differential equations, whereas the learning mechanism in an SNN is based on synaptic plasticity [8].

Spike train $s(t)$ can be defined as a sequence of k spike times $\{t^1, t^2, t^3, .., t^k\}$ at which a neuron fires during the neural activities. Spikes can be modeled using the Dirac delta function as $\delta(t - t^i)$ where t^i represents the spike time of ith spike, and can be expressed as below:

$$s(t) = \Sigma_{i=1}^k \delta \left(t - t^i \right) \tag{1}$$

Spiking neurons serve as the computational unit in an SNN. Several models represent biological neurons in terms of spiking neurons, but a simple *Leaky Integrate & Fire (LI&F)* model can be used as a base model for sufficiently explaining the underlying principles [9]. LI&F model is typically represented using differential equations as below:

$$\tau_{\text{mem}} \frac{dV_i}{dt} = -(V_i - V_{\text{rest}}) + RI_i - S_i(t)(V_i - V_{\text{rest}}) \tag{2}$$

$$\tau_{\text{syn}} \frac{dI_i}{dt} = -I_i(t) + \tau_{\text{syn}} \Sigma_j W_{ij} S_j(t) \tag{3}$$

where V_i represents the membrane potential of ith neuron, V_{rest} is the resting potential, τ_{mem} and τ_{syn} are the time constants, R is the input resistance, and $I_i(t)$ is the synaptic current. Equation (2) shows that V_i is a leaky integrator of the current $I_i(t)$, also, $I_i(t)$ is a weighted leaky integration of spike train S_j. In simpler terms, synaptic current goes up every time an incoming spike is experienced, where the magnitude of the spike in current is W_{ij}, otherwise, it experiences an exponential decay with a rate of τ_{syn}. The firing of neurons takes place when the membrane voltage attains the threshold value V_{th}, and once the spike is observed, voltage reduces back to the resting potential V_{rest}. This reduction is caused by $S_i(t)(V_i - V_{\text{rest}})$ in Eq. (1).

One of the simplest SNN architectures can be created using a single input layer and a single output layer (with single output), as shown in Fig. 5. Each input neuron j is connected with the single output neuron through some weight w_j also known as synaptic weight. Input neuron j carries and transmits input spike trains S_i^j to the output neuron, which in turn outputs the observed output spike train S_o^k ($k = 1$ for single layer SNN) as a result of some processing between the set of input spike trains and the corresponding synaptic weights whose function can be represented by $S_o = \text{synwt}\_\text{func}(S_i, W)$. Let us assume that there exist some real output spike trains S_r^k. Learning in an SNN is defined in terms of an error function E to minimize the total error between a pair of observed output spike trains and real spike trains $\mid S_o^k - S_r^k \mid$ for each output neuron where the total number of output neurons is N_o as shown in Eq. (4):

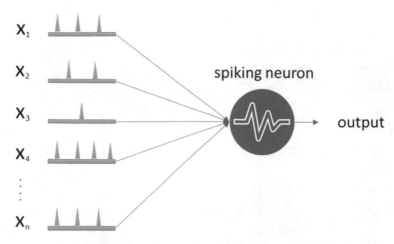

Fig. 5 An illustration of a single-layer SNN architecture. (Figure adapted from Roy et al. [7])

$$\min E\,(S_o, S_r) = \min \Sigma_{k=1}^{N_o} \mid S_o^k - S_r^k \mid \qquad (4)$$

The complete learning process in an SNN starts with random initialization of the synaptic weights W and then follows the below three steps:

(i) *Information Encoding and Decoding*: SNNs work with inputs and outputs in the form of spike trains, but the actual stimulus signal in the brain is mostly analog in nature. Therefore, a mechanism is needed to convert the analog signals into spike trains. Conversion of analog signals to spike trains is called encoding, and converting back to analog signals is called decoding. Various encoding and decoding algorithms have been developed over the past several years [10, 11]. However, we consider a viral encoding and decoding algorithm called Ben's spiker algorithm (BSA) to describe the process of information encoding and decoding [12]. A BSA-based encoder utilizes an FIR filter where two error values are computed at each time stamp. The first error value is calculated by subtracting the filter coefficients from the successive signal values, and the second error value is calculated using the unchanged signal value. If the first error value is lesser than the second error value minus a threshold value, a spike (positive) is produced, and eventually, the filter coefficients are subtracted from the signal values. The BSA-based decoder is a reasonably simple routine where the signals are reconstructed from the output spike trains by convoluting it with the FIR filter.

(ii) *SNNs Simulation*: Due to the SNN being a composite system of both continuous (analog signals) and discrete (spike trains) elements, the network simulation is different from that of ANN. There are various simulation strategies developed over the past several years that can typically be categorized as either a clock-driven strategy or event-driven strategy. We describe the Event-Driven Lookup

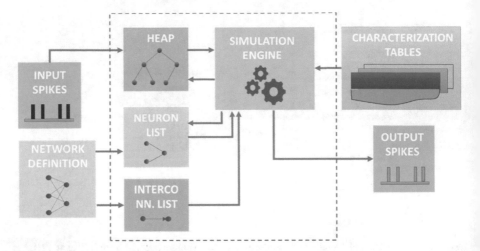

Fig. 6 An illustration of a step-by-step mechanism for simulating an SNN. (Figure adapted from Mohemmed et al. [13])

Table (ED-LUT) simulation strategy. Figure 6 describes the overall steps involved in the simulation.

Once the network is defined (initialized), a neuron list and an interneural connection list are generated. Next, multiple lookup tables corresponding to the neuronal and synaptic dynamics are computed that including one table for recording the exponential decay of the conductance across synapses; one table for storing the next spike time predictions when the next input spike time is absent; and a table for storing the membrane potential V_i for each input neuron i. Moreover, each neuron i keeps track of the state variables (membrane potential, excitation conductance G_{ex}, and inhibition conductance G_{ih}) at the last event as well as the time at which the last event occurs. Heap (Event Heap) is used for ordering the events (spikes), so they can be processed chronologically. Once a neuron receives a spike, it uses the lookup tables to regulate its response, and the newly generated spikes are recorded in the event heap. Typically, two kinds of events coincide, one when a neuron emits a spike (firing event) and the other when the emitted spikes reach the target neuron (propagated events). Since the firing events are a form of a prediction, therefore, the event handler must check the correctness of those firing events in a heap before being processed.

(iii) *Error Minimization and Learning*: A wide variety of learning algorithms exist for training an SNN. Some popular classes are perceptron-based algorithms, synaptic plasticity algorithms, and spike train convolution algorithms. However, we describe the ReSuMe algorithm in this section as it is based on the concept of synapse plasticity, which the animal brain utilizes to perform learning [13]. ReSuMe algorithm performs adjustment of the synaptic weights associated with the input spiking neurons to minimize the error function expressed in Eq. (7). The adjustment of the weights in the ReSuMe algorithm is made using Eq. (8).

$$\Delta w_i(t) = [s_r(t) - s_o(t)] \left[c + \int_0^\infty W(s) s_i \, (t - s) \, ds \right]$$ (5)

where c is a factor used for facilitating the convergence of the algorithm, and $W(s)$ is a factor considering the weight changes which occur due to the temporal correlation between the input spikes $S_i(t)$ and real output spikes $S_r(t)$. Therefore, $W(s)$ has been defined as:

$$W(s) = \begin{cases} A^+ \exp\left(\frac{-s}{\tau^+}\right) & \text{if } s \geq 0 \\ A^- \exp\left(\frac{s}{\tau^-}\right) & \text{if } s < 0 \end{cases}$$ (6)

Once the observed spike trains are obtained, the error is calculated using the function $E(S_o, S_r)$, as described in Eq. (7). If the error has reached a minimum threshold value, then the network is considered to be trained; the learning algorithm is run iteratively by updating the weights using Eq. (10) until the minimum threshold is reached.

$$W_i = W_i + \Delta w_i(t)$$ (7)

2.2 Memristors

Three basic circuit elements, that is, resistors, capacitors, and inductors, form the basis of all electronic systems, including semiconductors. They can be mathematically defined in terms of four fundamental circuit variables: current (i), voltage (v), charge (q), and magnetic flux (ϕ), as shown in Fig. 7. In theory, it is possible to set up a total of five one-to-one mathematical relationships among the four circuit variables, therefore, there was a possibility for the existence of a fourth circuit element that could validate the relationship between the voltage and magnetic flux. This element was discovered in 2008 and is known as a memristor [14]. It is a two-terminal resistive device that can adjust its internal resistance depending on the applied voltage, allowing it to store its internal state value.

This ability to store emulates the memory property and is thus called memory + resistor = memristor. Mathematically, they belong to a general class of dynamical systems called memristive systems that can be defined by Eqs. (11) and (12) [15].

$$v = R(x)i$$ (8)

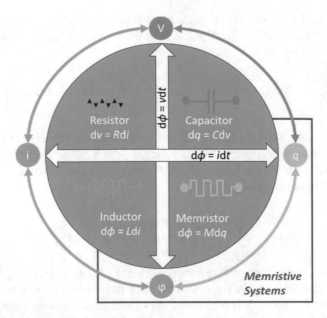

Fig. 7 An illustration of the fundamental circuit elements and their relationship with circuit variables. (Figure adapted from Prodromakis and Toumazou [14])

$$\frac{dx}{dt} = f(x, i) \tag{9}$$

where v denotes the voltage, i is the current, and $R(x)$ is the instantaneous resistance dependent on an internal state variable x. Figure 7 illustrates the pair of circuit variables used to define the mathematical relationship for the corresponding circuit element. It suggests that $\phi = f_M(q)$ defines a charge-dependent memristor. Solving the previous equation by performing differentiation and integration gives:

$$\frac{d\phi}{dt} = \frac{df_M(q)}{dq}\frac{dq}{dt} \tag{10}$$

(it can also be written as) $\quad \dfrac{dq}{dt} = \dfrac{dg_M(\phi)}{d\phi}\dfrac{d\phi}{dt}$ \hfill (11)

Equations (13) and (14) can be rewritten in terms of v and i as,

$$v(t) = \frac{df(q)}{dq}i(t) \tag{12}$$

$$i(t) = \frac{dg_M(\phi)}{d\phi}v(t) \tag{13}$$

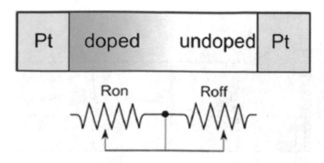

Fig. 8 An illustration of HP memristor with an equivalent circuit diagram. (Figure adapted from Kavehei et al. [11])

where, $M(q) = \frac{dg_M(\phi)}{d\phi}$ is the memristance for a charge-dependent memristor and $W(\phi) = \frac{df(q)}{dq}$ is the memductance for the flux-dependent memristor having the units of Ohm (Ω) and Siemens (S), respectively.

One of the first memristors was built at HP labs. It was developed by squeezing a thin film of titanium oxide (TiO_2) between two Platinum (Pt) surfaces where one end of the film was doped with oxygen and the other was left undoped. Doing so created two different resistances such that the doped region was at higher resistance while the undoped region was at a lower resistance, as shown in Fig. 8. The resistance of a device consisting of such a memristor can be modified by applying a voltage bias. HP memristors used TiO_2 as the dielectric, but ever since, various materials such as silicon oxide (SiO_2), copper oxide (CuO), nickel oxide (NiO), zinc oxide (ZnO), and aluminum oxide (Al_2O_3) have experimented with the memristors. Different materials show different resistive switching characteristics that can be either abrupt or gradual. An abrupt change in the resistance is more suitable for binary data storage/processing, whereas a gradual type is more suitable for analog computing. Set and reset operations of memory state in a memristor are performed by passing a voltage/current pulse on the memristor electrodes.

Due to the limitations of von Neumann architectures, the physical implementation of neuromorphic computing requires a different architecture; where one of the most widely used such architectures is known as the crossbar array. It provides a mechanism to represent axons and neurons in a parallelly connected design to emulate neurosynaptic processes in the human brain. Figure 9 shows a synaptic array circuit that is composed of the single crossbar array of M-($g_{j,k}$) and the constant-term circuit of 1/RB [16]. Here $g_{j,k}$ is the memristor's conductance at the crossing point between the jth row and kth column. $V_{IN,j}$ is an input voltage applied to the jth row. $V_{C,k}$ is a column-line voltage on the kth column. The column line, $V_{C,F}$ is connected to all the applied input voltages from $V_{IN,1}$ to $V_{IN,m}$ through R_B. In Fig. 9, $V_{C,F}$ enters G_F that constitutes the inverting OP amp with the negative feedback resistor R_{F1}. The output voltage of G_F is V_F that is connected to all column lines

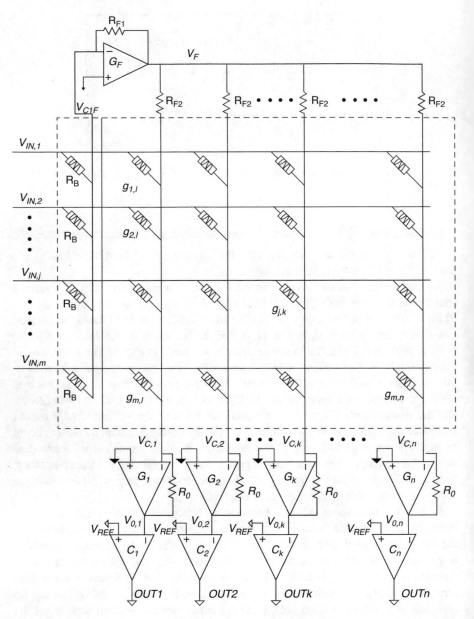

Fig. 9 An illustration of memristor-based crossbar array architecture. (Figure imported from Kim et al. [16])

$$V_{\mathrm{F}} = -\Sigma_{j=1}^{m} \frac{R_{\mathrm{F1}}}{R_{\mathrm{B}}} V_{\mathrm{IN},j} \qquad (14)$$

From $V_{C,1}$ to $V_{C,n}$ via R_{F2}, as shown in Fig. 9. By applying Kirchhoff's current law to the column line, $V_{C,F}$, we can calculate V_F with Eq. (14).

For the column lines, as you can see in Fig. 9, each column line is connected to its inverting amplifier, from G_1 to G_n. For example, $V_{C,1}$ enters G_1 with the negative feedback resistor, R_0. $V_{O,1}$ is the output voltage of G_1. Similarly, $V_{C,k}$ goes into G_k, and $V_{O,k}$ is the output voltage of G_k. $V_{O,k}$ can be calculated with Eq. (15).

$$V_{O,k} = -\left[\Sigma_{j=1}^{m}\left(R_0 g_{j,k} V_{IN,j}\right) + \frac{R_0}{R_{F2}} V_F\right] \qquad (15)$$

Assuming that RF1 = RF2 and combining Eq. (14) with Eq. (15), the following Eq. (16) can be obtained as follows:

$$V_{O,k} = -\left[\Sigma_{j=1}^{m}\left(R_0 g_{j,k} - \frac{R_0}{R_B}\right) V_{IN,j}\right] \qquad (16)$$

Adding the comparator, C_k, to the output voltage, $V_{O,k}$, we can decide if the neuron's output of the kth column, OUT_k, should be activated or not. V_{REF} is the reference voltage for the comparators from C_1 to C_n. When $V_{O,k}$ is larger than V_{REF}, OUT_k becomes 1. On the other hand, if $V_{O,k}$ is smaller than V_{REF}, OUT_k is decided to be 0, as indicated in Eq. (17).

$$OUT_k = \begin{cases} 1 & \text{if } V_{o,k} \geq V_{REF} \\ 0 & \text{if } V_{o,k} \leq V_{REF} \end{cases} \qquad (17)$$

3 Neuromorphic Hardware Accelerators

3.1 Overview

With the basic building blocks for neuromorphic computing already being discussed, the focus can be shifted toward the hardware implementation of neuromorphic computing systems. At the lowest tangible level of hardware implementations, Neuromorphic Hardware Accelerators (NHAs) come into the picture for practical implementation of the neuromorphic architectural design and the execution of brain-inspired neural network algorithms. Due to its huge potential in revolutionizing computing systems, many researchers and engineers in academia and industries are attempting to realize NHAs. This section discusses various popular NHAs, including their architectures and performance evaluation over some experimental applications.

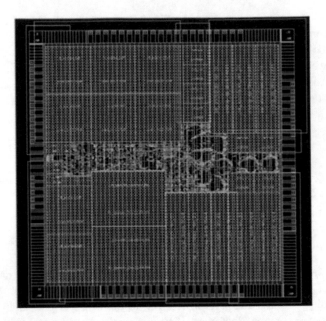

Fig. 10 A pictorial representation of a die in Darwin NPU chip. (Figure imported from Ma et al. [17])

3.2 Darwin Neural Processing Unit

Darwin is a Neural Processing Unit (NPU) for low-resource embedded technologies such as mobile phones, drones, and IoT devices [17]. Its application-specific integrated circuit (ASIC) prototype was produced using SMIC's 180 nm fabrication technique after being designed on an FPGA. Darwin NPU's design makes most of its neurons, synapse locations, and synaptic delays customizable parameters, enabling its overall configuration to be changed to meet the application's needs. Also, it utilizes time-multiplexing of physical neuron units to subside the computational cost, and its reconfigurable memory subsystem assists in reducing the memory resource cost during the processing (Fig. 10).

3.2.1 Darwin NPU Architecture

The entire microarchitecture of the Darwin NPU chip is depicted in Fig. 11. Eight physical neuron units on the chip use time multiplexing to simulate logical neurons. The sum of the physical neuron count and the level of time-division multiplexing determine the total number of logical neurons that can exist.

Address Event Representation (AER) format encodes the information for input and output spikes. Based on two pieces of information—the ID of the spike-generating neuron and the timing when the spike is generated—an AER packet is

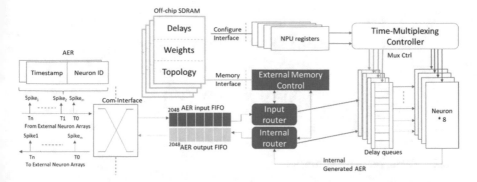

Fig. 11 Overall microarchitecture of Darwin NPU. (Figure adapted from Ma et al. [17])

used to specify and distinguish each spike. NPU is designed to activate only when an input AER packet arrives at the FIFO queue. The arriving packets are served online instead of in batch mode, improving the response speed to input stimuli. During no arrival of any packets, NPU enters into the sleep mode, thereby reducing the standby power consumption in the chip.

Dataflow in the NPU consists of the following steps:

- *Step 1*: NPU operates in an even-trigger fashion, that is, NPU is activated when a spike arrives at the input queue. Once arrived, NPU fetches the AER packet and matches the timestamp with the current time. If the timestamp matches successfully, the packet moves to step 2 (spike routing process); otherwise, it skips step 4.
- *Step 2*: In step 2, AER packets are examined for the information they contain, that is, timestamp and pre-synaptic neuron ID (source). This information is utilized for routing them to their post-synaptic neurons (destination). In addition, synaptic attributes such as weights and delays are stored in the off-chip DRAM.
- *Step 3*: In this step, a weight-sum value stored in the weight-sum queue is sent to the neuron after a fixed delay in time.
- *Step 4*: In this step, the status update of each neuron is executed. Firstly, the current status of the neuron is obtained from local status memory, and then the weighted sum of the current step is obtained from weight-sum queue. When a spike generation is observed, it is sent to the spike router (input or internal router) in the form of AER packets. Status update of each neuron is done using Eqs. (18)–(20). It is important to note that AER packets are translated into weight-delay information by a spike router.

$$V_j(t) = V_j(t-1) * \left(1 - \frac{\Delta t}{\tau_{\mathrm{m}}}\right) + \Sigma_i S_{ij} V_{\max} w_{ij} \qquad (18)$$

$$V_j(t) = \begin{cases} 0, & \text{if } t \in \left[T_\text{f}, T_\text{f} + T_\text{ref} \right. \\ H\left(V_\text{th} - V_j(t) \right) * V_j(t), & \text{else} \end{cases} \tag{19}$$

$$S_i(t) = H\left(V_i(t) \right) - V_\text{th} \Big) \tag{20}$$

3.2.2 Performance Evaluation

Darwin NPU has experimented with MNIST handwritten digit recognition problem [18]. A four-layer fully connected SNN was used as the neural network architecture. The parameter values of the SNN are as follows: 784 neurons in the input layers, both hidden layers contain 500 neurons each, whereas the output layer contains 10 neurons. Each input neuron represents a 28 × 28 pixel input image emitted. Each input neuron emits a spike train, which encodes the pixel intensity using a firing rate coding. The output for classification is chosen to be the output neuron with the highest firing rate. Instead of waiting for the whole spike train that encodes the input image, the NPU begins processing as soon as the first spike of the input spike train comes since the input spikes are processed in a streaming and online manner. The Darwin NPU is set up to operate at a clock rate of 25 MHz, with an average digit recognition delay of 0.16 s and overall classification accuracy of 93.8%. It has been compared with Minitaur 7, another SNN hardware accelerator, which utilized a Xilinx Spartan- 6 FPGA to create the same SNN [19]. Minitaur, in contrast, has a clock speed of 75 MHz, an average latency of 0.152 s, and a classification accuracy of 92%. With a slower clock speed and marginally higher classification accuracy, the Darwin NPU achieved a similar average latency as Minitaur.

3.3 Neurogrid System

Large-scale neural models incorporate experimental results from several levels of research to describe how intelligent behavior originates from bioelectrical processes at six orders of magnitude smaller in space and time (from nanometers to millimeters and from microseconds to seconds). Only a few models can bridge the technological gap due to unreasonably high computing costs yet fail to provide inferences comparable to human behavior. While consuming 40,000 times more power, a personal computer mimics a mouse-scale cortex model (2.5×10^6 neurons) 9000 times slower than a real mouse brain (400 W vs. 10 mW). By sharing synaptic and dendritic tree circuits, the Neurogrid project, which also employs an analog approach, seeks to reduce transistor count further [20].

3.3.1 Neurogrid Architecture

The Neurogrid software stack comprises a user interface (UI), a hardware abstraction layer (HAL), and driver components (driver). The user interface enables the user to engage with the simulation, evaluate the outcomes in real-time, and display the simulated neural network model. HAL maps the description of the analyzed model to the Neurogrid circuit. The driver uses the Neurogrid package to program this card into Neurocores via USB. The user interface, hardware abstraction layer, and driver components (driver) form the Neurogrid software stack. A daughterboard, a Lattice ispMACH CPLD, a Cypress EZ-USB FX2LP, and 16 Neurocores connected in a binary tree make up the hardware. The FX2 handles communications through USB, and the CPLD connects the FX2 and Neurocores. With the help of eight Cypress 4 MB SRAMs and a Xilinx Spartan-3E FPGA, the daughterboard implements primary axon-branching. A Neurocore (Fig. 12) has a 256×256 silicon-neuron arrays, a transmitter, a receiver, a router, and two RAMs.

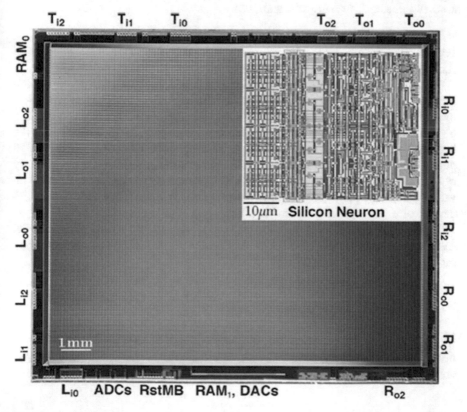

Fig. 12 A pictorial representation of Neurocore shown in Neurogrid chip. (Figure imported from Benjamin et al. [20])

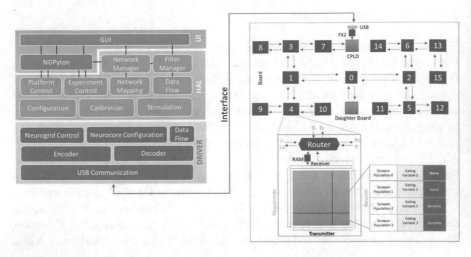

Fig. 13 An illustration of overall Neurogrid architecture for both software and hardware components. (Figure adapted from Benjamin et al. [20])

A neuron has a soma, a dendrite, four gating-variable, and four synapse-population (i.e., shared synapse and dendrite) circuits.

The Neurogrid neuron somatic, dendrite, synaptic, and ion channel population schemes use MOS devices to implement the dimensionless morphology of typical biological neuron models. Dimensionless models have few free parameters and can be implemented on various hardware platforms. However, MOS devices offer the highest densities. The Neurogrid neuron somatic, dendrite, synaptic, and ion channel population schemes use MOS devices to implement the dimensionless morphology of typical biological neuron models [21]. Dimensionless models are accessible on various hardware platforms and have fewer available parameters. However, MOS devices offer the highest densities (Fig. 13).

For a given passive membrane model with a capacitor C, a reversible potential E_{leak}, and a conductor G_{leak} with a current source I_{in}, this circuit can be described by

$$CV' = -G_{\text{leak}}(V - E_{\text{leak}}) + I_{\text{in}} \tag{21}$$

where V is the voltage across C. C, G_{leak}, E_{leak}, and I_{in}, are the four parameters in Eq. (21), however, the model has only two degrees of freedom. The reference voltage can be changed to E_{leak} and normalized with $G_{\text{leak}}V_n$ to give

$$\tau v' = -v + u \tag{22}$$

where $\tau = C/G_{\text{leak}}$, $v = (V - E_{\text{leak}})/V_n$, and $u = I_{\text{in}}/(G_{\text{leak}}V_n)$; and v is the voltage in a unit of V_n, and u is the current in a unit of $G_{\text{leak}}V_n$. Equation (22) has two parameters: τ and u, to match the model's two degrees of freedom. This strategy is general. By altering the reference voltage to E_{leak}, normalizing the voltage by V_n, the conductance by G_{leak}, and the current by $G_{\text{leak}}V_n$, any electrical model of the membrane can be made dimensionless. Henceforth, dimensionless equivalents of voltages can be denoted with v, conductances with g, and currents with i. The electrical model of neuron components (soma, dendrite, synapse, and ion channel population) is described below:

- *Soma*: The soma's dimensionless model is given by

$$\tau_s v_s' = -v_s + i_{\text{sin}} + \frac{1}{2}v_s^2 - g_K v_s - g_{\text{res}} v_s p_{\text{res}}(t) + v_d \tag{23}$$

where τ_s is the membrane time constant, i_{sin} is the input current, and v_d is the dendritic input. A second-order positive feedback $v_s^2 = 2$ shapes the spike producing a sodium current, refractory period modeling, active conductivity g_{res} recovery for unit amplitude pulse p_{res}, pressure duration t_{res}, and peak frequency tuning of the high-threshold potassium conductance g_K pattern. g_K is given by

$$\tau_K g_K' = -g_K + g_{K\infty} p_{\text{res}}(t) \tag{24}$$

where τ_K is the decay time constant and g_{K1} is the saturation value. The soma may also receive synaptic inputs.

- *Dendrite*: The dendrite's dimensionless model is given by

$$\tau_d v_d' = -v_d + i_{\text{din}} + i_{\text{bp}} p_{\text{res}}(t) - g_{\text{ch}}(e_{\text{ch}} - v_d) \tag{25}$$

where v_d is the membrane time constant, i_{din} is the input current, i_{bp} is the backpropagating input, and g_{ch} is the channel population's conductance, with reversal potential e_{ch}.

- *Synapse Population*: The synapse population's dimensionless model is given by

$$\tau_{\text{syn}} g_{\text{syn}}' = -g_{\text{syn}} + g_{\text{sat}} p_{\text{rise}}(t) \tag{26}$$

where g_{syn} is the synaptic time constant and g_{sat} is the saturation conductance for the population. The input spike triggers a single amplitude pulse $p_{\text{rise}}(t)$. t_{rise} simulates how long a neurotransmitter is available at regular intervals. The conductance g_{syn} decreases spatially in the common dendritic tree and provides inward currents $\zeta(n)g_{\text{syn}}(e_{\text{syn}} - v_s)$ and $\zeta(n)g_{\text{syn}}(e_{\text{syn}} - v_d)$ to the soma and dendrites, respectively.

$$\zeta(n) = \frac{1}{4\sqrt{\pi}}\left(1 + \left(\frac{1}{1-\gamma^2}\right)^{\frac{1}{4}}\right)^2 \frac{\gamma^n}{\sqrt{n}}$$

where γ is the damping coefficient of the silicon dendritic tree and n is the distance traveled in the number of neurons.

- *Ion-Channel Population*: The ion-channel population's conductance g_{ch} is obtained by scaling a maximum conductance g_{max} with a gating variable c. c is modeled as

$$\tau_{gv}c' = -c + c_{ss} \tag{27}$$

where c_{ss} is its steady-state activation or inactivation and τ_{ch} is its time constant. Where c_{ss} is given by

$$c_{ss} = \frac{\alpha}{\alpha + \beta} \text{ or } \frac{\beta}{\alpha + \beta}$$

where α and β model a channel's opening and closing rates, whose voltage dependence is modeled as

$$\alpha, \beta = \pm\frac{1}{2}(v_d - v_{th}) + \frac{1}{2}\sqrt{(v_d - v_{th})^2 + \frac{1}{4s^2}}.$$

Here, v_{th} is the membrane potential at which $c_{ss} = 1/2$ and s is the slope at this point. α and β satisfy a difference relation $\alpha - \beta = v_d - v_{th}$ and a reciprocal relation $\alpha\beta = 1/(16s^2)$, resulting in a sigmoidal dependence of c_{ss} on v_d. The gating variable's time constant is given by

$$\tau_{gv} = \frac{\tau_{max} - \tau_{min}}{2s(\alpha + \beta)} + \tau_{min} \tag{28}$$

τ_{gv} is bell-shaped with a maximum value of τ_{max} when $v_d = v_{th}$ and a minimum value of τ_{min} when $|v_d - v_{th}| \gg 1/(2s)$, to avoid unphysiologically short time constants.

The neuron tip is transmitted from the matrix and passed through a router to the parent Neurocore. All these digital circuits are event-driven, activated only when a spark occurs, and synthesize logic following Martin's synthesis for asynchronous circuits. The transmitter transmits multiple stimuli in series, and the receiver transmits multiple stimuli in sequence. A router sends packets from a source to multiple receivers in two phases: a point-to-point phase and a branching phase. In the point-to-point phase, routers forward packet up/down or left/right based on the bits in the first word of the packet. In the branching phase, the router copies the packets to the left and right ports. These overflow packets are either passed to the local neural array or filtered using the Neurocore 256 × 16-b SRAM location information marked as the second word in the packet. This method delivers outstanding throughput by sending packets to all potential receivers without running a memory scan. Unlike the network, which decides how packets are routed, these inquiries, which are the slowest operation, operate concurrently. The row address receives an addition of 2 bits of SRAM when a packet is transmitted. Which of the four shared receiver synapses will be triggered is specified by this bit.

3.3.2 Performance Evaluation

A 15-layer recurrent inhibitory network that was mapped onto a separate Neurocore was simulated by Neurogrid. The recurrent synaptic connections between each Neurocore and all of the others (including itself) were revealed by multicasting spikes. Each Neurocore was configured to accept the spikes from its three neighbors on either side, which blocked neighboring neurons via the common dendrite. Consequently, each model's 983,040 neurons received 50% of their inhibition from 7980 neurons that were located in a cylinder surrounding them that was 19 neurons in radius and seven layers deep. Globally synchronous spike activity is what is predicted and observed to result from such recurrent inhibitory connection patterns; 3.7 Hz was the frequency of the rhythmic synchronized activity, and an average of 0.42 spikes per second were emitted by the neurons. 2.7 W of power was consumed by Neurogrid during the simulation. The daughterboard was not required because interlayer connections were made between corresponding sites (i.e., columnar).

3.4 SpiNNaker System

SpiNNaker machine was developed primarily to support the kind of communication mechanism in the human brain [22]. SpiNNaker, created with massively parallel computing in mind, can support up to one million microprocessor cores; the upper limit is determined by cost and architectural convenience rather than predetermined standards. The communication infrastructure of the SpiNNaker architecture is built to transfer enormous quantities of tiny packets, in contrast to typical high-performance computer communications systems, which are ideal for huge data packets. A 40-bit packet is used to encode a single neural spike event. These 40 bits are made up of 8 control bits that indicate the packet type and other related information and 32 bits that serve as the spike neuron's AER identity. The sub-millisecond packet transmission times required for real-time neural simulations are achievable by communications networks. Regarding crucial performance requirements, SpiNNaker's design is willing to make some concessions. The asynchronous nature of communication systems governs the sequence in which packets are received, and occasionally packets may be dropped to prevent communication failures. One of the compromises the SpiNNaker architecture recognizes is the necessity for predictable performance. The stochastic order in which packets are received results from the asynchronous nature of communication networks, and occasionally packets may be lost to prevent communication deadlocks.

3.4.1 SpiNNaker Architecture

A single two-dimensional instruction, 96 KB of local memory, 128 MB of shared memory, a multi-node data array powered by an 18-core ARM968 processor, packet routers, and system-wide peripherals make up the SpiNNaker machine's architecture. Arithmetic operations are usually carried out in fixed-point because each processor core is a 200 MHz 32-bit general-purpose processor that lacks floating-point hardware. Leaky integrate-and-fire or Izhikevich's model can be modeled in each core [23]. A core can have fewer than 1000 of these model neurons, but no single neuron can have more than 1000 input synapses. Various factors can affect how many neurons a processor core can handle in real-time, but computational budgets frequently incur costs for input connections, input spikes, and average input rates (number of inputs per neuron). The CPU core can theoretically support up to 10 million connections per second, even if current software implementations saturate at roughly half the total bandwidth, which the plastic synaptic model further lowers. A lightweight multicast packet routing system that permits extremely high connectivity in the actual brain is a significant breakthrough in the SpiNNaker design. This mechanism is merely a generalization of AER.

The processor software transmits a packet containing just the neuron's ID when it decides that a neuron should spike. Only the routing structure is in charge of understanding where a packet is routed, not the processor issuing it. Each multicast-using node in a 2D triangular mesh has a packet router that evaluates each packet to determine its source before forwarding it to a group of six nearby nodes, a subset of 18 local processors, or both. The router table selects the route, which is initialized when the application is loaded into the computer.

Due to the limited length of packet source IDs (only 32 bits), it is not feasible to build a comprehensive routing table for every plausible source; hence numerous optimizations are applied to preserve the right table size. The "monitor processor" is one of the 18 computing cores on each SpiNNaker node. This decision is controlled for fault tolerance reasons, and the monitor serves as the operating system's secondary after selection. The 18th CPU may fail on specific nodes, but the other 17 processors are still available as fault-tolerant backups.

SpiNNaker nodes are included in single 300-ball 19 mm square grid array packs. A custom multiprocessor system-on-chip integrated circuit with 18 ARM968 processors, each with 32 KB of local instruction memory and 64 KB of data memory, is included in the package. These processors are connected to different on-chip shared resources via on-chip automatic networking. A second chip contains 128 MB of mobile SDRAM with a low power double data rate (DDR). The two chips are stacked on a package substrate, and their connections are made via gold wire soldering. It was determined that the SDRAM's overall bandwidth was 900 MB/s. After that, the package is attached to a printed circuit board (PCB, see Figs. 14 and 15). The circuit board's chip-to-chip connection sends a 4-b symbol with a 2-wire transition and a 1-wire connection for acknowledgment answers using an automatic 2 of 7 non-zero protocol. The gadget comprises a 48-node PCB, and the connection between the PCBs is made via a high-speed serial connection. Xilinx

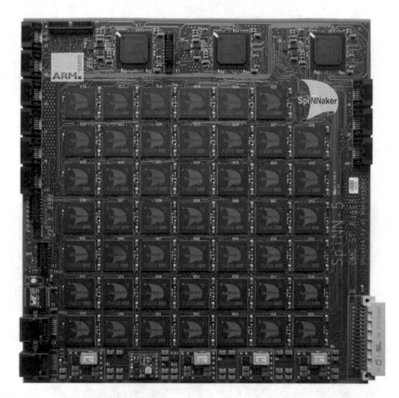

Fig. 14 A pictorial representation of SpiNNaker printed circuit board. (Figure imported from Furber et al. [22])

Spartan6 field programmable gate arrays are used to multiplex eight chip-to-chip links over each serial link. Theoretically, we could construct any size SpiNNaker (FPGA) machine using these straight connections. The 48-node PCB can then be utilized to construct different-sized SpiNNaker systems. A smaller four-node board is also available, which is excellent for training, development, and mobile robotics. The largest machine will have 1200 48-node boards in ten machine room cabinets and more than a million ARM CPU cores and will consume up to 75 kW of power (peak) (Fig. 16).

There are two types of software for SpiNNaker: programs that run directly on the SpiNNaker system and programs that use other systems, some of which may be able to connect with SpiNNaker. Operating systems for SpiNNaker chips are typically created in C. This software can also be divided into control and application software, which performs the user's computations (a simple operating system). The main communication channel between the SpiNNaker system and the outside world is Ethernet and IP-based protocols. Every SpiNNaker chip includes an Ethernet port, and a chip on the PCB typically uses this interface. The software uses it to load SpiNNaker code and gather data. We are investigating how to better use the FPGA-

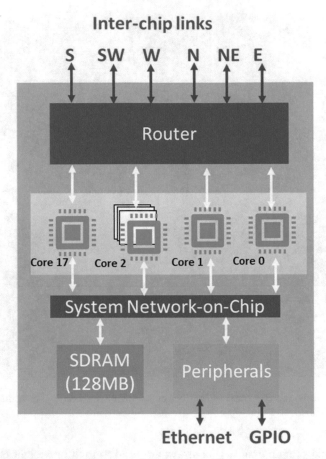

Fig. 15 An illustration of a SpiNNaker node architecture. (Figure adapted from Furber et al. [22])

provided gigabit connections to SpiNNaker systems to lessen some applications' data transfer bottleneck.

The SpiNNaker system's control program is known as SpiNNaker Control and Monitoring Program (SC&MP). The SpiNNaker chip contains an initial load code that may be loaded by an Ethernet interface or a chip-to-chip link to initially load SC&MP into a single chip over an Ethernet interface. Using chip-to-chip links, SC&MP may be disseminated across the system and loaded onto the additional 16 or 17 application cores on each chip. As a monitor processor, it continuously runs on a few cores and offers a range of services to the outside world. The SpiNNaker system uses a simple packet protocol called the SpiNNaker Datagram Protocol (SDP). SC&MP's role as a router allows SDP packets to be forwarded over Ethernet to and from any system core as well as to remote destinations. This protocol serves as the basis for advanced communication and application loading between the SpiNNaker chip and/or external machines. SDP is transmitted between chips as a series of point-

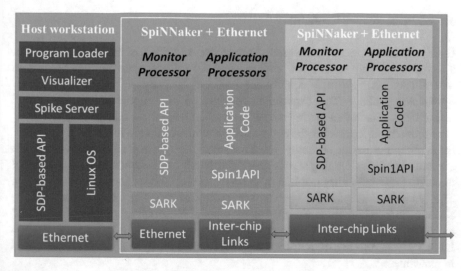

Fig. 16 An illustration of software components required for executing the SpiNNaker system. (Figure adapted from Furber et al. [22])

to-point packets using a chip-to-chip interface. For external SDP transport, packets are sent over an Ethernet interface to the remote endpoint.

SpiNNaker "application" is software that runs on one or more application cores of a system. It is usually written in C and uses SDP or multicast packets for communication. Since SpiNNaker's chip memory only provides a small amount of code and data storage, the operating system does not have much capacity to support it, so the amount of secondary code loaded with the program is small. The SpiNNaker Application Runtime Kernel Support Library is a requirement for each program (SARK). SARK offers runtime code for the application core to be used in creating an application runtime.

Additionally, it offers a library of functions tailored to particular applications, such as memory, and interrupts management. SARK also provides a communication link with the SC&MP program running on the monitor's CPU, allowing applications to communicate with and control other SpiNNaker chips or other systems. An SDP-based protocol is used to implement this feature. The ARM cross-compiler is used to generate applications for SARK and additional runtime libraries. The Application Load and Run (APLX) format created from the configuration step output file can be understood by the primary loader, which is a component of SC&MP. The SC&MP loader is then used to load the APLX files into SpiNNaker and the relevant memory locations on each application core. Most SpiNNaker programs make use of the event-handling library called Spin1 API, which enables the management of event queues and the connection of common interruptions to event handling code. When not processing events, the processor is in a low-power sleep mode.

3.4.2 Performance Evaluation

Using the hardware and software infrastructure described in the previous section, a network with up to 250,000 neurons and 80 million synapses was simulated in real-time on a 48-node SpiNNaker board with a power budget of 1 W per SpiNNaker package. This experiment showed 1.8 billion connections per second using a few nanojoules per event and neuron. The system consistently achieved maximum throughput using the current software infrastructure in terms of maximum throughput and power consumption. The flexibility of the platform allows for the exploration of new learning algorithms. Additionally, a biologically plausible model of cortical microcircuitry inspired by earlier work has shown good power efficiency. The platform's adaptability enables the testing of novel learning algorithms. In addition, a cerebral microcircuit model with a biological foundation motivated by earlier research showed outstanding energy efficiency. In this model, 10,000 Izhikevich neurons and 40 million synapses simulate the dynamics of cortical spikes in real-time.

3.5 IBM TrueNorth

The TrueNorth chip, the latest achievement of the DARPA's (Defense Advanced Research Projects Agency) SyNAPSE project, is composed of 4096 neurosynaptic cores tiled in a 2-D array, containing an aggregate of 1 million neurons and 256 million synapses [24]. It attains a peak computing performance of 58 GSOPS (Giga-Synaptic Operations per Second) and a computing power efficiency of 400 GSOPS/W (per watt). TrueNorth chips have been deployed on 1-, 4-, and 16-chip systems and small 2-inch × 5-inch boards for mobile applications. To develop applications for the TrueNorth architecture, the native TrueNorth Corelet language, the Corelet Programming Environment (CPE), and an extensive library of configurable algorithms are used [25].

3.5.1 TrueNorth Architecture

TrueNorth architecture is fundamentally different from traditional von Neumann architecture. Unlike von Neumann machines, it does not employ sequential programs that pull instructions into linear memory. Spike neurons are implemented in the TrueNorth architecture and are networked together. We can program the chip by defining the neurons' actions and interactions. Neurons exchange pulses in order to communicate. The frequency, timing, and spatial distribution of spikes can all be used to encrypt the data being conveyed. With the neurosynaptic core as a fundamental component, TrueNorth architecture is created to be highly parallelizable, event-driven, energy-efficient, and scalable. The left side of Fig. 17 shows a bipartite graph of neurons, which is a small section of a whole neural network.

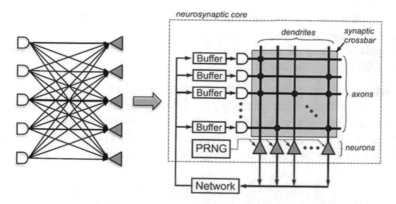

Fig. 17 An illustration of a bipartite neural network graph (*left*) and the corresponding logical representation of a neurosynaptic core (*right*). (Figure imported from Akopyan et al. [24])

A TrueNorth core, shown on the right side, is a hardware representation of this bipartite graph of neurons, with arbitrary connectivity between the input and output layers of the graph. The neural synaptic core includes neurons, memory for storing neural connections and parameters, and computing components for calculating membrane potentials using various neuron models. By physically locating the memory and computing units close to each other, the data movement is reduced thereby improving the efficiency and also lowering the power consumption in chips. In order to scale to a highly parallel architecture with 256 neurons and 64 thousand synapses per core, it incorporates 4096 neural synaptic cores on the TrueNorth chip.

As shown in Fig. 17 (right), a single neurosynaptic core consists of input buffers that receive spikes from the network, axons represented by horizontal lines, dendrites represented by vertical lines, and neurons (represented by triangles) that send spikes into the network. Synapses are the connectors between axons and dendrites, and black dots identify them. Transverse synaptic bands create each nucleus' synapses. Each neuron's output is linked to the corresponding axon's input buffer through an axon input buffer. Communication between these axons and the neuron they connect to may occur directly or indirectly through a routing network, depending on whether they are in the same or a distinct nucleus.

The computation of a neurosynaptic core proceeds according to the following steps:

1. The neuronal synaptic core receives spikes from the network and stores them in an input buffer.
2. When a 1 kHz trigger signal called "tick" arrives, the spike for the tick signal is read from the input buffer and propagated to the corresponding transverse axon.
3. The neuron's dendrite transports a spike of the axon when there is a synaptic connection between a horizontal axon and a vertical dendrite.
4. Each neuron adjusts the membrane potential after integrating an input spike.

5. When all spikes are integrated into a neuron, the leak value is subtracted from the membrane potential.
6. A spike is produced and sent to the network when the updated membrane potential exceeds the threshold.

Although network delay may affect the order of spikes leaving the network, the input buffer controls the computation inside the probe. The TrueNorth chip mimics the neuron's processing, so the operation corresponds to the way of a software neural network simulator. The membrane potential $V_j(t)$ of the jth neuron at tick t is computed according to the following simplified equation:

$$V_j(t) = V_j(t-1) + \Sigma_{i=0}^{255} A_i(t) w_{i,j} s_j^{G_i} - \lambda_j \tag{29}$$

where $A_i(t)$ is 1 if axon i has an input spike (temporary buffer) at time t, $w_{i,j}$ is 1 if axon i is connected to dendrite j, otherwise, the value is 0 I do not see. Assign one of the four types to each axon and denote the ith axon type as G_i, an integer between 1 and 4. The integer number s^{Gj} is used to specify the synaptic weight between each G-type axon and the jth dendrite. The membrane potential subtracts the leak λj for each tick equals the integer value j. The jth neuron produces a spike and injects it into the network when the revised membrane potential $V_j(t)$ exceeds the threshold j, restoring the state $V_j(t)$ to R. The TrueNorth chip's actual brain calculations are very complicated. For instance, in order to provide peak integration, drains, and thresholds, it employs a pseudo-random number generator (PRNG). The membrane potential can be recovered after jumps, and the leakage signal is programmable, dependent on the membrane potential's sign. By directly introducing the restrictions into the learning process, axonal and synaptic weight constraints are effectively managed (e.g., backpropagation). It has been demonstrated that the learning algorithm may use the available free parameters to identify the optimal solution given the architectural limitations.

The entire chip comprises 64 × 64 arrays of peripheral logic and neural synaptic cores, as shown in Fig. 18. A single core comprises a scheduler block, token controller, SRAM block, and neural and router block. The routers connect with the core and four other routers in the east, west, north, and south, creating a 2D mesh network. Each spike packet has a target core address, a target axon index, a target check box, and several flags for debugging purposes. When the spike reaches the router on the destination core, the router forwards it to the scheduler, marked by block A in Fig. 18. Prior to confirming the fixed objective indicated in the peak packet, the scheduler's primary function is to queue the input nodes. The scheduler saves each spike as a binary value in SRAM made up of 16 × 256-bit entries, or 16 labels and 256 axons. The scheduler reads all nodes $A_i(t)$ for the current tick when the neuronal core receives a tick and delivers them to the marker controller (B). The neural synaptic core performs a number of calculations, which the token controller controls. It processes each of the 256 neurons one at a time after getting a spike from the scheduler. Using Eq. (29), for neuron j, the synaptic connection $w_{i,j}$ is sent from the main SRAM to the symbol controller (C), and the parameters of

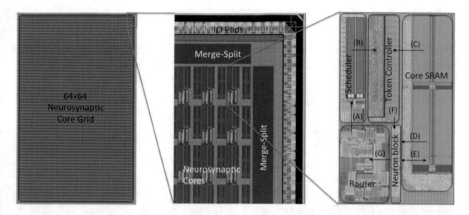

Fig. 18 An illustration of a TrueNorth Chip with magnified details of its neurosynaptic cores. (Figure imported form Akopyan et al. [24])

the remaining neurons (D) and the current membrane potential (E) are sent to the neuron block. If the bit-wise AND of the synaptic connectivity $w_{i,j}$ and the input spike $A_i(t)$ are 1, the signal controller sends a clock signal and an instruction (F) indicating the type of axon Gi to the neuronal unit to adjust the membrane potential sG_{ij}. In other words, a neural unit will receive commands and clock pulses only if two conditions are met: an active input spike and the corresponding synapse is active. At the same time, neural devices make calculations based on events. When the token controller has consolidated all the stimuli for the neuron, it sends some additional commands to the neuron block: one is to decrease the leakage $\lambda_{i,}$ and the other is to check the peak condition. The neural device sends a spike (G) packet to the router if the resulting membrane potential $V_i(t)$ is higher than the configurable threshold. New spike packets are then introduced into the network by the router. The updated membrane potential (E) value is written back to the primary SRAM, and a single neuron finishes the processing. The token controller stops providing clock pulses to individual neurons once the core has finished processing all of the neurons for the current token. The token controller commands the scheduler to advance the time slot to the following tick. The neuro-synaptic core is now silent and waits for the next tick, besides continuously relaying the input spike to the scheduler. The 2D grid of neurosynaptic cores can also be extended outside the chip thanks to its peripheral interface. The token controller then instructs the scheduler to advance its time pointer to the next tick. Due to the restricted number of I/O pins available, the merge split blocks located at the chip's edges combine spikes originating from many buses into a single stream of spikes that exits the chip. In contrast, this peripheral component also disperses an incoming stream of off-chip spikes to various buses in the core array.

3.5.2   Performance Evaluation

TrueNorth's performance has been evaluated in various applications under different studies. Table 1 shows the power consumption by TrueNorth for each of those applications [24].

3.6   BrainscaleS-2 System

The brain-inspired computing platform BrainScaleS-2 creates a physical representation of the most popular reductionist perspective of the biological brain to implement neuromorphic computing: a collection of neurons joined together by flexible synapses [26]. It differs from most other modeling techniques used by the computational neuroscience field in this respect. In particular, no differential equation is solved while the network model operates. Biological processes are not represented by discrete-time changes of many bits that approximate molecular biology in some way.

BrainscaleS-2 Architecture   Each neuromorphic BrainScaleS-2 core is made up of two digital control and plasticity processors, an event routing network for spike transmission, a synaptic crossbar, neuron circuits, analog parameter storage, and a whole custom analog core, as shown in Fig. 19. Each of the four quadrants of the physical structure has a synaptic crossbar with 256 rows and 128 columns. This division also applies to the neuron circuitry, with each of the 512 neuron circuits being assigned to a column. The two digital processors control the analog core's upper and lower halves at the design's top and bottom, respectively. Vector multiple instruction data (SIMD) extensions can be used to read and write digital states of the synaptic crossbar halves on parallel tracks, and analog traces can be read using a 512-channel analog-to-digital converter (ADC). (There are 256 channels per quadrant paired for correlation and anticorrelation measurements).

The event routing line divides the four quadrants and occupies a cross-shaped space that extends in all directions. It operates on address-coded packets; connects digital outputs, external inputs, and on-chip Poisson sources of neural circuits; and sends them to synaptic rows or off-chip destinations in one of the quadrants. The external systems' digital event and data interface implement a custom hardware communication protocol that supports reliable data transmission and efficient event communication. Figure 20 shows a BrainScaleS-2 system incorporated into hardware for commissioning and analog measurements.

Emulation of neuron and synapse dynamics takes place in specialized mixed-signal circuits that, together with other fully customizable components, form an analog neuromorphic core. A full-size application integrated circuit (ASIC) contains 512 neural sections with varied and rich dynamics. They evolve on a time scale accelerated 1000 times compared to the biological time domain, considering the time constants characteristic of semiconductor substrates. At its core, the neuron

Table 1 Power consumption by TrueNorth chip under various applications [24]

Applications	Network size				Communication power (Active)			Total power (Watts)
	#chips	#cores	#nets	#pins	Default	CPLACE	Improvement	
Saccade generator	1	2297	17523	35046	0.40 mW	0.11 mW	3.6×	0.049 W@0.75 V
Local Binary Patterns	1	3520	22032	44064	4.41 mW	2.33 mW	1.9×	0.064 W@0.75 V
Haar-like features	1	3875	45822	91644q	7.57 mW	1.81 mW	4.2×	0.058 W@0.75 V
K-means classifier	4	14832	445380	890760	20.03 mW	6.44 mW	3.1×	0.653 W@1.0 V
Grid classifier A	7	27644	461788	923576	589.90 mW	130.88 mW	4.5×	1.274 W@1.0 V
Grid classifier B	16	62375	960013	1920030	353.95 mW	70.06 mW	5.1×	2.515 W@1.0 V

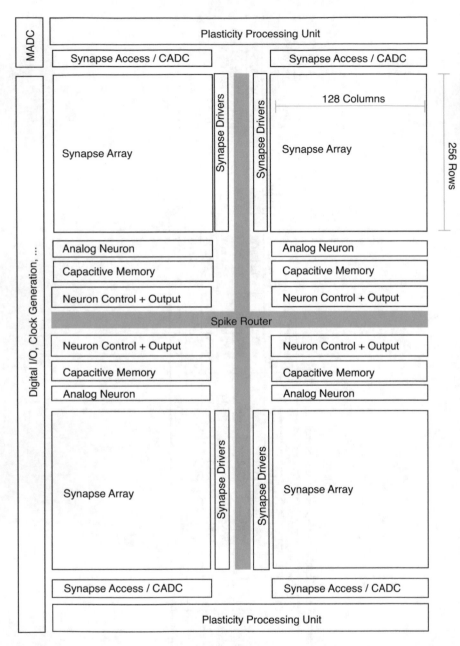

Fig. 19 A schematic diagram of BrainscaleS-2 chip. (Figure imported from Pehle et al. [26])

Fig. 20 A photo of a hardware setup embedded with BrainscaleS-2 chip (covered by white plastic). (Figure imported from Pehle et al. [26])

circuits faithfully implement the adaptive exponential integrate-and-fire (AdEx) model,

$$C_{\mathrm{m}} V' = -g_l (V - E_1) + g_l \Delta \exp \left(\frac{V - V_{\mathrm{T}}}{\Delta \mathrm{T}} \right) - w + I \tag{30}$$

$$\tau_{\mathrm{w}} w' = a (V - E_{\mathrm{L}}) - w \tag{31}$$

The first differential equation describes the evolution of the membrane potential V on capacitance C_{m} [27]. The current $I = I_{\mathrm{syn}} + I_{\mathrm{stim}}$ accumulates in the membrane containing both direct external stimuli and currents of synaptic interactions. Moreover, g_l denotes the leakage conductance, which pulls the membrane toward the leakage potential E_{L}. The exponential term implements vital positive feedback that mimics the natural shape of action potentials in biological neurons and is controlled by an exponential slope Δ_T and a soft threshold V_T. An output spike is emitted when the membrane potential exceeds the threshold V_{th}. In this case, the membrane is fixed at the recovery potential V_r and maintained during the refractory period t_r. The second differential equation captures the dynamics of the adaptation current w allowing the neuron to adapt to its previous activation and firing activity. The adaptation state decays back to zero with a time constant τ_w and is driven by the deflection of the membrane potential, scaled with the subthreshold adaptation strength a. In case of an action potential, w is incremented by b implementing spike-triggered adaptation.

The system has two digital plastic and control processors called the plastic processing unit (PPU). The primary purpose of this part of the system is to complement the flexible and configurable architecture of neurons and synapses with an equally flexible digital control architecture.

3.7   Intel Loihi

Loihi is a 60-mm^2 chip (Fig. 21) fabricated in Intel's 14-nm process that advances the state-of-the-art modeling of spiking neural networks in silicon [28]. It incorporates a variety of cutting-edge features new to the discipline, including programmable synaptic learning rules, dendritic compartments, hierarchical connectivity, and synaptic delays.

Loihi Architecture With three embedded ×86 processing cores, 128 neuromorphic cores, and off-chip communication ports that hierarchically extend the mesh in four planar directions to additional chips, Loihi comprises a manycore mesh. Asynchronous network-on-chip (NoC) transports packetized messages for all communication between cores. The NOC provides spike messages for SNN computation, barrier messages for temporal synchronization across cores, write, read request, and read response messages for core management, and ×86-to-×86 messaging. All message types may be addressed to any on-chip core and may be sourced externally by the host CPU or on-chip by the x86 cores. The hierarchical encapsulation of messages for off-chip communication through a second-level network is possible. The mesh protocol scales to 4096 on-chip processors and up to 16,384 chips using hierarchical addressing. Basic spiking brain units of 1 K (compartments) organized into sets of trees make up each neuromorphic core. The compartments share configuration and state variables in ten architectural memory along with their fan-in and fan-out connectivity. Every algorithmic time step's state variables are updated in a time-multiplexed, pipelined fashion. When a neuron's activation rises over a certain threshold, it produces a spike message that is sent to several fan-out compartments in the destination cores. Supporting a variety of workloads necessitates having SNN connectivity characteristics that are adaptable and well-equipped. Some desirable networks may require dense, all-to-all connectivity, while others require sparse connectivity. Some may have uniform

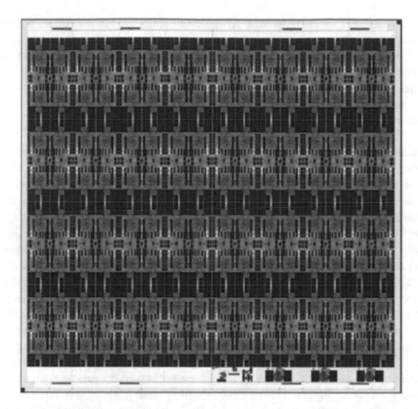

Fig. 21 An illustration of the Loihi chip plot. (Figure imported from Davies et al. [28])

graph degree distributions, while others may have power law distributions. Some may need synaptic weights, such as those required to support learning, whereas others can get by with binary connections. Algorithmic performance often scales with network size, as determined by both neurons counts and neuron-to-neuron fan-out percentages. This law is evident down to biological levels (1:10,000). Supporting networks with high connectivity with current integrated-circuit technology is extremely difficult because of the $O(N^2)$ scaling of connection state in the number of fan-outs.

- *Sparse network compression.* Loihi offers a dense matrix connectivity model along with three sparse matrix compression models. In these models, the fan-out neuron indices are calculated using the stored index state with each synapse's state variables.
- *Core-to-core multicast.* Every neuron can send a single spike to as many cores as the network connectivity may demand.
- *Variable synaptic formats.* Any weight precision between one and nine bits, whether signed or unsigned, is supported by Loihi, and weight precisions may

be combined (with scale normalization) even within the fan-out distribution of a single neuron.

• *Population-based hierarchical connectivity.* It is possible to define and map connection templates to certain population instances during operation as a generalized weight-sharing mechanism, for example, to enable convolutional neural network types. This capability can significantly reduce the number of connectivity resources needed for a network.

Loihi is the first fully integrated SNN chip that supports the above features. All earlier chips, such as the previously most synaptically dense, store their synapses as dense matrices, severely limiting the range of networks that can be supported effectively. A programmable learning engine is included in every Loihi core, allowing it to change synaptic state variables over time in response to past spike activity. The learning engine uses filtered spike traces to support the broadest class of feasible rules. Learning rules can be programmed into microcode and accommodate a wide range of input terms and synaptic target variables. Each synapse that needs to be adjusted has a learning profile connected to it that is associated with specific sets of these rules. Pre-synaptic neurons, post-synaptic neurons, or synaptic classes map profiles. The learning engine can execute straightforward paired STDP rules and far more intricate ones, including triplet STDP, reinforcement learning with synaptic tag assignments, and complex rules that use both averaged and spike-timing traces. The chip's whole logic implementation is asynchronous bundled data, functionally deterministic, and digital. This chip's logic enables event-driven spike generation, routing, and consumption with maximum activity gating during idle periods. This method of implementation works well for SNNs, which by their very nature exhibit highly sparse activity both in space and time.

Figure 22 shows the operation of the neuromorphic mesh as it executes an SNN model. Any neurons that reach a firing state create spike messages that the NoC sends to all cores, including their synaptic fan-outs, as each core individually iterates across its collection of neuron compartments. The second box shows the spike distributions for two model neurons, $n1$, and $n2$, located in cores A and B. The third box shows the NoC traffic as well as additional spike distributions from other firing neurons. Spike (and all other) messages are distributed by the NoC using a dimension-order routing mechanism, and only unicast distributions are supported by the NoC itself. The output process of each core transmits one spike to each core on a list of destination cores for a firing neuron's fan-out distribution. The mesh employs two separate physical router networks for deadlock prevention purposes related to reading and chip-to-chip message transactions. The cores alternately send their spike messages across the two physical networks to maximize bandwidth efficiency.

At the end of the time-step, we need to ensure all spikes have been delivered and that the cores can safely proceed to time-step $t + 1$. They employ a barrier synchronization method, as shown in the fourth box, instead of a globally distributed time reference (i.e., clock) that must plan for the worst-case chip-wide network activity. Each core exchanges barrier messages with the cores next to it as soon as it has finished servicing its compartments for time-step t. The barrier messages clear

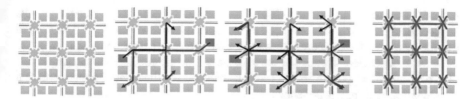

Fig. 22 An illustration of the step-by-step mesh operation in Loihi chip. (Figure adapted from Davies et al. [28])

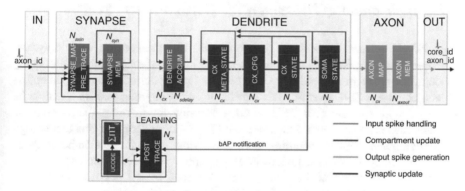

Fig. 23 An illustration of the overall microarchitecture of the Loihi neuromorphic core. Figure imported from Davies et al. [28]

any spikes currently in flight and then spread a time-step advance notification to every core in a subsequent phase. Cores advance their time-step and start to update compartments for time $t + 1$ as soon as they receive the second phase of barrier messages. The Loihi mesh is provably deadlock-free as long as management activity is limited to a particular "preemption" phase of the barrier synchronization process that any embedded $\times 86$ core or off-chip host may introduce on demand.

Figure 23 shows the internal structure of the Loihi neuromorphic core. The primary memories that store the connectivity, configuration, and dynamic state of all neurons mapped to the core are represented in this diagram by the colored blocks. With ECC overhead, the core's total SRAM capacity is 2 Mb. The memory and dataflow arc coloring depict the four main operating modes of the core: input spike processing (green), synaptic updates (blue), purple neuron for compartment updates, and synaptic update (red). Depending on the state and configuration of the core, each of these modes can run independently at various frequencies with little synchronization.

The adaptable learning engine is shown as a black structure with the letters UCODE. Each memory's number of logical addresses is indicated by the values it has annotated, which correspond to the primary resource limitations of the core. Network connectivity restrictions are imposed by the number of input and output axons (N_{axin} and N_{axout}), the quantity of synaptic memory (N_{syn}), and the overall number of neuron compartments (N_{cx}). The minimal number of synaptic delay units supported, eight in Loihi, is indicated by the option Nsdelay. When a given mapped network requires fewer neuron compartments, larger synaptic delay values, up to 62, may be supported. Sections of the core's pipeline are subjected to varying levels of serialization and parallelism in order to balance the throughput constraints that typical workloads may experience.

3.8   Applications of Neuromorphic Hardware Accelerators

- *Self-Driving Cars*: Success in developing a fully autonomous vehicle system requires solving various problems superior to human-level perception modeling, cognitive world model building, and power reduction during real-time perception and cognition [29]. Event-driven neuromorphic computing is envisioned to be much more suitable than GPU-based processing for learning spatio-temporal semantics of the real-world environment [30]. Current deep learning technology for computer vision is becoming increasingly complex and relies on humongous data training to perform human-level perception; however, humans consume only a tiny fraction of the total power consumed by the deep learning-based computer vision models [31]. Therefore, it is essential to discover technologies that can emulate the dynamics of the human brain to help develop power-efficient computer vision applications, and neuromorphic computing holds the promise to achieve the same.
- *Robotics*: The main components in which robotics is traditionally categorized— perception, decision-making, and control—are strictly connected and influence each other [32]. Exploring the interactions and understanding how neural codes represent sensory information and decision-making processes and how these are affected by behavior and vice versa represent a neuromorphic level of complexity that allows robots to interact effectively with humans. It is an upcoming research mission to bring agents. Togo Advances in understanding the roles, working principles, and interactions of different brain regions with other regions at different temporal and spatial scales have led to the design of artificial architectures that use neural models, synaptic plasticity mechanisms, and connectivity structures for specific functions. It is crucial to find the appropriate level of detail and abstraction for each neural computational primitive and develop a principled methodology for combining them. The community starts with detailed models of brain regions to realize basic functions and finds reduced models that can be implemented on neuromorphic hardware.
- *Smell Sensors*: The theoretical performance capacities of machine olfaction are not clearly constrained by anything short of their deployed environment's fundamental signal-to-noise limits [33]. The essential ability to distinguish similar odors can be permanently increased by deploying an array with many other chemical sensors if the chemical sensors respond to the appropriate set of ligands and discriminate some of them. Proportional scaling of network size is

controlled in terms of execution time and power consumption in neuromorphic devices. Mechanical olfaction is an early and successful adopter of neural morphology sampling and computational strategies due to the detailed description of olfactory system networks through experimental and computational neurobiology. Continued development of large-scale sensory networks and post-selection neural circuits for signal conditioning and context-based category learning is an essential focus for near-term development. Overall, comparing and analyzing different computational motifs in these functional biomimetic architectures could answer a wide range of theoretical and task-dependent questions, such as when to turn off plasticity in selective plasticity architectures and the transformative potential of learning rules locally implemented in specific network contexts.

- *Audition Sensors*: Neuromorphic audition technology is inspired by the amazing capability of human hearing [34]. Even in challenging hearing situations, humans can interpret speech using only a little portion of their 10 W brain. A key objective of creating artificial hearing algorithms, hardware, and software is to match human hearing capabilities. The natural temporal encoding carried by the asynchronous events (pulses, or sets of pulses, within a specified time window), the event-driven form of brain computing, and other neuromorphic bio-inspired features can enable more energy-efficient solutions for hardware-friendly models that solve an auditory task. It is desirable to extract interaural time differences (ITD) from the asynchronous events of the spiking silicon cochleas since ITD is helpful for a spatial audition at lower latencies [35].

4 Analog Computing in NHAs

4.1 Overview

In Sect. 3, we discussed a plethora of NHAs developed in recent years. The current advancement in NHA is primarily based on SNN simulation on CMOS circuits consisting of transistors arranged in a crossbar array to perform Multiply-Accumulate (MAC) operation. However, further reducing power consumption and improving computational speed requires processing in a physical or analog paradigm rather than a digital one. Existing developments are digital processors in the sense that they utilize transistors to store data. Digital processors are bound by the need to move data between compute and storage units and thus incur the data movement cost. Therefore, it is necessary to incorporate analog techniques as they can perform MAC operations on the go inside the memory. Analog implementations of deep learning algorithms for image recognition tasks have been demonstrated to perform sufficiently well in the past [36]. Analog MAC operation can be performed by exploiting the physical properties of memristive devices and arranging them in a crossbar array structure [37].

4.2   Magnetoresistive Random Access Memory

Magnetoresistive Random Access Memory (MRAM) is a family of storage circuits capable of storing data in magnetic states in magnetoresistive devices such as spin valves and magnetic tunneling junctions (MJTs). Stored data is read by measuring the device resistance to determine their magnetic states [38]. Among various MRAM categories, spin-transfer torque MRAM (STT-MRAM) can provide high endurance and scalability without sacrificing any significant performance [39].

4.3   MRAM-Based In-Memory Computing

Analog implementation of STT-MRAM has been a very challenging idea due to the low-resistance property of MRAM, which would cause high power consumption in crossbar arrays for MAC operations because conventional crossbar arrays rely on electrical current-based summation. However, in one of its recent works, Samsung utilized resistance-based summation to achieve low-power consumption [40]. Figure 24 shows a 64×64 MRAM crossbar array used in this work where each cell consists of two field-effect transistors (FETs) and two MJTs.

Figure 25 shows a detailed illustration of a single-bit cell extracted from the crossbar array. In each bit-cell, left and right FETs consist of gates that are operated via voltage IN (either high ($R_H = 1$) or low ($R_L = -1$)) and its complementary. Each side of MJT-FET paths stores weight W (either high (1) or low (-1)) and its complementary value. RH and RL resistances add FET switch resistance to the MJT resistance. There are a total of four possible configurations for IN and W values (IN,W): $\{(-1,-1), (-1,1),(1,-1),(1,1)\}$ and an output of either R_H or R_L. Each column in the crossbar array has bit-cells arranged in a series combination, which adds up each bit cell's output to produce a total output of R_i in the ith column.

This crossbar array architecture was tested on the traditional MNIST dataset to perform an image recognition task [41]. A two-layer neural network was used to assess the performance. It consists of the first layer with 128 neurons and the second layer with 10 neurons corresponding to the 10 digits. Figure 26 illustrates the classification accuracy results obtained by the analog implementation versus digital implementation of a neural network to perform an image recognition task on the MNIST classification dataset. The experiment was repeated three times, and the overall accuracy was approximately 93.4%, whereas its digital counterpart achieved an accuracy of 95.7%. The decrease in analog implementation arises from the analog noises in the overall architecture. This MRAM crossbar array achieved a power efficiency of 262 tera-operations per second per watt (TOPS/W).

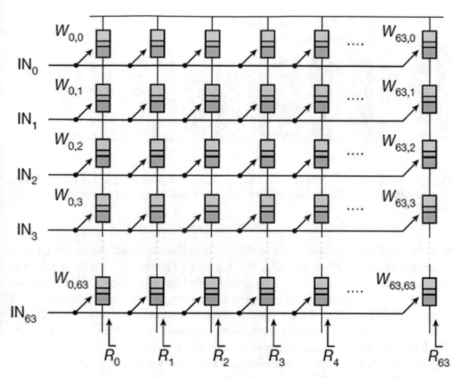

Fig. 24 An illustration of the 64 × 64 crossbar array architecture of MRAM. (Figure imported from Jung et al. [40])

Fig. 25 An illustration of a bit cell for the MRAM crossbar array. (Figure imported from Jung et al. [40])

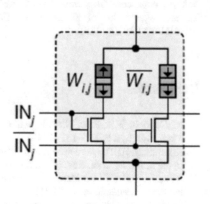

5 Conclusions

Increasing demand for high-performance computing at every scale ranging from embedded systems to cloud centers has led researchers to explore this new dimension of computing called neuromorphic computing. It is heavily inspired by the structure and function of the human brain. A new domain of AI algorithms

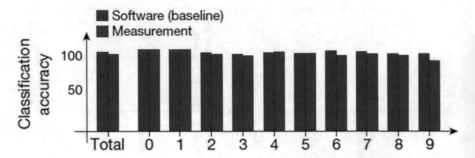

Fig. 26 A comparison graph of classification accuracy of the proposed approach (measurement) vs. a baseline software-based neural network architecture (baseline). (Figure imported from Jung et al. [40])

is under constant development to simulate the actions of neurons and synapses to perform learning mechanisms, with Spiking Neural Networks serving as the basis for such implementations. Various Neuromorphic Hardware Accelerators have been developed over the years by emulating the neuro-synaptic behaviors using a crossbar array architecture. Despite the progress, these hardware accelerators are still built with CMOS transistors at their base. The von Neumann bottleneck needs to be eliminated by exploring other devices, such as magnetoresistive, capable of in-memory computing. This research domain is relatively new and might open a new perspective on how we think about computers in the future.

Acknowledgments This work was supported by the Brain Pool Program through the National Research Foundation of Korea (NRF), funded by the Ministry of Science and ICT (NRF-2019H1D3A1A01071115).

References

1. McCulloch, W.S., Pitts, W.: A logical calculus of the ideas immanent in nervous activity. Bull. Math. Biophys. **5**, 115–133 (1943)
2. Rosenblatt, F.: The perceptron: a probabilistic model for information storage and organization in the brain. Psychol. Rev. **65**(6), 386–408 (1958)
3. Kim, S., Deka, G.C.: Hardware accelerator systems for artificial intelligence and machine learning. Academic Press (2021)
4. Hebb, D.O.: The organization of behavior: a neuropsychological theory. Psychology Press (2005)
5. Jeong, D.S., Kim, I., Ziegler, M., Kohlstedt, H.: Towards artificial neurons and synapses: a materials point of view. Royal Society of Chemistry (RSC). RSC Adv. **3**(10), 3169 (2013)
6. Thompson, S.E., Parthasarathy, S.: Moore's law: the future of Si microelectronics. Mater. Today. Elsevier BV. **9**(6), 20–25 (2006)
7. Roy, K., Jaiswal, A., Panda, P.: Towards spike-based machine intelligence with neuromorphic computing. Nature. **575**, 607–617 (2019)
8. Davidson, S., Furber, S.B.: Comparison of artificial and spiking neural networks on digital hardware. Front. Neurosci. Frontiers Media SA. **15** (2021)

9. Tuckwell, H.C., Wan, F.Y.M.: Time to first spike in stochastic Hodgkin–Huxley systems. Phys. A. Elsevier BV. **351**(2–4), 427–438 (2005)
10. Nadasdy, Z.: Information encoding and reconstruction from the phase of action potentials. Front. Syst. Neurosci. Frontiers Media SA. **3**, 6 (2009)
11. Kavehei, O., Iqbal, A., Kim, Y.S., Eshraghian, K., Al-Sarawi, S.F., Abbott, D.: The fourth element: characteristics, modeling, and electromagnetic theory of the memristor. Proc. R. Soc. A. The Royal Society. **466**(2120), 2175–2202 (2010)
12. Schrauwen, B., Van Campenhout, J.: BSA, a fast and accurate spike train encoding scheme. Proc. Int. Joint Conf. Neural Netw. IEEE. **4**, 2825–2830 (2003)
13. Mohemmed, A., Schliebs, S., Matsuda, S., Kasabov, N.: Span: spike pattern association neuron for learning Spatio-temporal spike patterns. Int. J. Neural Syst. **22**(04), 1250012 (2012)
14. Prodromakis, T., Toumazou, C.: A review on memristive devices and applications. In: IEEE International Conference on Electronics, Circuits and Systems, pp. 934–937 (2010)
15. Strukov, D.B., Snider, G.S., Stewart, D.R., Williams, R.S.: The missing memristor found. Nature. **453**(7191), 80–83 (2008)
16. Kim, K.H., Gaba, S., Wheeler, D., Cruz-Albrecht, J.M., Hussain, T., Srinivasa, N., Lu, W.: A functional hybrid memristor crossbar-array/CMOS system for data storage and neuromorphic applications. Nano Lett. **12**, 389 (2012)
17. Ma, D., Shen, J., Gu, Z., Zhang, M., Zhu, X., Xu, X., Xu, Q., Shen, Y., Pan, G.: Darwin: a neuromorphic hardware co-processor based on spiking neural networks. J. Syst. Archit. Elsevier BV. **77**, 43–51 (2017)
18. LeCun, Y., Bottou, L., Bengio, Y., Haffner, P.: Gradient-based learning applied to document recognition. Proc. IEEE. **86**(11), 2278–2324 (1998)
19. Neil, D., Liu, S.C.: Minitaur, an event-driven FPGA-based spiking network accelerator. IEEE Trans. Very Large Scale Integr. (VLSI) Syst. **22**(12), 2621–2628 (2014)
20. Benjamin, B.V., Gao, P., McQuinn, E., Choudhary, S., Chandrasekaran, A.R., Bussat, J.M., Alvarez-Icaza, R., Arthur, J.V., Merolla, P.A., Boahen, K.: Neurogrid: a mixed-analog-digital multichip system for large-scale neural simulations. Proc. IEEE. **102**(5), 699–716 (2014)
21. Sodini, C.G., Ko, P.K., Moll, J.L.: The effect of high fields on MOS device and circuit performance. IEEE Trans. Electron Devices. **31**(10), 1386–1393 (1984)
22. Furber, S.B., Galluppi, F., Temple, S., Plana, L.A.: The spinnaker project. Proc. IEEE. **102**(5), 652–665 (2014)
23. Izhikevich, E.M.: Simple model of spiking neurons. IEEE Trans. Neural Netw. **14**(6), 1569–1572 (2003)
24. Akopyan, F., Sawada, J., Cassidy, A., Alvarez-Icaza, R., Arthur, J., Merolla, P., Imam, N., Nakamura, Y., Datta, P., Nam, G.J., Taba, B.: Truenorth: design and tool flow of a 65 mW 1 million neuron programmable neurosynaptic chip. IEEE Trans. Comp-Aided Des. Integr. Circuits Syst. **34**(10), 1537–1557 (2015)
25. Sawada, J., Akopyan, F., Cassidy, A.S., Taba, B., Debole, M.V., Datta, P., Alvarez-Icaza, R., Amir, A., Arthur, J.V., Andreopoulos, A., Appuswamy, R.: Truenorth ecosystem for brain-inspired computing: scalable systems, software, and applications. In: SC'16: Proceedings of the International Conference for High Performance Computing, Networking, Storage and Analysis. IEEE, pp. 130–141 (2016)
26. Pehle, C., Billaudelle, S., Cramer, B., Kaiser, J., Schreiber, K., Stradmann, Y., Weis, J., Leibfried, A., Müller, E., Schemmel, J.: The BrainScaleS-2 accelerated neuromorphic system with hybrid plasticity. Front. Neurosci. **16** (2022)
27. Brette, R., Gerstner, W.: Adaptive exponential integrate-and-fire model as an effective description of neuronal activity. J. Neurophysiol. **94**(5), 3637–3642 (2005)
28. Davies, M., Srinivasa, N., Lin, T.H., Chinya, G., Cao, Y., Choday, S.H., Dimou, G., Joshi, P., Imam, N., Jain, S., Liao, Y.: Loihi: a neuromorphic manycore processor with on-chip learning. IEEE Micro. **38**(1), 82–99 (2018)
29. Schwarting, W., Alonso-Mora, J., Rus, D.: Planning and decision-making for autonomous vehicles. Annu. Rev. Control Robot. Auton. Syst. **1**(1), 187–210 (2018)

30. Moreira, O., Yousefzadeh, A., Chersi, F., Cinserin, G., Zwartenkot, R.J., Kapoor, A., Qiao, P., Kievits, P., Khoei, M., Rouillard, L., Ferouge, A.: NeuronFlow: a neuromorphic processor architecture for live AI applications. In: 2020 Design, Automation & Test in Europe Conference & Exhibition (DATE). IEEE, pp. 840–845 (2020)
31. Alzubaidi, L., Zhang, J., Humaidi, A.J., Al-Dujaili, A., Duan, Y., Al-Shamma, O., Santamaría, J., Fadhel, M.A., Al-Amidie, M., Farhan, L.: Review of deep learning: concepts, CNN architectures, challenges, applications, future directions. J. Big Data. 8(1), 1–74 (2021)
32. Garcia, E., Jimenez, M.A., De Santos, P.G., Armada, M.: The evolution of robotics research. IEEE Robot. Autom. Mag. 14(1), 90–103 (2007)
33. Rozas, R., Morales, J., Vega, D.: Artificial smell detection for robotic navigation. In: Fifth International Conference on Advanced Robotics' Robots in Unstructured Environments. IEEE, pp. 1730–1733 (1991)
34. Vanarse, A., Osseiran, A., Rassau, A.: A review of current neuromorphic approaches for vision, auditory, and olfactory sensors. Front. Neurosci. 10, 115 (2016)
35. Klumpp, R.G., Eady, H.R.: Some measurements of interaural time difference thresholds. J. Acoust. Soc. Am. 28(5), 859–860 (1956)
36. Yao, P., Wu, H., Gao, B., Tang, J., Zhang, Q., Zhang, W., Yang, J.J., Qian, H.: Fully hardware-implemented memristor convolutional neural network. Nature. 577, 641 (2020)
37. Xia, Q., Yang, J.J.: Memristive crossbar arrays for brain-inspired computing. Nat. Mater. 18, 309 (2019)
38. Heinonen, O.: Magnetoresistive materials and devices. In: Introduction to Nanoscale Science and Technology (2004)
39. Apalkov, D., Dieny, B., Slaughter, J.M.: Magnetoresistive random access memory. Proc. IEEE. 104, 1796 (2016)
40. Jung, S., Lee, H., Myung, S., Kim, H., Yoon, S.K., Kwon, S.-W., Ju, Y., Kim, M., Yi, W., Han, S., Kwon, B., Seo, B., Lee, K., Koh, G.-H., Lee, K., Song, Y., Choi, C., Ham, D., Kim, S.J.: A crossbar array of magnetoresistive memory devices for in-memory computing. Nature. 601, 211–216 (2022)
41. Deng, L.: The MNIST database of handwritten digit images for machine learning research. IEEE Signal Process. Mag. 29(6), 141–142 (2012)

Hardware Accelerators for Autonomous Vehicles

Junekyo Jhung, Ho Suk, Hyungbin Park, and Shiho Kim

1 Introduction

1.1 Overview

Autonomous vehicles (AVs) are massive systems composed of many sensors, electrical/electronic (E/E) hardware devices, and software technologies to improve driving performance to prevent accidents, provide optimized driving conditions to drivers, or automate driving. Recent developments in hardware capability have led to advancements in real-time implementation for software applications based on deep learning (DL) algorithms. These require complex computation, efficient power consumption, and many other elements to be considered. As deep learning-based applications become eligible for real-time implementation, the vehicle industry has adopted the applications to enhance monitoring systems to driving control systems in vehicles. Even though different manufacturers implement various strategies for autonomous vehicles, they share common factors to enable autonomous driving. Autonomous vehicles are commonly composed of sensing, perception, and decision systems. Sensors are integrated to acquire raw data of environments around a vehicle, such as global navigation satellite systems and inertial measurement units (GNSS/IMU), cameras, light detection and ranging (LiDAR), radars, and ultrasonic sensors. Perception systems leverage the raw data gathered from the sensing systems to various modules such as localization, object detection, or object tracking to detect and predict vehicle environments during driving. The perception results are fed into applications of autonomous driving systems to make appropriate decisions for lateral/longitudinal controls of autonomous vehicles to ensure driving quality

J. Jhung · H. Suk · H. Park · S. Kim (✉)
School of Integrated Technology, Yonsei University, Incheon, South Korea
e-mail: ejhung@yonsei.ac.kr; sukho93@yonsei.ac.kr; phb88@yonsei.ac.kr; shiho@yonsei.ac.kr

© The Author(s), under exclusive license to Springer Nature Switzerland AG 2023   269
A. Mishra et al. (eds.), *Artificial Intelligence and Hardware Accelerators*,
https://doi.org/10.1007/978-3-031-22170-5_9

and safety. Once, advanced driver assistance systems (ADAS) were implemented through conventional engineering models. Still, as deep learning-based autonomous driving systems are widely installed in vehicles, the demand for fast, accurate, and efficient AI accelerators is also increasing in the automotive industry. This section introduces sensors, technologies, and E/E architecture for autonomous vehicles for better understanding. A detailed explanation of the AI accelerator is covered in later sections.

1.2 Sensors

Autonomous vehicle systems and technologies are composed of hardware and software systems that conduct like a human driver. Autonomous vehicles sense the raw data of environments around them, conduct perception based on the raw data to interpret the circumstances, make decisions and behaviors to drive appropriately, and avoid accidents to ensure safety. Types of system architecture are countless according to strategies built by manufacturers, but we introduce prevailing systems and technologies generally used for autonomous vehicles around 2022.

Sensors perceive raw data of environments based on their mechanical methodologies to provide geographical or visual information [1, 2]. GNSS/IMU, camera, LiDAR, radar, and ultrasonic sensors are frequently integrated onboard sensor modalities with different properties and purposes. Features of each sensor and acquired data differentiate the roles to automate vehicle functionalities.

GNSS/IMU The global navigation satellite system and inertial measurement unit (GNSS/IMU) system provide global positions and inertial updates at a high rate (e.g., 200 Hz) to figure out the localization information of autonomous vehicles. However, GNSS has fundamental limitations because it only works well in clear sky conditions away from skyscrapers blocking satellite signals. Independent of external factors, IMU is proper to obtain the motion information in real-time, but its accuracy becomes lower as the error diverges with time. Therefore, by combining the advantages of GNSS and IMU to overcome each disadvantage, the fused system can provide accurate real-time data updates for the precise localization of autonomous vehicles. Not only the IMU but RTK also could support the GNSS by rectifying the position error with the aid of a ground base station.

Camera Cameras provide colored visual data (images) as human eyes that can be leveraged into countless vision-based deep learning applications such as object tracking and object detection (lane, traffic signal, vehicle, pedestrian, drivable space, etc.), and many others. The sensor runs in real-time (e.g., 60 Hz) and generates gigabytes of high-definition raw data per second, which are essential for the performance of autonomous vehicles. Since color information is one of the most critical factors in driving, cameras are the primary modality in sensor integration for autonomous vehicles. Furthermore, the passive sensor does not emit any signal like

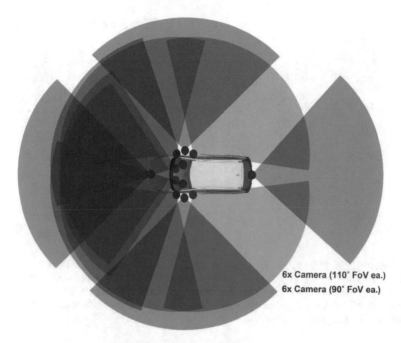

6x Camera (110° FoV ea.)
6x Camera (90° FoV ea.)

Fig. 1 Example of camera sensors integration in an autonomous vehicle of Seamless Transportation Lab (STL), Yonsei University [3]: Numerous cameras are installed on a vehicle to obtain various regions of interest (ROI) according to its purpose. The integration shows that the installation strategy focuses on avoiding blinding areas around the vehicle

lasers or radio waves. It is extensible to increase the number of cameras to extend a region of interest (ROI) to reduce blinding areas.

For this reason, cameras are installed at a vehicle's front, rear, left, and right sides to obtain comprehensive visibility as illustrated in Fig. 1. Rich image data is used in various vision-based deep learning applications such as depth estimation, object detection and tracking, segmentation, and many others, so most manufacturers integrate camera sensors into their vehicles. A stereo camera has multiple lenses apart and can provide disparity information to derive depth information by leveraging binocular vision like humans. Thermal cameras use infrared radiation, far-infrared (FIR) cameras capture heat sources, and near-infrared (NIR) cameras penetrate haze and fog. High dynamic range (HDR) cameras acquire multiple images with different shutter speeds for various brightness to create a brightness-optimized image.

LiDAR Light detection and ranging (LiDAR) emits infrared light waves and calculates reflected waves based on the time of flight (TOF) to measure the distance of reflected points, referred to as point clouds. The sensor provides three-dimensional (3D) positional data (x, y, z) and 1D intensity data (I) of a reflected point and has a detection range of up to 200 meters with guaranteed accuracy of centimeter-level, so it is widely applied to scan static and dynamic objects around

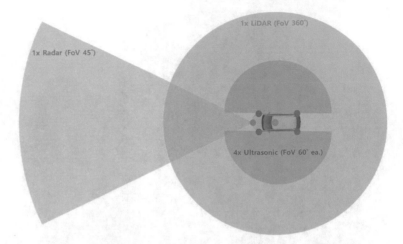

Fig. 2 Example of LiDAR, radar, ultrasonic sensors integration in an autonomous vehicle of Seamless Transportation Lab (STL), Yonsei University [3]: Each sensor is installed in a different location of the vehicle according to its purpose. A LiDAR is located on the rooftop to obtain 360 degrees of FoV around the vehicle, a radar measures front object information, and ultrasonic sensors scan near objects to the vehicle

AVs. The horizontal field of view (FoV) can be extended to 360 degrees which can sense the distance of the environment surrounding AVs. Rotating LiDARs are generally installed on the rooftop of vehicles, while solid-state LiDARs are attached to the side of the vehicle body as shown in Fig. 2. FMCW (frequency modulated continuous wave) LiDARs use continuous waves to apply coherent detection and measure instantaneous velocity based on Doppler shift within a detection range of 300 meters [4]. LiDAR's more extensive sensing range and capability of performing in environments with a lack of illumination, such as nighttime, make the sensor competitive with a camera. The advantages of the sensor are leveraged into implementing localization, 3D object detection, and high-definition (HD) map generation with deep learning-based technologies.

Radar and Ultrasonic Radar and ultrasonic sensors emit radio waves and ultrasonic waves to measure the distance to objects and retrieve 3D position information by calculating the time of each bounce back from objects in environments. Radars have longer detection distances than LiDARs but are less accurate and so do ultrasonic sensors. However, they are cost-effective, and the generated data can be leveraged directly into autonomous driving applications with less processing. Radars commonly measure the relative speeds of objects in front of the vehicle, and ultrasonic sensors measure close objects around the vehicle for obstacle avoidance as depicted in Fig. 2.

The sensors above have different capabilities that can compensate for others, as illustrated in Table 1. A camera is a cost-effective sensor capable of providing color and contrast information but unable to detect speed and is easily affected

Table 1 Comparison of sensor capabilities [5]: Each sensor has different strengths and weaknesses compared to others. Depending on the purpose of the autonomous vehicle's module, a single sensor can be used or sensor fusion can be applied to enhance performance or efficiency. For instance, if a camera and a radar are fused, a cost-effective module can be built that outputs object detection results with accurate distance measurement

Capabilities	Sensors		
	Camera	LiDAR	Radar
Sensor Cost	5	1	5
Sensor Size	5	1	5
Speed Detection	1	4	5
Resolution	5	4	3
Range	5	4	4
Proximity Detection	2	2	4
Color/Contrast Information	5	0	0
Functionality in Snow/Fog/Rain	2	3	5
Functionality in Bright	4	5	5
Functionality in Dark	1	5	5

∗ Score: 1 to 5 (higher is better)

by illumination. A LiDAR can cover what a camera cannot obtain or is weak, such as accurate depth information. Still, the former sensor is less cost-efficient and limited in certain adverse weather conditions, such as fog or snow, that light reflection is affected. A radar does what a LiDAR is limited to but has a weakness in range, resolution, and acquiring color/contrast information which a camera can do. Therefore, building a strategy for an autonomous vehicle in which sensors will be integrated is crucial. The sensor integration will affect not only the performance of autonomous driving but also the hardware integration strategies.

1.3 Technologies

During the last decades, developments in hardware and software for AI applications have advanced the functional automation of vehicles. AVs are combined with various sensors to obtain more environmental data around the vehicles to provide more precise detection and perception results. Nowadays, AV technologies leverage rich sensor information and promising performance by AI/DL-based methods within actual vehicles. Autonomous driving tasks such as lane-keeping, adaptive cruise control, lane changing, automatic emergency braking systems, and others are adapting various applications, including object detection, classification, tracking, and so on, to take advantage of AI/DL technologies.

Localization is a critical and fundamental task that finds where the ego-vehicle is located relative to a reference environment [6–8]. Not only does localization correct the vehicle position within lanes, but it is also a mandatory requirement

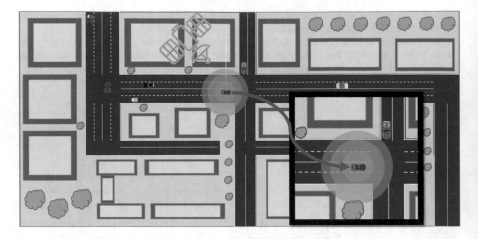

Fig. 3 An example of state estimation and localization for autonomous vehicles.

for global navigation so that vehicles can accurately plan paths to a destination as shown in Fig. 3. Therefore, autonomous vehicles can perform maneuvers based on the perception (object detection, segmentation) or decision (path planning, lateral and longitudinal control) applications depending on the location resulting from the localization task in the given environment. Generally, GNSS and IMU are utilized to find the global coordinates of the ego-vehicle. However, satellite signals of GNSS are vulnerable to being blocked by tall obstacles like skyscrapers, so relying solely on satellite-based localization is not recommended. For implementing feature-based localization, cameras and LiDARs are used in the mapping task to locate the relative position of the ego-vehicle in the road environment.

Object detection is a computer vision task to detect objects in input data such as images, videos, or point clouds acquired from the camera, radar, or LiDAR sensors, whose outputs can be seen in Fig. 4. According to target domain dimensions, there is 2D and 3D object detection in the research area. Primarily, 2D object detection [9–13] is based on image data from camera sensor. In contrast, 3D object detection [14–18] adapts camera, LiDAR, or their fusion data to estimate more accurate spatial information in metric space. Advanced deep learning-based methods extract features to classify objects and estimate their location within the input data by providing information on bounding boxes. In cooperation with the localization task, object detection is widely leveraged into ADAS applications, such as adaptive cruise control (ACC), lane-keeping assist (LKA), lane change assist (LCA), and autonomous emergency braking (AEB) system, for recognizing existence with the location of objects such as vehicles, pedestrians, cyclists, traffic lights, and road signs around the ego-vehicle.

Segmentation provides denser pixel-wise classified information of an image that can be used for applications that require precise classification results to understand scenic features around the ego-vehicle. Semantic segmentation [9] is a single entity

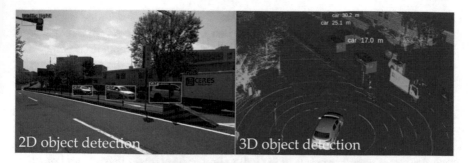

Fig. 4 Examples for 2D object detection (*left*) and 3D object detection (*right*) [19]

Fig. 5 Examples for semantic (*left*) and instance segmentation (*right*) [19]

when multiple objects are in the same class, such as the sky, building, car, or person. In contrast, instance segmentation [20] further analyzes distinct individual objects over multiple objects of the same class as depicted in Fig. 5. Panoptic segmentation [21] combines the features of semantic and instance segmentation that identifies objects concerning classification and distinguishes all instances individually on a given input image data. In addition to the localization and object detection, segmentation could be utilized in the multiple AV functionalities that require rich pixel information of environments around the AVs.

Object tracking estimates multiple objects' heading (yaw angle) and velocity to track the object in terms of sequential time (frame) and to predict their future trajectory so that AVs can determine proper perception functions to avoid collisions. Target objects for multiple objects tracking (MOT) include various and usual dynamic objects in on-road environments such as pedestrians, vehicles, animals, and so on [22]. Multiple camera systems or fusing modality systems with the camera, LiDAR, and radar are frequently used to obtain spatial information about objects and track them in a wide range [23, 24]. Usually, the tracking prediction results are leveraged into autonomous driving tasks so that the ego-vehicle can make appropriate perceptions and decisions for local path planning.

Numerous technologies leverage sensor modalities in AVs to make the vehicle safer and execute a better driving performance. Deep learning networks based on CNN, GAN, and Transformers are recently applied as backbone networks to achieve more accurate performance that can be eligible for AV implementation. Even though a network can perform with great accuracy, computation complexity depends on many factors, including network architecture, the configuration of input data, algorithms for pre−/post-processing, and many others. Not only for the network itself but hardware specifications for AVs implementation and integration also have constraints compared to ideal experimental environments. Implementation systems with hardware accelerators should also satisfy conditions such as frame rate, computing power, and power efficiency based on the limited spatial and technical resources. Therefore, it is challenging to implement feasible and practical technologies for AVs considering integrated hardware systems on the vehicles, such as electrical/electronic architecture, sensors, power, and computation systems.

1.4 Electrical/Electronic (E/E) Architecture

Electrical/Electronic architecture represents the integrated technology system of hardware/software applications, wiring, network communications, and others that contribute to various vehicle functions such as vehicle lateral and longitudinal controls, body, chassis, power train, security, safety, and infotainment. The different features were incrementally implemented into a vehicle with electronic control units (ECUs) which communicate with each other through in-vehicle networks such as controller area network (CAN) and gateway. As technologies for vehicle functionality have been developed with more complex features like ADAS, recent vehicles require up to a hundred ECUs to leverage many implemented devices, including sensors and software-based applications. Over the evolution of vehicle technology, the E/E architecture of a vehicle has been built in various structures to optimize spatial and power efficiencies, as shown in Fig. 6.

Autonomous vehicles (AVs) can be integrated with various sensor and software algorithm strategies. The vehicle E/E architecture should inevitably be able to manage a large amount of raw data and complex computation for better automated ADAS or driving control performance while preserving various vehicle factors. As larger systems of AVs require more power supply, it is essential to handle limited power sources efficiently. The power modes are hosted on an ECU to selectively manage the supply into several domains of E/E according to the vehicle's situations and purposes. For functional safety of E/E architecture, ISO 26262 [26] has been adopted as the standard in the industry, which categorizes functions as automotive safety integrity levels (ASIL) [26] and quality management. The standard pursues addressing any hazards caused by systematic malfunctioning of E/E architecture to detect, avoid, and control the failures to ensure an acceptable level of safety is being achieved. E/E architecture of AVs communicates internally and externally via onboard diagnostics (OBD), high-speed backbone bus (FlexRay, Ethernet, CAN), USB, cellular, Bluetooth, or Wi-Fi that provide potential threats from malicious attacks against vehicle systems. Therefore, robust cybersecurity

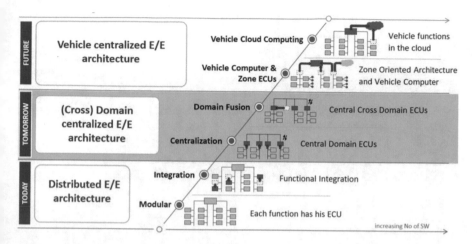

Fig. 6 Evolution of E/E architecture [25]: From the first vehicle CAN network in the 1980s, many OEMs have constructed various E/E architectures in terms of integrated vehicle functionalities. The main goals for the future are to improve functional safety, security, and efficiencies according to increasing applications leveraging enormous data from AVs

can be achieved with a layered approach with mechanisms repeated at critical points in the architecture, and integrated hardware security modules (HSM) may be required. Especially AVs contain more sensors and computing modules for automating functions to monitor, manage, and control the vehicles.

Distributed (decentralized) E/E architectures have been constructed and evolved over a few decades since the advent of basic cruise control in the late 1950s. Decentralized architectures can support separation between ECUs, allowing for one or a few related functions to be implemented on a single ECU. In addition, replacing malfunctioning ECUs is simple since limited functions are deployed on the ECUs. However, the one-to-one mapping between vehicular functions and ECUs of decentralized E/E architectures has drawbacks in implementing technological enhancements for modern vehicles with many functionalities. More ECUs are required for more functionalities, increasing wiring harness and constraints in optimizing the architectures. Also, technological enhancements enlarge the scale of a vehicle system and the complexity of E/E architecture by considering functional stability, safety, fuel efficiency, and so on. Manufacturers replace the decentralized architecture with centralized architecture with high computing power and large storage capacity to improve the scalability of E/E architectures for AVs. Automotive original equipment manufacturers (OEMs) are experimenting with three types of centralized architectures: domain centralized, cross-domain centralized, and zone-oriented architectures.

Each domain in a domain centralized architecture has a corresponding domain cluster that includes one domain control unit (DCU) and sub-domain ECUs. A DCU has multi-core control processing units (CPUs) to host significant domain functionalities, while sub-domain ECUs require less powerful CPUs to handle

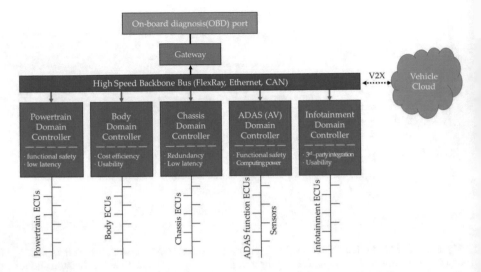

Fig. 7 Domain-centralized E/E architecture [27]: As recent vehicles include various electronic-based domains with complex functions, many ECUs and their collaboration are required. Each domain controller focuses on specific goals depending on its role. For example, the ADAS domain controller pursues functional safety and computing power since it leverages many sensors to perform complicated autonomous driving tasks

auxiliary domain functions. As depicted in Fig. 7, the new architecture is a hybrid architecture of distributed ECUs and domain controllers, which can be integrated and operated independently in terms of encapsulated domains such as powertrain, body, chassis, infotainment, and ADAS. The domain-wise integration with domain control units (DCUs) reduces ECUs in a vehicle and enables the logical grouping of software related to each domain. The key feature of the architecture is that each software-based function can be modified independently without affecting other software. Even though centralized domain architectures address the drawbacks of decentralized architectures, they still face some challenges in implementing functionalities for AVs. Heavy interaction among DCUs because of high-end functions (e.g., ADAS) makes boundaries between the DCUs unclear, and current communication bandwidths are insufficient as such functions arise.

To compensate for the drawbacks mentioned above, the cross-domain centralized architecture contains cross-domain ECUs (CDCUs) whose functions of different domains are consolidated onto single ECUs. For instance, functions for the powertrain and chassis domains can be consolidated into a single-vehicle motion control CDCU [25]. Still, cross-domain centralized architectures require more wiring for more complex and vehicular functions, and the issue can be resolved in a zone-oriented (vehicle-centralized or zonal) architecture [28].

Zone-oriented architectures divide a car into zones such as "front right" or "rear left," as illustrated in Fig. 8. Each zone includes zone ECUs (ZCUs), which are connected to sensors and actuators similar to sub-domain ECUs in centralized

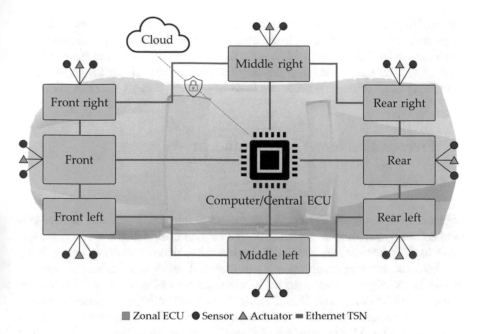

■ Zonal ECU ● Sensor ▲ Actuator ▬ Ethernet TSN

Fig. 8 Zone-oriented (vehicle-centralized or zonal) E/E architecture: Generalized architecture with divided zones, corresponding zone ECUs (ZCUs), and a computer/central ECU in a vehicle

domain architectures. However, they collect and transfer information of their zone to the central server, which processes complex functions with powerful computing assets. The architectures can reduce wiring harness weight and eliminate potential threats in terms of functional safety by connecting sensors to the central server for data processing [25]. The connection ensures faster and more secure data communication internally within the vehicle and externally to the cloud. The type of architecture is also optimized for implementing over-the-air (OTA) so that the vehicle can update firmware or software to improve functionalities.

As domain controllers deal with enormous data from vehicle sensors, they require significant processing power, and often hardware accelerators are leveraged to fulfill the necessity. Graphics processing units (GPUs), digital signal processors (DSPs), and field programmable gate arrays (FPGAs) are frequently used as hardware accelerators for implementing various applications for ADAS and ECUs such as object detection, sensor fusion, security gateways, or parallel acquisition of the engine data [29, 30].

Recently, BMW and Volkswagen adopted a three-domain E/E architecture that includes vehicle controller, intelligent cockpit, and driving domains. At the same time, Tesla has quasi-central architecture (central computing platform and regional controllers), which includes a central computing module (CCM), body control module left (BCMLH), and body control module right (BCMRH) with the self-developed Linux [31]. The main common goal of the approaches is to reduce the wiring harness, satisfy software/hardware safety standards, and process real-time high-level automated technologies for autonomous driving with optimization of E/E architecture. In addition, to enable large-scale integration of many different software and allow OTA updates of individual functions while maintaining the performance and safety of a vehicle, it is necessary to satisfy the AUTOSAR standard [32].

2 Prerequisites for AI Accelerators in AVs

2.1 Overview

Autonomous vehicles (AVs) are complex systems integrated with various hardware and software technologies. Building design strategies for autonomous vehicle systems has been challenging because the process should also ensure power efficiency, performance accuracy, mobility, cost, safety, and many other things. Above all, safety is the highest priority factor to consider for constructing AV systems no matter what sensors, detection, classification, localization, perception, decision-making, or control algorithms comprise the systems. This chapter describes various requirements for AI accelerators in AV systems to ensure performance and safety. Then, several international standards in the field of automotive E/E to which the AI accelerator belongs are covered.

2.2 Requirements of AI Accelerators in AVs

The main goal of autonomous vehicles (AVs) is to maintain and guarantee safety. Many hardware sensors (camera, LiDAR, radar, ultrasonic, etc.) are integrated into AVs to gather data around the ego-vehicles to provide information for various tasks of the autonomous driving system, such as object detection, path planning, and many others. The vehicles generate approximately 1GB/s datastream, which enlarges as more sensors are integrated. As a result, the embedded processing systems are required to digest the enormous amount of data in real-time so that the autonomous driving tasks operate in purposed performances. The refresh rate needs to be as fast as the frame rate to cope with any unexpected situation, even though AVs are cruising at high speed. In contrast to in-lab environments, the best-performing hardware for processing, computing, and inferencing data from AVs have limitations in being integrated into the vehicles due to cost and power inefficiencies. Therefore, AI accelerators must satisfy the following factors for safer AVs that perform their proposed autonomous driving tasks, including ADAS.

Computing power is one of the critical factors in advancing AVs toward higher SAE automation levels [33] by enabling the implementation of more complex computation tasks. The automotive industry is engaging in transforming developed

Table 2 Computational
requirements for different
SAE levels and sensors

Computational requirements		
SAE levels	#Sensors	TOPS
L2	17	2
L3	24	24
L4	28	320
L5	32	4000+

AI and DL applications into practical systems for automation based on enormous amounts of resource data from various sensors integrated into AVs. Therefore, the computing power of AI accelerators should meet the computational requirements of AVs to process all algorithms with ensuring functional safety and performance in real-time. Generally, hardware accelerators are required to handle two tera operations per second (TOPS) for L2, 24 TOPS for L3, 320 TOPS for L4, and 4000+ TOPS for L5, as shown in Table 2. The magnitudes can be varied, but it is clear that the computation power is needed more for higher SAE levels because of the increasing complexity of driving tasks.

Low latency in AV systems is necessary to deal with the massive volume of collected sensor data to ensure safety because missing a single data frame can cause failures of autonomous driving tasks and cause severe threats. Therefore, lowering and ultimately getting rid of latency is a must factor for AVs to compute AI algorithms (detection, classification, and prediction) faster and to make proper control decisions of the vehicles in real-time. Processing tasks at the local level using AI accelerators at the local level in AV system architecture not only lowers transfer latency but also distributes computation cost by operating independently. The acceleration in processing speed ensures the performance of AVs at high speed and perceives more environment information to enable algorithms to output more accurate results. The quantitative requirements vary depending on the role of each driving task, but it is clear that faster processing drives safer AV systems.

Energy efficiency is one of the vital factors for AVs because many hardware devices, including sensors, are generally electric-powered. Higher power consumption implies inefficiencies in weight and cost by demanding additional power supply and thermal management systems. Recent Level 1 to Level 3 AVs demand 30 W or more, but hundreds and thousands of watts are needed above Level 3(L4 and L5) to flawlessly process fully automated driving based on enormous data from sensors in Table 3. Instead of just adding a supplementary power supply such as a battery, leveraging embedded systems with energy-efficient processors within a limited energy budget is more practical and feasible for AVs. Many AVs are using GPUs for processing AI-based autonomous driving tasks. Despite general purpose, GPUs are less energy efficient than FPGAs and ASIC accelerators that are purpose-designed and optimized chip architecture to distribute workloads. As AI algorithms become advanced to solve more complex driving tasks, high computation costs could be indispensable, so energy-efficient AI accelerators are essential to maintain power consumption in AVs.

Table 3 Power requirements
in different applications for
autonomous vehicles

Power requirements	
Application	Power (Watts)
Individual sensors	0.5–5
L1–L3	10–30
L4–L5	100+

Scalability is also an important requirement for AI accelerators. AVs are adopting various approaches to design their systems to cope with more complicated driving tasks to automate their vehicles and achieve higher SAE levels. Updates for both hardware and software should be flexible, and accelerator scalability should support the changes. Furthermore, *reliability* and *security* are critical to ensuring the safety of AVs. The functional safety of accelerators must be guaranteed while processing tremendous amounts of data and algorithms to avoid failures that violate any standards related to AVs such as ISO26262 or the safety of the intended functionality (SOTIF). *Cost* needs to make the whole AV system with a reasonable budget to implement practical and feasible AVs.

2.3   Standards for AI Accelerators in AVs

2.3.1   IEC 61508

Before demonstrating the development process for AI accelerators applied in autonomous vehicles, international standards for automotive electric/electronic systems should be explained first as described in Fig. 9 for clear comprehension. As electronic systems affecting safety were adopted in vehicles in the 1980s and 1990s, *Development Guidelines for Vehicle Based Software* was published by Motor Industry Software Reliability Association (MISRA) in 1994 [34]. It was the first consensus for the safety of vehicle electronic systems [35–37].

After that, in 1998, International Electrotechnical Commission (IEC), which had more than 50 member nations at that time, published the first international standard for the safety of electronics [38]. *IEC 61508*, titled *Functional Safety of Electrical/Electronic/Programmable Electronic(E/E/PE) Safety-related Systems*, is an international standard that presents overall requirements to guarantee the safety of equipment under control (EUC), EUC control system, and safety-related system of E/E/PE [39].

The safety-related system of E/E/PE is allocated a probability called safety integrity that indicates how well the safety functions are performed under all given conditions within a given period. Based on the safety integrity, in phase 4 of the safety lifecycle, a discrete rank called safety integrity level (SIL) is determined [40]. The three stages of hazard identification, hazard analysis, and risk assessment apply to draw a target SIL. SIL defined in either quantitative or qualitative way is shown in Figs. 10 and 11, respectively. The figures are described as the risk matrix method.

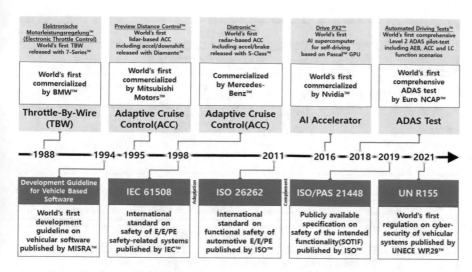

Fig. 9 Milestones of vehicular E/E technologies (*top*) and guidelines, international standards, and regulations (*bottom*)

Safety Integrity Level	Low Demand Mode		High Demand or Continuous Mode	
	Average Probability of Failure on Demand per Year	Risk Reduction Factor (Fail Every f Years)	Average Probability of Failure on Demand per Hour	Risk Reduction Factor (Fail Every f Hours)
SIL 1	$10^{-2} \le p < 10^{-1}$	$10 < f \le 100$	$10^{-6} \le p < 10^{-5}$	$100{,}000 < f \le 1{,}000{,}000$
SIL 2	$10^{-3} \le p < 10^{-2}$	$100 < f \le 1{,}000$	$10^{-7} \le p < 10^{-6}$	$1{,}000{,}000 < f \le 10{,}000{,}000$
SIL 3	$10^{-4} \le p < 10^{-3}$	$1{,}000 < f \le 10{,}000$	$10^{-8} \le p < 10^{-7}$	$10{,}000{,}000 < f \le 100{,}000{,}000$
SIL 4	$10^{-5} \le p < 10^{-4}$	$10{,}000 < f \le 100{,}000$	$10^{-9} \le p < 10^{-8}$	$100{,}000{,}000 < f \le 1{,}000{,}000{,}000$

Fig. 10 Safety integrity level (SIL) defined in a quantitative way in the IEC 61508

Consequence of Damage	Frequency and Exposure Time	Possibility of Avoidance	Probability of Unwanted Occurrence		
			W1: Very Low	W2: Low	W3: Relatively High
C1: Minor Injury			No Requirements	No Requirements	No Special Requirements
C2: Serious Permanent Injury or One Dead	F1: Rare	P1: Possible	No Requirements	No Special Requirements	SIL 1
		P2: Impossible	No Special Requirements	SIL 1	SIL 2
	F2: Frequent	P1: Possible	No Special Requirements	SIL 1	SIL 2
		P2: Impossible	SIL 1	SIL 2	SIL 3
C3: Several Dead	F1: Rare	P1: Possible	No Special Requirements	SIL 1	SIL 2
		P2: Impossible	SIL 1	SIL 2	SIL 3
	F2: Frequent	P1: Possible	SIL 1	SIL 2	SIL 3
		P2: Impossible	SIL 2	SIL 3	SIL 4
C4: Disaster	F1: Rare	P1: Possible	SIL 1	SIL 2	SIL 3
		P2: Impossible	SIL 2	SIL 3	SIL 4
	F2: Frequent	P1: Possible	SIL 2	SIL 3	SIL 4
		P2: Impossible	SIL 3	SIL 4	Multiple System Required

Fig. 11 Safety integrity level (SIL) defined in a qualitative way in the IEC 61508

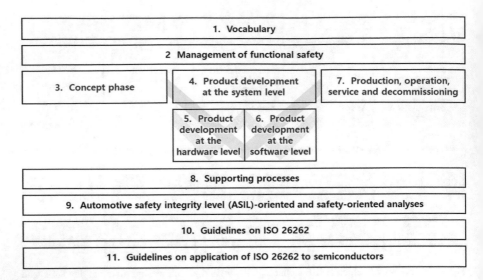

Fig. 12 Overview of the ISO 26262. Part 3, 4, 5, 6, and 7 belong to the V-model

As a target probability of failure, SIL specifies safety requirements for safety functions. SIL is decided among 1, 2, 3, and 4. On top of that, the SIL 1 has the lowest reliability and the SIL 4 has the highest, on the contrary. It means that a higher level of SIL requires more risk reduction for the system [41]. In accordance with the SIL, safety functions should be designed to meet the safety requirements with the objective of risk reduction [42]. Ultimately, by the safety functions, EUC risk becomes lower than the tolerable risk target and the system could be regarded as safe.

2.3.2   ISO 26262

IEC 61508 has served as a traditional standard for system designing and developing. As many parts of vehicles have become more electrified, a need for a new standard specialized in vehicles has also been raised. In response to the needs, in 2011, International Organization for Standardization (ISO) published *ISO 26262*, titled *Road Vehicles — Functional Safety*, as the new international standard specialized in the automotive industry [26]. ISO 26262 covers the whole safety lifecycle of vehicles, and it mainly focuses on product development phases with the V-model. The V-model is the core of the ISO 26262 and it contains the *Part 3: Concept phase*, *Part 4: Product development at the system level*, *Part 5: Product development at the hardware level*, *Part 6: Product development at the software level*, and *Part 7: Production, operation, service and decommissioning* as shown in Fig. 12.

   ISO 26262 is an adaptation of the IEC 61508 for automotive electrical/electronic (E/E) [43]. Likewise, a concept of the safety integrity level (SIL) from the IEC

Severity of Possible Injury	Probability of Exposure	Controllability by Driver		
		C1: Simple	C2: Normal	C3: Difficult
S1: Light and Moderate	E1: Very Low	Quality Management	Quality Management	Quality Management
	E2: Low	Quality Management	Quality Management	Quality Management
	E3: Medium	Quality Management	Quality Management	ASIL A
	E4: High	Quality Management	ASIL A	ASIL B
S2: Severe and Possibly Life Threatening	E1: Very Low	Quality Management	Quality Management	Quality Management
	E2: Low	Quality Management	Quality Management	ASIL A
	E3: Medium	Quality Management	ASIL A	ASIL B
	E4: High	ASIL A	ASIL B	ASIL C
S3: Life Threatening and Fatal	E1: Very Low	Quality Management	Quality Management	ASIL A
	E2: Low	Quality Management	ASIL A	ASIL B
	E3: Medium	ASIL A	ASIL B	ASIL C
	E4: High	ASIL B	ASIL C	ASIL D

Fig. 13 Automotive safety integrity level (ASIL) defined in qualitative way in the ISO 26262

61508 is inherited to the ISO 26262 as an automotive safety integrity level (ASIL), which determines a class of automotive-specific risk. Though SIL and ASIL share similar concepts, ASIL should take account of the distinctiveness of the vehicle domain. IEC 61508 generally targets various industrial equipment, and it assumes that trained engineers operate the equipment in designated situations. On the other hand, the ISO 26262 targets vehicles, so manufacturers should consider that users of the vehicles are ordinary folks with a driver's license as a minimum qualification and drivers are exposed to various road environments. Since the maintenance of the vehicle fully depends on its owner, automotive manufacturers should present stricter and robust criterion for the functional safety.

According to the *ISO 26262 Part 3: Concept phase – Clause 6: Technical safety concept*, hazard analysis and risk assessment (HARA) are required to determine the ASIL. In the HARA process, there are three factors that affect the ASIL classification: severity of possible injury, probability of exposure, and controllability by driver. Through the combinations of these factors, results of the ASIL for all individual identified hazardous events are drawn out. Like the SIL, ASIL is chosen among A, B, C, and D levels. ASIL A represents the least stringent level, whereas ASIL D represents the most stringent level. Described as a risk matrix, ASIL is shown in Fig. 13.

As a guidance to present some methods and examples for ASIL determination, in 2015, Society of Automotive Engineers International (SAE International) issued *SAE J2980*, titled *Considerations for ISO 26262 ASIL Hazard Classification* [44]. In this way, automotive manufacturers set the specification of individual automotive E/E systems by adjusting factors for ASIL classification at their discretion [45].

Based on the results of HARA, manufacturers find the highest ASIL from the results and define it as a safety goal that could be decomposed to functional safety requirements like Fig. 14. In other words, after the concept phase by determining

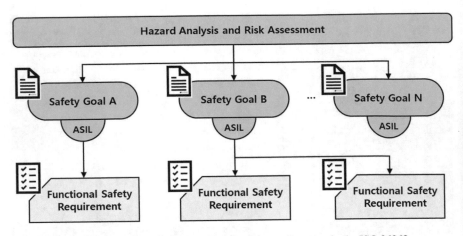

Fig. 14 Hierarchy of safety goals and functional safety requirements in the ISO 26262

the ASIL, a development phase specifying the requirements proceeds. Its details are described in the *ISO 26262 Part 3: Concept phase – Clause 7: Functional safety concept*.

So far, the concept phase including HARA to derive ASIL based on SIL has been described. Hereafter, descriptions will focus on a development phase of the AI accelerator. As vehicles equipped with the ADAS achieving not only passive safety but also active safety are widely available in the 2010s, Euro NCAP executed the world's first comprehensive ADAS test in 2018. That same year, ISO published the second edition of the ISO 26262 including new parts, *Part 11: Guideline on application of ISO 26262 to semiconductors*, in a response to the widespread adoption of semiconductors supporting ADAS. Today, vehicles are equipped with dozens to hundreds of semiconductors. Especially, it is about to commercialize the autonomous driving systems of Level 3 or higher that goes beyond ADAS. Since the release of NVIDIA Drive PX2, the world's first AI accelerator for self-driving car platform, in 2016, AI accelerators that can process high bandwidth signals from various sensors in real-time have been spotlighted these days [46]. Then, how should AI accelerators for autonomous vehicles be developed to comply with the standards and guidelines?

To achieve the safety goal specified by HARA, safety measures are required to prevent the violation caused by failure. The safety measure (or safety mechanism) is a technical solution to maintain the functional safety (or safe state) by controlling systematic failures and random hardware failures in order to prevent unreasonable risk resulting from hazards caused by malfunctioning behavior occurring in E/E systems [47, 48]. As mentioned earlier, to comply with the safety goals, the safety measures should be implemented according to the specification in the functional safety requirements. From the functional safety requirements, technical safety requirements defined in the *ISO 26262 Part 4: Product development at the system level – Clause 6: Technical safety concept* are derived. From the technical safety

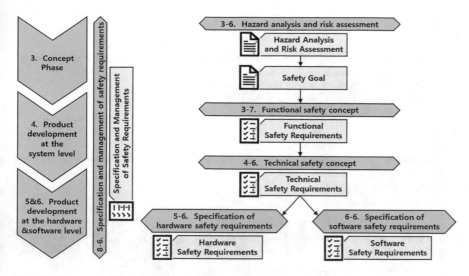

Fig. 15 Structure of safety requirements in the ISO 26262

requirements, hardware safety requirements defined in the *ISO 26262 Part 5: Product development at the hardware level – Clause 6: Specification of hardware safety requirements* are derived. The structure of safety requirements described so far is illustrated in Fig. 15. But, the hardware safety requirements are applied only in case the safety goal corresponds to one of ASIL B, C, and D. AI accelerators should be developed satisfying the hardware safety requirements in accordance with the *ISO 26262 Part 5: Product development at the hardware level* since AI accelerators are semiconductor hardware. *ISO 26262 Part 11: Guidelines on application of ISO 26262 to semiconductors* could also be a good reference in the development process.

After establishing the specification of safety requirements is done, safety measures are implemented to satisfy the specification. According to the *ISO 26262 Part 1: Vocabulary – Clause 3.61: Fault tolerant time interval*, when a fault occurs, the implemented safety measure activates within the fault tolerant time interval (FTTI) to turn the fault into detected multiple point fault or perceived multiple point fault for the purpose of preventing the occurrence of hazardous events [47]. Otherwise, additional emergency operation is carried out within the emergency operation time interval (EOTI) to complete a transition from the fault to a safe state. See Fig. 16 for better understanding.

Then what is the detailed procedure to implement the safety measures in compliance with ISO 26262? In the implementation process of safety measures, to quantitatively analyze fulfillment of the requirements for the ASIL of the safety goal, failure modes effect analysis (FMEA) method, which Ford adopted after the Pinto affair (Grimshaw v. Ford Motor Co. case) in the 1980s, could be utilized [49, 50]. The FMEA process proceeds as follows. According to the *ISO 26262 Part 5: Product development at the hardware level – Annex B: Failure mode classification of a hardware element*, failure modes that are failures in intended behavior are defined

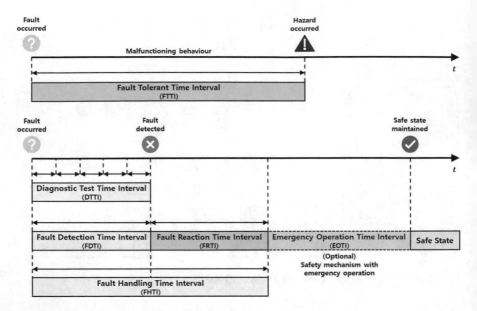

Fig. 16 Safety relevant time intervals in the ISO 26262. One without safety mechanism (*top*) and the other with an implemented safety mechanism including emergency operation (*bottom*)

depending on the use cases and safety measures. There are several types of failure modes, but they are not all dangerous and only latent multiple-point fault, residual fault, and single-point fault are risky. Classification of failure modes is described in Fig. 17. After classifying failure modes, it is time to estimate a diagnostic coverage (or failure mode coverage), which is a percentage of each failure rate detected or controlled by the implemented safety measure. However, the terms "failure mode" and "diagnostic coverage" seem unfamiliar. Let us clarify the details of the terms.

First, what is the failure mode? Descriptions related to the failure modes of safety measure for digital components like AI accelerators can be found in the *ISO 26262 Part 11: Guidelines on application of ISO 26262 to semiconductors – Clause 5: Specific semiconductor technologies and use cases*, and *Annex A: Example on how to use digital failure modes for diagnostic coverage evaluation*. According to the clauses, failure modes are classified into four types of models: (1) function omission (FM1) which is function not delivered when needed, (2) function commission (FM2) which is function executed when not needed, (3) function timing (FM3) which is function delivered with incorrect timing, and (4) function value (FM4) which is function that provides incorrect output. An example of failure modes for the accelerator is shown in Fig. 18.

Second, the diagnostic coverage of safety measure could be determined as in the example of *ISO 26262 Part 5: Product development at the hardware level – Annex D: Evaluation of the diagnostic coverage*. For instance, the diagnostic coverage ranking can be classified into three categories: low (more than 60%), medium (more than 90%), and high (more than 99%). The claimed diagnostic coverage could be

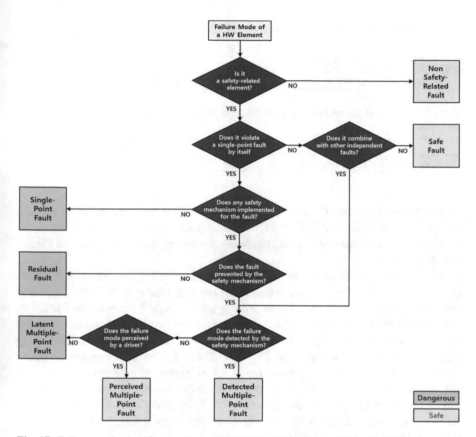

Fig. 17 Failure mode classification in the ISO 26262. Faults marked in red are dangerous, and faults marked in blue are considered safe

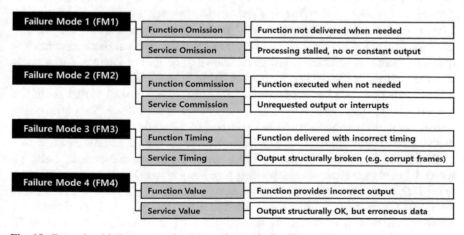

Fig. 18 Example of failure modes for the accelerator in the ISO 26262

Table 4 Example of the target SPFM and the LFM values in the ISO 26262

	ASIL B	ASIL C	ASIL D
Single point fault metric (SPFM)	≥90%	≥97%	≥99%
Latent fault metric (LFM)	≥60%	≥80%	≥90%

Table 5 Example of the random hardware failure target value for the evaluation of PMHF in the ISO 26262

	ASIL B	ASIL C	ASIL D
Random hardware failure target values	$<10^{-7}$/h	$<10^{-7}$/h	$<10^{-8}$/h

supported by dependent failure analysis (DFA) method and fault injection method described in the *ISO 26262 Part 11: Guidelines on application of ISO 26262 to semiconductors – Clause 4.7: Semiconductor dependent failure analysis* and *Clause 4.8: Fault injection*.

Then, after calculating the single point fault metric (SPFM), latent fault metric (LFM), and probabilistic metric for random hardware failures (PMHF) regarding the failure mode case of each component, a table should be drawn for FMEA to check whether those metrics meet requirements or not. Both SPFM and LFM, which are metrics related to random hardware failure that cause violation of the safety goal, must be met according to the *ISO 26262 Part 5: Product development at the hardware level – Clause 8: Evaluation of the hardware architectural metrics*. To determine the hardware part failure rate data, it is possible to use acknowledged industry sources like *IEC 61709*, Siemens *SN 29500*, and *FIDES 2009 EdA*, which are regarded as conservative figures [51]. Example target values for the SPFM and the LFM are presented in Table 4.

After which, quantitative target values for random hardware failure rate are evaluated, expressed as average probability per hour over the operational lifetime, to assure whether the values reach the target or not according to the PMHF method defined in the *ISO 26262 Part 5: Product development at the hardware level – Clause 9: Evaluation of safety goal violations due to random hardware* [52, 53]. Through the evaluation result, it can be proved that the residual risk due to random hardware failure is sufficiently low to be known to be safe to tolerate. An example of target values for PMHF is presented in Table 5. In addition to the PMHF, there is also a method called evaluation of each cause of safety goal violation (ECC). For evaluation, either PMHF or ECC is freely chosen to perform. Examples of the FMEA result embracing the process described so far could be found in *ISO 26262 Part 5: Product development at the hardware level – Annex E* and *Annex F*.

Evaluation to ensure that the performance of the intended functional behavior served by hardware elements meets safety requirements can be performed according to the *ISO 26262 Part 8: Supporting processes – Clause 13: Evaluation of hardware elements*. Based on the difficulty of verifying the safety-related functionality, a hardware element is classified as one of class I (simple), class II, or class III (intricate). Since the AI accelerator corresponds to class III, tests, analyses, and arguments should be carried out in combination to evaluate it. During the ISO 26262 process, the whole process including evaluation is executed iteratively until requirements for the safety goal are met [54].

Descriptions so far cover the process based on the ISO 26262 for developing AI accelerators for autonomous vehicles in the aspect of hardware. Not only the hardware, software that performs safety-related functionality to mitigate hardware faults of the AI accelerator must also be developed, and the development process should follow the description in the *ISO 26262 Part 6: Product development at the software level* [55]. But since the description of *Part 6* is less specific than *Part 5*, this chapter would not deal with details for the software development process. While developing the hardware and software constituting the AI accelerator, specification of hardware-software interface (HSI) such as memory or controller area network (CAN) bus that provides interaction between the hardware and software should also be iteratively refined. Through the integration of hardware, software, and HSI, a system that achieves functional safety could be developed, and ultimately, a vehicle could be completely configured by integrating items [56].

2.3.3 ISO 21448

Although the ISO 26262 deals with arrangement of safety measures to secure the functional safety of a system that integrates hardware and software, it mainly focuses on controlling the random hardware failure. However, an autonomous driving system utilizing machine learning–based AI may make wrong decisions on maneuver of the vehicle due to the temporary decrease in the recognition performance of algorithms expressed in the deep neural network or due to the hitch in the processing performance of AI accelerators. For convenience, in order to supplement the lack of the ISO 26262, which assumes that the intended functionality is safe, in 2019, *ISO/PAS 21448 (publicly available specification)*, titled *Road Vehicles – Safety of the Intended Functionality*, was published, and in 2021, *ISO/DIS 21448 (draft international standard)* was published successively [57].

ISO 21448 deals with the guidance on a series of implementation phases for achieving safety of the intended functionality (SOTIF), in order to provide measures to prevent hazards that the intended functionality cannot correctly perform dynamic driving task (DDT), within the operation design domain (ODD), due to functional insufficiency. The ultimate goal of the SOTIF principle is to avoid the situation where hazardous behavior resulting from insufficiencies of specification or performance limitation occurred by a triggering condition lead to harm. To satisfy the SOTIF, define vehicle-level safety strategy (VLSS) encompassing functional requirements, and consider SOTIF HARA, ODD, and automation level of *SAE J3016* to implement VLSS as driving policy eventually. SOTIF process consists of design, hazards evaluation, functional insufficiencies evaluation, hazardous scenarios evaluation, verification and validation, and SOTIF release in order. The process to guarantee the SOTIF in the ISO 21448 is carried out in parallel with ISO 26262, and the parallel process could be aligned in the V-model to complement each

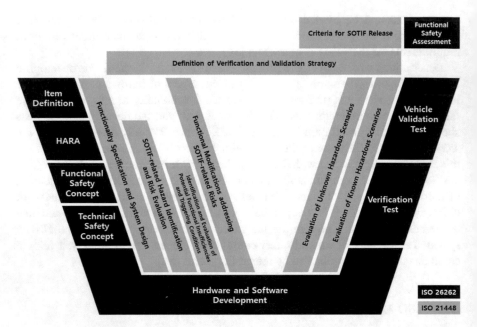

Fig. 19 V-Model in the ISO 21448. Parts marked in black belong to the ISO 26262 and parts marked in gray belong to the ISO 21448

other. For instance, identification and evaluation of hazards phase in the ISO 21448 is aligned with the HARA process in the ISO 26262. However, unlike ISO 26262, ASIL is not determined for the SOTIF-related hazardous event. See Fig. 19 for more details.

In the SOTIF, software that performs functions is a main issue, but hardware executing the software cannot be ignored either. According to the *ISO 21448 – Annex D: Guidance on specific aspects of SOTIF*, autonomous vehicle architecture could be divided into three subsystems as described in Fig. 20. Each subsystem consists of three parts: sensing, planning, and actuation part and AI accelerator belongs to the planning subsystem that receives data from the perception subsystem composed of various sensors sensing the surrounding environment, then decides the next maneuver of ego-vehicle by machine learning–based algorithm like deep learning, and finally delivers the decision to the actuator.

Unlike in general domains, AI accelerators in autonomous vehicles do not require versatility for various tasks, but a design that specialized in autonomous driving functions to meet maximum time efficiency for real-time requirements and also meet robustness for safety requirements is expected to make maneuver decisions. For example, to achieve the purpose of real-time inferencing, it could be possible to design an AI accelerator to match the architecture to the size and shape of datastream provided by sensors, and to process repeatedly performed computing operations in parallel as quickly as possible. Furthermore, in consideration of functional requirements, ODD, automation level, etc., it is recommended to guarantee the intended

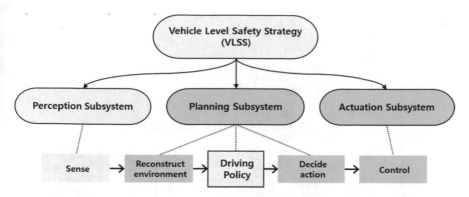

Fig. 20 Simplified architecture for autonomous vehicles. Under the vehicle level safety strategy (VLSS), the system consists of three subsystems: perception (*left*), planning (*middle*), and actuation (*right*)

functionality by preventing the performance limitation due to insufficient computing power. After the design process, identify hazards and evaluate risks. If the risk of harm is unacceptable, evaluate insufficiencies of subsystems encompassing the AI accelerator and functions first, and modify the specification and design, if necessary, to start the whole process loop again. See Fig. 21 for details of ISO 21448 activities.

After the SOTIF activities, when residual risk becomes reasonable to accept, then operation phase activities such as the field monitoring process continue. If a potential SOTIF issue is discovered in this operation phase, functional insufficiencies related with algorithms could be addressed relatively easily through over the air (OTA) update. But, when a semiconductor such as an AI accelerator needs to be improved, a recall may occur for hardware replacement. Recalling vehicles that have already been marketed is burdensome, so it is essential to make a vehicle with guaranteed safety by complying with ISO 26262 and ISO 21448 before launch. However, ISO 21448 deals with reducing SOTIF risks through functional modification of sensors, actuators, and algorithms, but hardly deals with the modification loop for the decision subsystem to which the AI accelerator belongs. Moreover, the aforementioned standards only provide minimum required processes and guidelines to remain technology neutral, and authorities generally do not specify the requirements in detail. In conclusion, since it is entirely up to development parties, including OEMs, to determine the specification of the AI accelerator, those responsible for development should structure a tight development process with the goal of guaranteeing the safety of autonomous vehicles.

ANSI/UL 4600, titled *Evaluation of Autonomous Products*, was published in 2020 by American National Standards Institute (ANSI), in collaboration with Underwriters Laboratories (UL). ISO 21448 is geared toward Level 3 and lower autonomous vehicles, whereas *ANSI/UL 4600* serves as an international standard for Level 4 and Level 5 autonomous vehicles that do not require driver's attention [58]. Unlike the ISO 26262 and ISO 21448 that utilize technical approach, ANSI/UL 4600 aims to protect the safety of fully autonomous vehicles, based on a safety

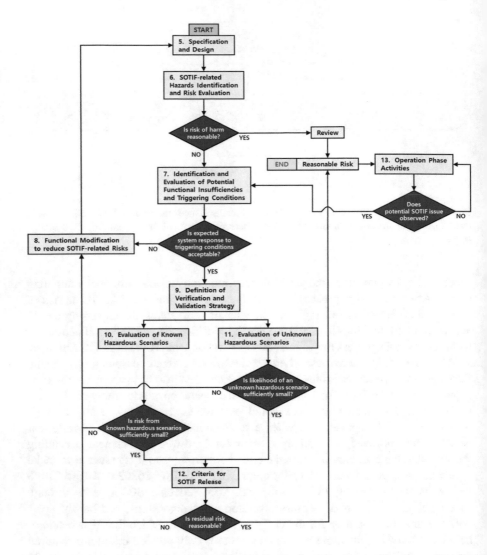

Fig. 21 Dependency of activities in the ISO 21448. Starting from the specification and design phase (marked in *red START*), the whole activity process is repeated until the risk reaches a reasonable level to accept (marked in *blue END*)

case approach consisting of claims, arguments, and evidence. Since ANSI/UL 4600 is compatible with ISO 26262 and ISO 21448, it is recommended to apply the ANSI/UL 4600 to the development process if the objective is to develop an autonomous driving system of Level 4 or reflection of views from US industry and academia is needed [59, 60].

2.3.4 Standards Compliance for the Products Liability

Then why is compliance with standards necessary for automotive manufacturers? Generally, market-leading manufacturers in the automotive industry voluntarily follow the international standard. For instance, in the United States, manufacturers perform self-test to validate their vehicle and if they find any defects, decide to recall their products. However, fully delegating the validation process to the manufacturers could incur moral hazard problems. To prevent those problems, the National Highway Traffic Safety Administration (NHTSA) under the U.S. Department of Transportation purchases vehicles released on the market with the purpose of testing the standards compliance. Based on the test results, the NHTSA has its authority to force recall and impose a fine to the manufacturer if necessary.

Products Liability Law also serves as an indirect regulation for standards compliance too. Manufacturers must prove that they did not commit negligence like design flaws when consumers file lawsuits against them in the event of any safety issues related to their vehicles. In this case, manufacturers could defend themselves based on a fact of the compliance with international standards such as ISO 26262 and ISO 21448 because it means that even the state-of-the-art technologies following those safety standards could not work as precautions for the safety failures [61–63]. Under the autonomous driving system of Level 3 or higher that goes beyond ADAS, the responsibility for decision failure within ODD lies not with the driver, but with the automotive manufacturer who failed to set the specifications of algorithm and AI accelerator properly, so the importance of compliance with standards for product liability is expected to grow.

2.3.5 UN R155

In the near future, vehicles will no longer depend on their drivers, and onboard AI accelerators in autonomous driving systems will be anticipated to play a primary role in taking charge of decisions for driving. In such a future, it is crucial to protect the cybersecurity of AI accelerator because an attack from a malicious user outside of the ego-vehicle to access the control of the AI accelerator could lead to catastrophic results. Despite such concerns, the ISO 26262 covers cybersecurity of embedded software, but the ISO 21448 does not care about attacks caused by vehicle security vulnerabilities.

To fill the gaps in existing standards, in 2016, SAE published *SAE J3061*, titled *Cybersecurity Guidebook for Cyber-Physical Vehicle Systems*, to provide recommended practices as a guidebook for cybersecurity [64]. After that, *ISO/SAE 21434*, titled *Road Vehicles – Cybersecurity Engineering*, was published in 2021 to supersede SAE J3061, and it specifies requirements for cybersecurity in vehicles [65]. While being compatible with ISO 21448 and ISO 26262, ISO/SAE 21434 covers the additional area as shown in Fig. 22. But ISO/SAE 21434 does not provide specific solutions, so it is up to manufacturers to determine the details. *SAE J3101*, titled *Hardware Protected Security for Ground Vehicles*, published in 2020 could be

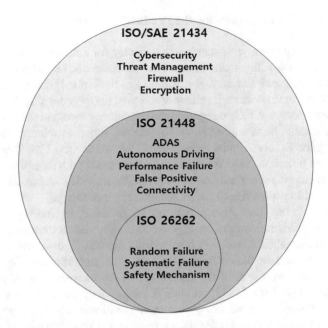

Fig. 22 Scope of the ISO 26262, ISO 21448, and ISO/SAE 21434

a good reference during the development of AI accelerators because the document deals with hardware protected security [66].

The World Forum for Harmonization of Vehicle Regulations (WP.29), a working party belonging to the United Nations Economic Commission for Europe (UNECE), published *UN R155*, a regulation covering automotive cybersecurity [67, 68]. Unlike the previously published international standards that encourage voluntary compliance to manufacturers, UN R155 is the first legally binding regulation on the automotive industry, in the field of autonomous vehicles and connected cars. The regulation, which took effect in 2021, will be enforced in the EU first for newly released vehicles from July 2022. As WP.29 is a global organization with 64 participating countries that agree on type approval for vehicles and vehicular equipment, it is clear that the impact of UN R155 will be tremendous.

In addition to complying with international standards in the development process for product liability, it is expected that autonomous vehicles and AI accelerators mounted on them comply with regulations set by authorities in the future. As vehicles are in the dangerous domain that can cause casualties, safety is paramount above any other matters. To ensure the functional safety, safety of the intended functionality, and cybersecurity as much as possible, certification for compliance with standards or regulations should not be merely a pointless formal act to sell vehicles.

3   Recent AI Accelerators for AVs

3.1   Overview

With recent advancements in deep learning (DL), OEM automakers (Mercedes Benz [69], BMW [70], Volkswagen [71], Tesla [72], Hyundai [73], etc.), tier-1 automotive components suppliers (Bosch [74], Continental [75], etc.), and other autonomous driving technology development companies (Waymo [76], Uber [77], Zoox [78]) are adapting DL-based technologies to implement autonomous driving. Tesla's Autopilot has demonstrated well-performing automated ADAS and lately released FSD (full self-driving) package to pursue SAE Level 4 and Level 5 AVs. To achieve higher levels of automation in autonomous vehicles (AVs), a number of prerequisites must be met, including low latency, high energy efficiency, low cost, and adherence to standards. These capabilities are necessary for AVs to effectively navigate complex driving situations. Since AI accelerators have achieved low latency, energy, and cost-efficiency, their demand keeps growing as AI algorithm-based AVs adapt to decentralized E/E architecture. The performance is competitive with conventional and graphics processing units (GPUs), so practical products are being implemented into AVs. Manufacturers in the industry are designing and developing their hardware form factors with GPUs, field-programmable gate arrays (FPGAs), and application-specific integrated circuits (ASICs) to satisfy requirements and implement well-performing algorithms and applications of AVs within the given conditions. This chapter aims to provide an overview of the current trends and developments in AI accelerators for AVs, as well as an overview of the commercialized products that are currently available on the market.

3.2   Industrial Trend

The automotive industry engages in deep learning technology transformations into existing AV systems to proceed to higher automation levels. Since AVs acquire enormous amounts of data from different sensors, they require efficient and powerful computing AI processors to leverage AI-based tasks in ADAS. Most existing AI tasks for AVs, such as object detection, classification, and segmentation, rely on central processing units (CPUs), GPUs, FPGAs, or ASICs to simultaneously process multiple tasks. Generally, the magnitudes of computational requirements are approximately two Tera operations per second (TOPS) for Level 2, 24 TOPS for Level 3, 320 TOPS for Level 4, and 4,000+ TOPS for Level 5 [5]. Conventional CPUs tend to have flexibility but struggle with processing more complex AI tasks since they are linear, making them have limited computation abilities. GPUs compensate for the constraint of CPUs with multiple cores, which is why many autonomous vehicles have started using onboard processing systems with GPUs. However, computing power and power consumption are significant issues for

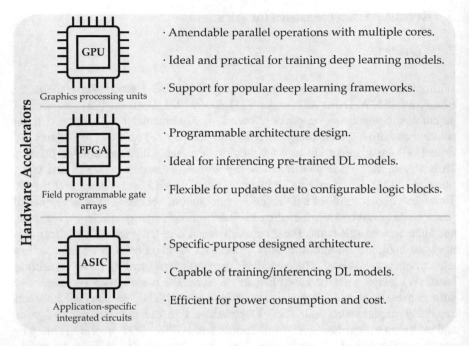

Fig. 23 Comparison between hardware accelerators: GPUs, FPGAs, and ASICs [79]

designing AVs to process more complicated AI algorithms at higher automation levels. Thus, different processors and AI accelerators, such as FPGAs and ASICs, are being used to satisfy the requirements, including performance, energy, and cost efficiencies (Fig. 23).

While the movement of data into and out of AI chips (data bandwidth) is a big challenge, making sure that linear algebra and matrix operations can be done efficiently at the highest level of throughput with the lowest amount of energy is another common challenge. Cutting-edge chips or dedicated co-processors are becoming the mainstream for on-device, edge, and even cloud AI processing. Autonomous vehicle technology is complicated because it requires processing massive data captured by the sensors (camera, LiDAR, Radar, and Ultrasound). And it has to provide real-time feedback, such as traffic conditions, events, weather conditions, road signs, traffic signals, and others [80–82]. This complicated functionality requires high trillions of operations per second (TOPS) to process multiple challenging tasks simultaneously (e.g., object extraction, detection, segmentation, tracking, and more). It also consumes high power depending on the operation. Lastly, high-speed processing, reliability, and accuracy are essential and must be better than humans.

Graphics Processing Unit (GPU) The NVIDIA Drive platform [83] was introduced in 2015 for the first time. It is aimed at providing deep learning-based functionality to vehicles and has been a market-leading company in GPU-based embedded systems for autonomous driving. The first chip Drive CX is based on a

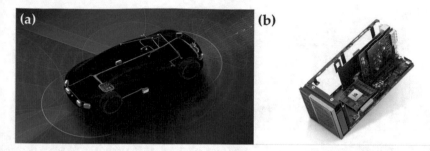

Fig. 24 (a) NVIDIA Drive Hyperion autonomous vehicle development platform. (b) NVIDIA Drive AGX Orin [84]

single Maxwell GPU architecture Tegra X1 SoC. This chip was marked as a digital cockpit computer due to its capability. Drive PX includes two Tegra X1 SoCs, and it is targeted at the initial development platform for autonomous driving. In 2016, NVIDIA released Drive PX 2, composed of Pascal architecture-based GPUs and Tegra X2 SoC processors for neural net processing to enable various AI algorithms. It is used as the vehicle computation unit (VCU) capable of installing multiple sensors, including cameras, radars, one LiDAR, GNSS, and DSRC antennas, to collect data of 360 degrees around vehicles. The first Volta GPU-based solution, Drive PX Xavier, delivers 30 TOPS of computation while consuming 30 W of power. Then Drive PX Pegasus stated its capability for L4+ autonomous driving with 320 TOPS computational performance and 500 W power consumption. The GPU specialist has been incorporating ASICs on the latest Ampere GPU-based system, where Drive AGX Orin delivers 254 TOPS through a combination of CPU, GPU, image signal processor (ISP), and accelerators such as deep learning accelerator (DLA) and programmable vision accelerator (PVA). Likewise, such computing capabilities of GPU-based systems are being leveraged into autonomous driving tasks and driver monitoring and visualization, AI cockpits, and other various tasks in AVs. However, the higher power consumption of GPUs is still a key challenge for aiming for an energy-efficient high level of AVs (Fig. 24).

Field-Programmable Gate Array (FPGA) While CPUs or GPUs are purposed to run general AI algorithms, FPGA is designed to be programmable and run specific algorithms in real-time. FPGA executes targeted tasks that require less computational resources, faster processing speed, and lower power consumption, so it becomes an appropriate option for building efficient and accelerating AV systems. Also, FPGA's scalability allows minimum changes for new updates since it is composed of configurable logic blocks such as digital signal processing (DSP). FPGA blocks can be connected using its spatial fabric, which enables the scale of optimized computing resources while GPU cores communicate through memory systems that are not directly connected. The structural feature reduces latency and brings more communication efficiency.

Fig. 25 (**a**) HEV/EV System with Intel FPGA-based applications [85]. (**b**) Intel's AI accelerators for IVE (in-vehicle experience) applications [87]

Intel's FPGAs (Agilex, Startix, Arria, Max, Cyclone) provide a solution that can aggregate large amounts of data from AV's multiple sensors and convert them into a unified format (e.g., MIPI CSI-2) for essential computing tasks in ADAS [85]. The re-programmable/re-configurable processor can be used to build custom security functions with secure boot, cryptographic acceleration, and other features because FPGA is able to manage secure data handling policy by its flexibility of adapting to changing requirements. Meanwhile, demand for IVE (in-vehicle experience) is increasing to interact with their vehicle's infotainment system as driving tasks become gradually automated. FPGAs are being leveraged into IVE applications such as HUD (head-up display), driver monitoring system (DMS), blind spot detection, gesture and voice recognition, and so on. Moreover, electric vehicle systems apply FPGAs to improve efficiency and reliability. Here, electrified vehicles need a DSP for control and maintenance functions such as an onboard charger, traction inverter, DC/DC converter, motor control, and battery management systems for improved performance, cost-efficiency, functional safety, etc. Intel claims that their AI soft processor called the NPU (neural processing unit), whose architecture is flexibly designed with one or multiple connected FPGAs (Intel Stratix), has more than ten times higher average performance than GPUs (NVIDIA T4 and V100). Xilinx's Zynq [86] FPGA accelerator, designed for AVs, includes multiple ARMs and accelerates CNN inference with nested-loop algorithms that deliver 14 fps/watt when running CNN tasks which outperforms Tesla K40 GPU's 4 fps/watt. Also, the solution can process in real-time that reaches 60 fps in a 1080p video stream for object tracking tasks. Like this, we need to design neural network accelerators on FPGA for AVs to take advantage of high performance and efficiency (Fig. 25).

Application-Specific Integrated Circuits (ASIC) ASICs have high flexibility in hardware implementation and can cope with specific requirements for achieving complex performance for specific AV algorithms. Recently, many manufacturers in the automotive industry are building their own custom ASICs that provide the highest cost efficiency and performance against an off-the-shelf component. In 2019, Tesla [88] developed their full self-driving (FSD) ASIC-based computer that runs on their automotive and aims for a fully autonomous driving system, L5. The company pursues camera-based autonomous driving; the FSD computer is adapted to vision-based processing.

The architecture includes integrated camera interfaces with 2.5G pixels per second serial input to accommodate multiple camera sensors surrounding the car and 128-bit wide LPDDR4 DRAM for other sensors like radars. An independent 24-bit ISP (image signal processor) enriches image data by applying noise reduction or tone mapping, and H.265 video encoder exports data. 12 ARM A72 CPUs run several light tasks at 2.2 GHz, and GPU supporting FP32 and FP16 achieves 600 GFLOPS at 1GHz for data processing. The FSD computer has the primary purpose as neural network accelerator (NNA), so two independent NNAs and power supplies are equipped to each device to minimize functional failure and ensure the safety of the AVs. Each NNA contains SRAM to store model parameters and temporary results, which delivers computation power of 144 TOPS. It can analyze input images at 1050 fps while consuming 15 W to 72 W. AI-specific designed ASICs show more performance, power consumption, and cost efficiencies and outperform GPUs and FPGAs to be more integrated into AVs for computing platforms these days. Similar to FPGAs, ASICs are being applied to various ADAS, IVE, and other applications in AVs, including DMS (driver monitoring system), TPMS (tire pressure monitoring systems), and BMS (battery management systems) to enhance AVs.

In conclusion, AV is a massive system with critical hardware and software requirements to ensure safety. AVs will face diverse environments that are never to be the same and, thus, must have robust hardware accelerators considering the capacity of the vehicles. Currently, most existing AV tasks rely on CPUs, GPUs, FPGAs, ASICs, or generic processors. The industrial trends show GPUs, FPGAs and ASICs are booming, but some constraints need to be resolved, such as cost inefficiency or time consumption on redesigning to be taken into safe functioning AVs. Furthermore, evolving AI technology would vary options for designing hardware accelerators to build optimized system architecture in AVs to leverage enhanced hardware and software performance to advance to L5 and beyond.

3.3 Commercialized Products

This part introduces representative products of companies designing and planning chips for autonomous vehicles such as Tesla, Mobileye, Nvidia, etc.

Tesla FSD (Full Self Driving) Hardware System The main feature of FSD computer is the configuration of two FSD chips, in which the power and control devices are completely independent systems and are composed of a stable system [88]. Additional parts include CPUs, ISP, GPU and video encoder, eight camera connectors, self-powered subsystem, DRAM, and flash memory for various preprocessing and post-processing needs. All data from sensors such as eight cameras, radar, GPS, ultrasonic sensor, wheel scale, steering angle, and map data

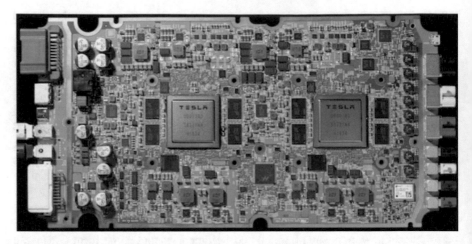

Fig. 26 FSD computer with two Tesla FSD chips in dual configurations including sensors like cameras [88]

are input/output through FSD computer, and 72 TOPS hardware accelerators for neural inference are provided (Fig. 26).

FSD chips are System-on-a-Chip (SoC) and autonomous driving chips designed by Tesla and introduced for its own vehicles in 2019. Made by Samsung's 14 nm process technology, the chip is designed around two NNAs each with 32 MB SRAM and a 96 × 96 MAC array with a performance of 36 TOPS at 2GHz. At the same time, the rest of the system consists of a cluster of three quad-core Cortex-A72 cores operating at 2.2GHz used for general-purpose processing. There are 12 64-bit ARM cores. There is also a Mali G71 MP12 GPU with 1GHz performance, which is mainly responsible for relatively light post-processing (Fig. 27).

In FSD SoCs, every few milliseconds, a new input frame is input through a dedicated image signal processor, pre-processed and stored in DRAM. When a new frame becomes available in main memory, the CPU instructs the NNA to start processing the frame. When the data and parameters of the processed frame are finished passing through SRAM and back to DRAM, it triggers the CPU complex again to generate an interrupt. The GPU is responsible for the post-processing of algorithms outside the scope of the NNA (Fig. 28).

The convolution loop has been refactored to focus on parallel processing of MAC operations while minimizing power and maximizing computational bandwidth. The convolution loop has been refactored to focus on parallel processing of MAC operations while minimizing power and maximizing computational bandwidth. Concatenating multiple images in parallel is unsuitable for stability reasons in driving situations as the images must start processing as soon as they arrive. They do a few other optimizations here. This chip can merge the output pixels in the X and Y dimensions of multiple output channels in parallel within the output channel.

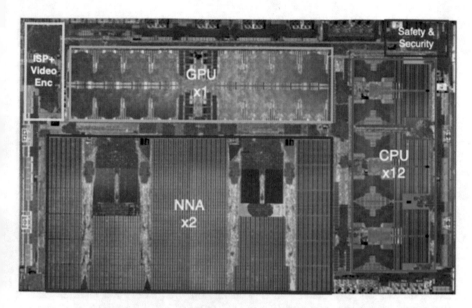

Fig. 27 FSD chip die photo with major blocks [88]

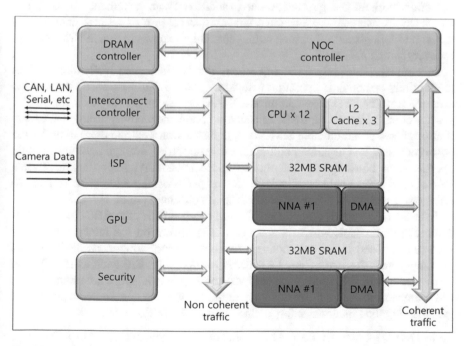

Fig. 28 SoC block diagram [88]

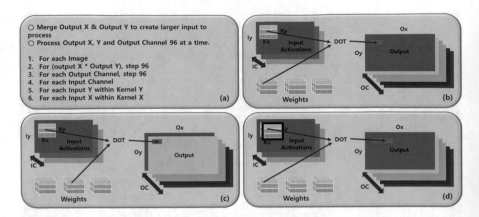

Fig. 29 Convolution refactoring and dataflow [88]

This process parallelizes the work and optimizes 96 pixels to work concurrently. All input weights are shared when working on all pixels in a channel, and it also swaps output channels and input channel loops. This parallelized computing capacity allows you to work sequentially on all output channels to share all input activations without losing resources due to additional data movement. To parallelize computations, we can use refactored convolution ropes to maximize data sharing across multiple pixels within each output channel, as shown in Fig. 29. This is done for every output channel.

In summary, the Tesla FSD chip adopts the own-designed NNA optimized in the multiply-accumulate operation instead of general multi-purpose GPUs as AI accelerators. The FSD chip minimizes DRAM access by sharing data for the dot product computation through SRAM caches to break the memory access bottleneck. It also utilizes programmable SIMD (single instruction multiple data) units for the post-processing and separate pooling unit for supporting average and max pooling operations to boost up the inferencing speed eventually. According to Tesla's presentation, FSD could handle 2300 frames per second, which is more than 20 times faster than the previous version of the self-driving system, HW 2.5.

In terms of standards compliance, the FSD chip satisfies the AEC-Q100 Grade-2 reliability standard, which is a stress test qualification for integrated circuits such as SoC used in the automotive environment. Furthermore, its design seems to comply with the international standards for vehicular E/E such as ISO 26262. Two independent FSD chips powered by separate power supplies are mounted on a FSD computer to secure the redundancy of the self-driving system in the purpose of guaranteeing the functional safety defined in the ISO 26262.

Tesla FSD D1 Chip At Tesla AI Day, the company announced its new D1 Chip, custom processors based on 7 nm process technology with 50 billion transistors [89, 90]. This chip has an area of 645 mm^2, smaller than both NVIDIA A100 (826 mm^2) and AMD Arcturus (750 mm^2). Specifications-wise, the chip has 354 training nodes

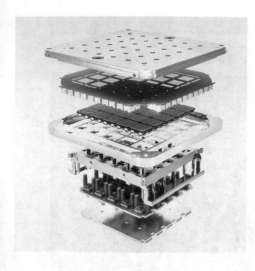

362 TFLOPs BF16/CFP8
22.6 TFLOPS FP32

10TBps/dir. On-Chip Bandwidth
4TBps/edge. Off-Chip Bandwidth

400W TDP

645mm²
7nm Technology
50 Billion
Transistors
11+ Miles
Of wires

Fig. 30 Tesla D1 Chip [89, 90]

based on a 64-bit superscalar CPU with four cores. Those are explicitly designed for
8×8 multiplications and support a wide range of instructions used for AI training,
including FP32, BFP16, CFP8, INT32, INT16, and INT8. According to Tesla, their
D1 chip offers 22.6 FLOPS of single-precision compute performance (FP32) and
up to 362 TFLOPS in BF16/CFP8. This performance is achieved within a TDP
of 400 W for a single D1 chip. For AI training, scalability is an important aspect,
which is why Tesla came up with high-bandwidth interconnects (low latency switch
fabric) with up to 10 TB/s. The I/O ring around the chip has 576 lanes, each offering
112 Gbit/s of bandwidth (Fig. 30).

NVIDIA Jetson AGX Orin The Jetson AGX Orin series can deliver up to
275 TOPS (INT8) of AI performance and includes 64 GB and 32 GB modules
[91]. The modules feature the NVIDIA Orin SoC integrated with an NVIDIA
Ampere architecture GPU, 12-core Arm® Cortex®-A78AE CPU, next-generation
deep learning (NVDLA v2.0) and vision accelerators (PVA v2.0), and a video
encoder and decoder while requiring 15 W to 60 W of power. Multiple concurrent
AI application pipelines are enabled with high-speed IO, 204 GB/s of memory
bandwidth, and 32 GB/64 GB of DRAM (Fig. 31).

An Ampere GPU is composed of 2 Graphics Processing Clusters (GPCs), up
to 8 Texture Processing Clusters (TPCs), up to 16 Streaming Multiprocessors
(SMs), 192 KB of L1-cache per SM, and 4 MB of L2 Cache. Jetson AGX Orin
64 GB contains 2048 CUDA cores and 64 Tensors with up to 170 Sparse TOPS
of INT8 Tensor compute, while Jetson AGX Orin 32 GB includes 1792 CUDA
cores and 56 Tensor cores with up to 108 Sparse TOPS of INT8 Tensor compute.
NVIDIA enhanced the Tensor cores in the Ampere GPU to support sparsity—
a fine-grained compute structure that can double throughput and reduce memory

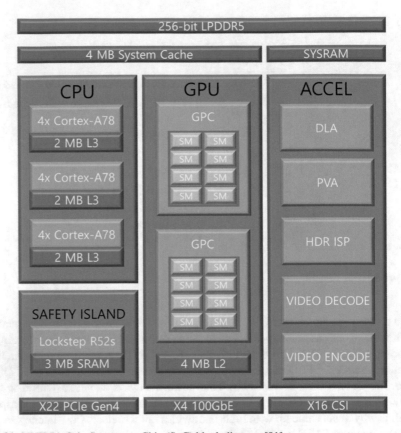

Fig. 31 NVIDIA Orin System-on-Chip (SoC) block diagram [91]

usage. The Tensor cores with programmable fused matrix-multiply-and-accumulate units execute concurrently alongside the CUDA cores. It implements floating-point HMMA (half-precision matrix multiply and accumulate) instructions and IMMA (integer matrix multiply and accumulate) instructions to accelerate dense linear algebra computations, signal processing, and deep learning inference (Fig. 32).

The NVIDIA Deep Learning Accelerator (NVDLA) architecture is designed to be scalable and highly simplified configuration, integration, and portability to accelerate deep learning inference operations [92]. NVDLA is composed of a convolution core, single data processor (SDP), planar data processor (PDP), channel data processor (CDP), data reshape, and dedicated memory engines. Convolution core supports sparse weight compression to save memory bandwidth, built-in Winograd convolution to improve efficiency on computing certain sizes of filters, and batch convolution to save additional memory bandwidth by reusing weights when running multiple inference operations in parallel. Convolution buffer, an internal RAM in the core, is for weight and input feature storage to avoid repeated access to system memory and improves memory efficiency. SDP is a single-point

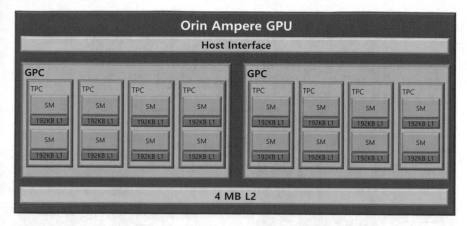

Fig. 32 NVIDIA Jetson AGX Orin 64GB's Ampere GPU block diagram [91]

lookup engine for activation functions that has a lookup table to implement the application of linear and nonlinear functions. It can support most common activation functions, element-wise operations, and nonlinear functions such as ReLU, PReLU, precision scaling, batch normalization, bias addition, and a sigmoid tangent. PDP is a planar averaging engine for pooling that provides specific spatial operations such as maximum pooling, minimum pooling, and average pooling. CDP is a multi-channel averaging engine for advanced normalization functions that operate on channel dimensions. Data reshape engine performs data format transformation such as splitting, slicing, merging, contraction, or reshape-transpose for performing inference on a convolutional network. The bridge DMA (BDMA) module provides copy operations between the system DRAM and the dedicated high-performance memory interface. Each block is separate and configurable, so a system can scale up the performance of a unit (e.g., convolution) without modifying other units in the accelerator.

Three major connections are implemented to the NVDLA: configuration space bus (CSB), interrupt, and data backbone (DBB) interfaces. CSB interface is a synchronous, low-bandwidth, low-power, 32-bit control bus used by a CPU to access the NVDLA configuration registers. An interrupt interface is asserted when a task is completed, or an error occurs. DBB interface is a synchronous, high-speed, and highly configurable data bus that connects NVDLA and the main system memory subsystems (Fig. 33).

Intel Mobileye EyeQ Intel Mobileye first announced EyeQ5 in 2016 and started mass production in 2021. EyeQ5 Mid, produced through a 7 nm process, offers 4.6 DL TOPS (int8) performance supporting a front autonomous driving system camera of the vehicle. With similar architecture to EyeQ5 Mid, EyeQ5 High offers 16 DL TOPS (int8) computing power covering the vehicle's surround cameras [93]. By supplying the EyeQ series, Intel Mobileye has maintained partnerships with automotive manufacturers such as Volkswagen, BMW, Hyundai, Nissan, and

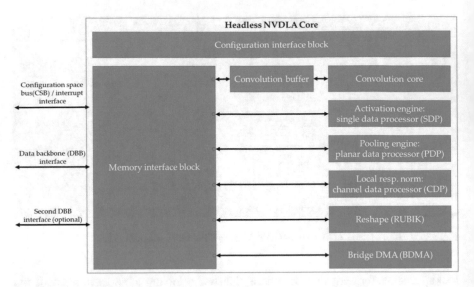

Fig. 33 Internal architecture of NVDLA core [92]

Volvo [94]. At CES 2022, Intel Mobileye introduced their new future line-ups, EyeQ6 and EyeQ Ultra [95]. EyeQ6 Light contains a CPU with two cores and eight threads, an LPDDR4 channel, general compute accelerators cluster, and a DL accelerators cluster. The general compute accelerators cluster has 1 PMA6 (programmable macro array), 2 VMP6 (vector microcode processor), as well as 1 MPC6 (multithreaded processing cluster), and the DL accelerators cluster has 1 XNN6 (deep learning accelerator). EyeQ6 Light will be produced in 2023 and replace the EyeQ4 Mid to cover Level 1 and 2 autonomous driving systems. Besides, EyeQ6 High offers 34 DL TOPS (int8) computing power to support the vehicle's surround cameras. A CPU with 8 cores and 32 threads, LPDDR5 channel, 2 general compute accelerators clusters with 1 PMA6, 2 VMP6, 2 MPC6, and 2 DL accelerators clusters with 2 XNN6 are mounted on the EyeQ6 High. EyeQ Ultra is the high-end model of the Mobileye EyeQ line-up. EyeQ Ultra could support Level 4~5 AV systems, including surround cameras, LiDARs, and radars. It will offer 176 DL TOPS (int8) provided by a 12-core 24-thread CPU, LPDDR5 channel, 2 general compute accelerators clusters with 4 PMA7, 8 VMP7, and 12 MPC7, as well as 2 DL accelerators clusters with 8 XNN7 [96, 97] (Fig. 34).

Qualcomm Snapdragon Ride Platform The Snapdragon Ride platform consists of hardware, software, open stacks, and development kits [98]. The hardware platform has a combination of SoC and accelerators. The software platform provides safe middleware, operating systems, and drivers. The Snapdragon Ride hardware platform supports a single/multiple safety system-on-chip (SOC) or safety SoC that combines safety accelerators to meet the different levels of autonomous driving requirements specified by the SAE. A single Snapdragon Ride SoC supports SAE Level 2–3 solutions, enabling autonomous highway driving up to 30 TOPS,

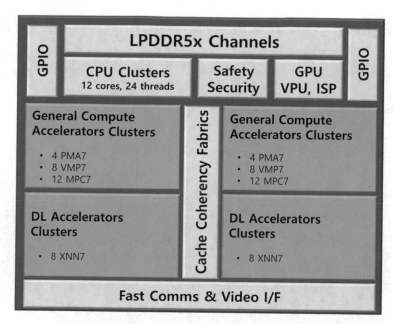

Fig. 34 Intel Mobileye EyeQ Ultra Floorplan [96]

requiring only passive cooling in a small form factor. In addition, the multi-SoC solution that combines autonomous driving system SoC and autonomous driving accelerator can implement autonomous driving solutions (ADS) for level 4 to level robot taxis. This ADS solution can be achieved with a relatively low power consumption level (Fig. 35).

Samsung Exynos Auto Samsung Exynos Auto T5123 communication chip and Samsung Exynos Auto V7 processor are integrated into Volkswagen ICAS 3.1 infotainment system [99]. The Exynos Auto T5123 is a 3GPP Release 15 telematics control device designed to provide fast and seamless 5G connectivity in standalone (SA) and non-standalone (NSA) modes for next-generation connected vehicles. It is a hardware accelerator for autonomous vehicles, delivering critical information to the vehicle in real-time at up to 5.1 Gbps, allowing passengers to access new services such as high-definition content streaming and video calls. The V7 processor integrates eight 1.5 GHz Cortex-A76 CPU cores and 11 G76 GPU cores, separated into three cores for cluster display and AR-HUD and eight "big" domains for the Central Information Display (CID). This separation allows the GPU to support multiple systems simultaneously and ensures safer operation by preventing one domain from interfering with another. The Exynos Auto V7 provides robust data protection with an isolated security processor for encryption operations and hardware keys using one-time programmable (OTP) or physical unclonable function (PUF). In terms of functional safety, the Exynos Auto V7 meets automotive safety integrity level B (ASIL B) requirements, including built-in safety islands and safety

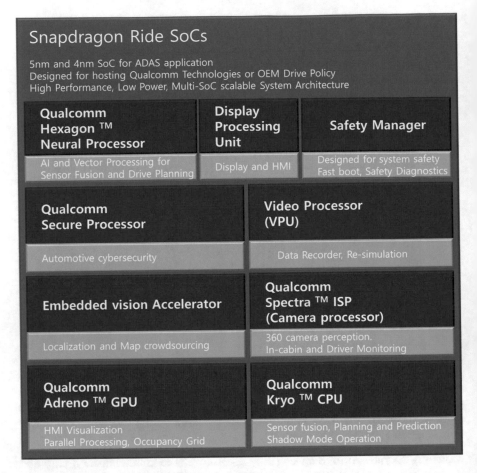

Fig. 35 Qualcomm Snapdragon Ride platform [98]

support for digital clusters that detect and manage faults to maintain safety through fault management units (FMUs) (Fig. 36).

NXP S32S The NXPS 32S processor uses the new Arm Cortex-R52 core family, which incorporates all Arm processors' highest level of safety [100]. This suite provides four independent processing paths at the ASIL D to support secure parallel computing. The S32S architecture also supports new "failure availability" features, allowing devices to detect errors and continue to operate after isolation. This feature is critical for future self-driving applications. NXP has partnered with OpenSynergy to develop a fully functional, real-time hypervisor that supports NXPS32S products.

Open Synergy's COQOS micro SDK is one of the first hypervisor platforms to leverage ArmCortext-R52's exceptional hardware capabilities, allowing multiple real-time operating systems to be integrated into microcontrollers that require up to ISO26262 ASIL D levels of security. Multi-OS and stack-independent

Fig. 36 Samsung Exynos Auto V7 [99]

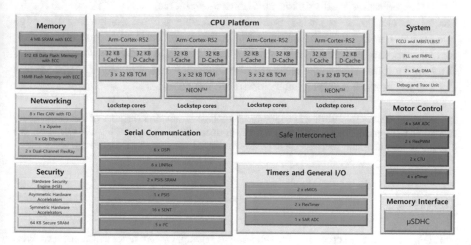

Fig. 37 NXP S32S processor block diagram [100]

COQOS micro SDKs provide secure and faster context switching than traditional microcontroller-only solutions. NXP's Companion ASIL D safety system-based chip (FS66 Functional Safety Multi-Output Power Supply IC) provides a comprehensive safety solution and high-capacity integrated flash memory (up to 64 M bytes) for real-time wireless updates without processor downtime. Advanced electric motor-controlled peripherals, including hardware security engines that support private and public keys, and PCIe and motor control software libraries for use in autonomous driving system domain monitoring applications, are available (Fig. 37).

4 Conclusion

In this chapter, we have discussed AI accelerators for autonomous vehicles in terms of prerequisites and recent trends in the industry. As autonomous driving systems are developed with advanced technologies, including deep learning-based methods, the demand for the enhanced capability of integrated hardware is increasing to supply higher computing power with better power and cost efficiencies. At the same time, the demand for E/E that guarantees proper functionality and the safety of AVs in failure-free without driver's monitoring is also growing. The AI accelerators that could meet these demands are drawing lots of attention from the automotive industry. GPUs, FPGAs, and ASICs are recent industry trends that have capabilities to execute complex computation tasks leveraging limited resources within vehicles to achieve prerequisites to ensure their proposed functionalities and safety.

Furthermore, to be free from product liability, standards for safety should comply in the development process of the AI accelerator. There are already plenty of commercialized AI accelerators in the auto market after NVIDIA released the first AI accelerator platform for autonomous vehicles, Drive PX2, in 2016. As the AI accelerator is regarded as an indispensable part of completing the edge computing–based fully autonomous vehicle, it is clear that it will constantly evolve, and its market will continue to grow. As a core technical huddle to prepare Level 5 fully autonomous vehicles, which will dominate as future mobility, AI accelerators deserve to get more and more attention in the future.

Acknowledgments This work was supported by Korea Evaluation Institute of Industrial Technology (KEIT) grant funded by the Korea government (MOTIE) (No.RS-2022-00154651, 3D semantic camera module development capable of material and property recognition).

References

1. Liu, S., et al.: Edge computing for autonomous driving: Opportunities and challenges. Proc. IEEE. **107**(8), 1697–1716 (2019)
2. Yurtsever, E., et al.: A survey of autonomous driving: Common practices and emerging technologies. IEEE Access. **8**, 58443–58469 (2020)
3. Seamless Transportation Lab. Yonsei University: Research: Intelligent Vehicles. Available online: https://sites.google.com/site/shihoyonsei/research. Accessed 3 Jan 2022
4. Li, Y., Ibanez-Guzman, J.: Lidar for autonomous driving: The principles, challenges, and trends for automotive lidar and perception systems. IEEE Signal Process. Magaz. **37**, 50–61 (2020)
5. GSA Global: Autonomous Driving and Sensor Fusion SoCs. Available online: https://www.gsaglobal.org/forums/autonomous-driving-and-sensor-fusion-socs/. Accessed 2 Jan 2022
6. Qin, T., Chen, T., Chen, Y., Su, Q.: Avp-slam: Semantic visual mapping and localization for autonomous vehicles in the parking lot. In: International Conference on Intelligent Robots and Systems (IROS) (2020), pp. 5939–5945
7. Nobis, F., Papanikolaou, O., Betz, J., Lienkamp, M.: Persistent map saving for visual localization for autonomous vehicles: An orb-slam 2 extension. In: International Conference on Ecological Vehicles and Renewable Energies (EVER) (2020), pp. 1–9
8. Lin, J., Zhang, F.: Loam livox: A fast, robust, high-precision LiDAR odometry and mapping package for LiDARs of small FoV. In: International Conference on Robotics and Automation (ICRA) (2020), pp. 3126–3131

9. Girshick, R., Donahue, J., Darrell, T., Malik, J.: Rich feature hierarchies for accurate object detection and semantic segmentation. In: Proceedings of the IEEE/CVF Conference on Computer Vision and Pattern Recognition (CVPR) (2014), pp. 580–587
10. Girshick, R.: Fast r-cnn. In: Proceedings of the IEEE/CVF Conference on Computer Vision and Pattern Recognition (CVPR) (2015), pp. 1440–1448
11. Redmon, J., Divvala, S., Girshick, R., Farhadi, A.: You only look once: Unified, real-time object detection. In: Proceedings of the IEEE/CVF Conference on Computer Vision and Pattern Recognition (CVPR) (2016), pp. 779–788
12. He, K., Zhang, X., Ren, S., Sun, J.: Deep residual learning for image recognition. In: Proceedings of the IEEE/CVF Conference on Computer Vision and Pattern Recognition (CVPR) (2016), pp. 770–778
13. He, K., Gkioxari, G., Dollár, P., Girshick, R.: Mask r-cnn. In: Proceedings of the IEEE/CVF Conference on Computer Vision and Pattern Recognition (CVPR) (2017), pp. 2961–2969
14. Zhou, Y., Tuzel, O.: Voxelnet: End-to-end learning for point cloud based 3d object detection. In: Proceedings of the IEEE/CVF Conference on Computer Vision and Pattern Recognition (CVPR) (2018), pp. 4490–4499
15. Lang, A.H., Vora, S., Caesar, H., Zhou, L., Yang, J., Beijbom, O.: Pointpillars: Fast encoders for object detection from point clouds. In: Proceedings of the IEEE/CVF Conference on Computer Vision and Pattern Recognition (CVPR) (2019), pp. 12697–12705
16. Zheng, W., Tang, W., Jiang, L., Fu, C.-W.: SE-SSD: Self-ensembling single-stage object detector from point cloud. In: Proceedings of the IEEE/CVF Conference on Computer Vision and Pattern Recognition (CVPR) (2021), pp. 14494–14503
17. Xu, Q., Zhou, Y., Wang, W., Qi, C.R., Anguelov, D.: Spg: Unsupervised domain adaptation for 3d object detection via semantic point generation. In: Proceedings of the IEEE/CVF International Conference on Computer Vision (ICCV) (2021), pp. 15446–15456
18. Mao, J., Niu, M., Bai, H., Liang, X., Xu, H., Xu, C.: Pyramid r-cnn: Towards better performance and adaptability for 3d object detection. In: Proceedings of the IEEE/CVF International Conference on Computer Vision (ICCV) (2021), pp. 2723–2732
19. Yurtsever, E., Lambert, J., Carballo, A., Takeda, K.: A survey of autonomous driving: Common practices and emerging technologies. IEEE Access. **8**, 58443–58469 (2020)
20. Liu, Z., Lin, Y., Cao, Y., Hu, H., Wei, Y., Zhang, Z., Lin, S., Guo, B.: Swin transformer: Hierarchical vision transformer using shifted windows. In: Proceedings of the IEEE/CVF International Conference on Computer Vision (ICCV) (2021), pp. 10012–10022
21. Kirillov, A., He, K., Girshick, R., Rother, C., Dollár, P.: Panoptic segmentation. In: Proceedings of the IEEE/CVF Conference on Computer Vision and Pattern Recognition (CVPR) (2019), pp. 9404–9413
22. Milan, A., Leal-Taixé, L., Reid, I., Roth, S., Schindler, K.: MOT16: A benchmark for multi-object tracking. arXiv preprint arXiv:1603.00831 (2016)
23. Hu, H.-N., Cai, Q.-Z., Wang, D., Lin, J., Sun, M., Krahenbuhl, P., Darrell, T., Yu, F.: Joint monocular 3D vehicle detection and tracking. In: Proceedings of the IEEE/CVF International Conference on Computer Vision (ICCV) (2019), pp. 5390–5399
24. Zhou, X., Koltun, V., Krähenbühl, P.: Tracking objects as points. In: European Conference on Computer Vision (ECCV) (2020), pp. 474–490
25. Navale, V.M., Williams, K., Lagospiris, A., Schaffert, M., Schweiker, M.-A.: (R)evolution of E/E architectures. SAE Int. J. Passeng. Cars-Electron. Electr. Syst. **8**(2), 282–288 (2015)
26. International Organization for Standardization (ISO): ISO 26262 - Road vehicles - Functional safety (2018)
27. McKinsey and Company: Automotive software and electrical/electronic architecture: Implications for OEMs. Available online: https://www.mckinsey.com/ industries/automotive-and-assembly/our-insights/automotive-software-and-electrical-electronic-architecture-implications-for-oems. Accessed 31 Jan 2022
28. Brunner, S., Roder, J., Kucera, M., Waas, T.: Automotive E/E-architecture enhancements by usage of ethernet TSN. In: 2017 IEEE 13th Workshop on Intelligent Solutions in Embedded Systems (WISES) (2017), pp. 9–13

29. Olmedo, I.S., Capodieci, N., Cavicchioli, R.: A perspective on safety and real-time issues for gpu accelerated adas. In: IECON 2018-44th Annual Conference of the IEEE Industrial Electronics Society (2018), pp. 4071–4077
30. Bandur, V., Selim, G., Pantelic, V., Lawford, M.: Making the case for centralized automotive E/E architectures. IEEE Trans. Vehic. Technol. **70**(2), 1230–1245 (2021)
31. ReportLinker: Intelligent Vehicle E/E Architecture and Computing Platform Industry Research Report, 2021. Available online: https://www.globenewswire.com/news-release/2021/09/01/2289809/0/en/Intelligent-Vehicle-E-E-Architecture-and-Computing-Platform-Industry-Research-Report-2021.html. Accessed 10 Feb 2022
32. AUTOSAR: AUTOSAR: Enabling continuous innovations. Available online: https://www.autosar.org/. Accessed 25 Jan 2022
33. SAE International: J3016: Taxonomy and Definitions for Terms Related to Driving Automation Systems for On-Road Motor Vehicles. Available online: https://www.sae.org/standards/content/j3016_202104/. Accessed 14 Feb 2022
34. Motor Industry Software Reliability Association (MISRA): Development Guidelines for Vehicle Based Software (1994)
35. Motor Industry Software Reliability Association (MISRA): A brief history of MISRA. Available online: https://www.misra.org.uk/a-brief-history-of-misra/. Accessed 16 May 2022
36. Ward, D.D., Kendall, I.R.: Automotive Software Engineering Using the MISRA Guidelines. SAE Trans. **109**, Section 7: Journal of Passenger Cars: Electronic and Electrical Systems (2000), 257–265
37. Ward, D.D.: MISRA standards for automotive software. In: 2nd IEEE Conference on Automotive Electronics (2006), pp. 5–18
38. International Electrotechnical Commission (IEC): International Electrotechnical Commission - Who we are - National Committees. Available online: https://www.iec.ch/nationalcommittees#nclist. Accessed 16 May 2022
39. International Electrotechnical Commission (IEC): IEC 61508 - Functional Safety of Electrical/Electronic/Programmable Electronic(E/E/PE) Safety-related Systems (2010)
40. Charlwood, M., Turner, S., Worsell, N.: UK Health and Safety Executive Research Report 216 - A methodology for the assignment of safety integrity levels (SILs) to safety-related control functions implemented by safety-related electrical, electronic and programmable electronic control systems of machines (2010). Available online: https://www.hse.gov.uk/research/rrhtm/rr216.htm
41. General Electric (GE): About SIS Management - SIL Assessment - Risk Reduction Factor. Available online: https://www.ge.com/digital/documentation/meridium/Help/V43050/Default/Subsystems/SISManagement/SISManagement.htm#RiskReductionFactor.htm
42. Marszal, E.M., Scharpf, E.W.: Safety Integrity Level Selection – Systematic Methods Including Layer of Protection Analysis (2002)
43. Society of Automotive Engineers International (SAE International): Application of ISO 26262 in Distributed Development ISO 26262 in Reality (2009)
44. Society of Automotive Engineers International (SAE International): SAE J2980 - Considerations for ISO 26262 ASIL Hazard Classification (2018)
45. Van Eikema Hommes, Q.: Assessment of the ISO 26262 Standard. Road Vehicles-Functional Safety (2012). Available online: https://www.volpe.dot.gov/infrastructure-systems-and-technology/advanced-vehicle-technology/assessment-iso-26262-standard. Accessed 16 May 2022
46. NVIDIA: NVIDIA CES 2016 Press Conference (2016). Available online: https://www.slideshare.net/NVIDIA/nvidia-ces-2016-press-conference. Accessed 16 May 2022
47. Greb, Karl and Seely, Anthony, "Design of Microcontrollers for Safety Critical Operation" (2009), Available online: https://web.archive.org/web/20150906100246/http://www.ti.com/ww/en/mcu/tms570/downloads/Design_of_Microcontrollers_for_Safety.pdf. Accessed on 16 May 2022

48. Freescale Semiconductor: Functional Safety and Safety Standards: Challenges and Comparison of Solutions (2007). Available online: https://www.nxp.com/files-static/training_presentation/TP_AUTO_FUNCT_SAFETY.pdf. Accessed 16 May 2022
49. International Electrotechnical Commission (IEC): IEC 60812 - Failure modes and effects analysis (FMEA and FMECA) (2018)
50. Carlson, C.S.: Understanding and Applying the Fundamentals of FMEAs. Annu. Reliab. Maintainab. Symp. **10**, 1–35 (2014)
51. International Electrotechnical Commission (IEC): IEC 61709 - Electric components - Reliability - Reference conditions for failure rates and stress models for conversion (2017)
52. Rausand, M., Barros, A., Hoyland, A.: System Reliability Theory: Models, Statistical Methods, and Applications, 3rd edn. Wiley Online Library (2020)
53. National Instruments: What is the ISO 26262 Functional Safety Standard? Available online: https://www.ni.com/en-us/innovations/white-papers/11/what-is-the-iso-26262-functional-safety-standard-.html#toc2. Accessed 16 May 2022
54. Palin, R., Ward, D., Habli, I., Rivett, R.: ISO 26262 safety cases: Compliance and assurance. IET (2011)
55. Salay, R., Queiroz, R., Czarnecki, K.: An analysis of ISO 26262: Using machine learning safely in automotive software. arXiv preprint arXiv:1709.02435 (2017)
56. Jeon, S.-H., Cho, J.-H., Jung, Y., Park, S., Han, T.-M.: Automotive hardware development according to ISO 26262. In: 13th international conference on advanced communication technology (ICACT) (2011), pp. 588–592
57. International Organization for Standardization (ISO): ISO/DIS 21448 - Road vehicles - Safety of the intended functionality (2021)
58. American National Standards Institute (ANSI) and Underwriters Laboratories (UL): ANSI/UL 4600 - Evaluation of Autonomous Products (2022)
59. Edge Case Research: An Overview of Draft UL 4600: Standard for Safety for the Evaluation of Autonomous Products (2019). Available online: https://edgecaseresearch.medium.com/an-overview-of-draft-ul-4600-standard-for-safety-for-the-evaluation-of-autonomous-products-a50083762591. Accessed 16 May 2022
60. Koopman, P., Ferrell, U., Fratrik, F., Wagner, M.: A safety standard approach for fully autonomous vehicles. In: International Conference on Computer Safety, Reliability, and Security (2019), pp. 326–332
61. Helmig, E.: ISO 26262 - Functional Safety in Personal Vehicles: Responsibilities and Liabilities of Functional Safety Managers (2021). Available online: https://www.ra-helmig.de/fileadmin/docs/publikationen/ISO_26262_Liability_Functional_Safety_Managers.pdf. Accessed 16 May 2022
62. ROHM Semiconductor: ISO 26262: Functional Safety Standard for Modern Road Vehicles. Available online: https://www.rohm.com/electronics-basics/standard/iso26262. Accessed 16 May 2022
63. Freescale Semiconductor: Addressing the Challenges of Functional Safety in the Automotive and Industrial Markets. Available online: https://www.nxp.com/docs/en/white-paper/FCTNLSFTYWP.pdf. Accessed 16 May 2022
64. Society of Automotive Engineers International (SAE International): SAE J3061 - Cybersecurity Guidebook for Cyber-Physical Vehicle Systems (2021)
65. International Organization for Standardization (ISO) and Society of Automotive Engineers International (SAE International): ISO/SAE 21434 - Road vehicles — Cybersecurity engineering (2021)
66. Society of Automotive Engineers International (SAE International): SAE J3101 - Hardware Protected Security for Ground Vehicles (2020)
67. United Nations Economic Commission for Europe (UNECE): WP.29 – Introduction. Available online: https://unece.org/wp29-introduction. Accessed 16 May 2022

68. United Nations (UN): UN Regulation No 155 – Uniform provisions concerning the approval of vehicles with regards to cybersecurity and cybersecurity management system (2021). Available online: https://unece.org/sites/default/files/2021-03/R155e.pdf. Accessed 16 May 2022

69. Mercedes Benz: Computer brains and autonomous driving. How artificial intelligence makes cars fit for the future. Available online: https://group.mercedes-benz.com/innovation/case/autonomous/artificial-intelligence.html?r=dai. Accessed 18 Mar 2022

70. BMW: BMW Group integrates AI into everyday work. Available online: https://www.bmw.com/en/events/nextgen/artificial-intelligence.html. Accessed 18 Mar 2022

71. Volkswagen Data Lab: Welcome to Data: Lab Munich. Available online: https://datalab-munich.com/. Accessed 18 Mar 2022

72. Tesla: Tesla Artificial Intelligence and Autopilot. Available online: https://www.tesla.com/AI. Accessed 18 Mar 2022

73. Hyundai Motor Group: Artificial Intelligence: Neural Networks and the Future of Mankind. Available online: https://tech.hyundaimotorgroup.com/mobility-service/ai/. Accessed 18 Mar 2022

74. Bosch: Bosch Center for Artificial Intelligence. Available online: https://www.bosch-ai.com/. Accessed 18 Mar 2022

75. Continental: Continental puts its own supercomputer for vehicle AI system training, powered by NVIDIA DGX, into operation. Available online: https://www.continental.com/en/press/press-releases/continental-puts-its-own-supercomputer-into-action/. Accessed 18 Mar 2022

76. Waymo: Waymo Driver. Available online: https://waymo.com/waymo-driver/. Accessed 18 Mar 2022

77. Uber: Uber AI. Available online: https://www.uber.com/uberai/. Accessed 18 Mar 2022

78. Zoox: The Full-stack Behind Autonomous Driving. Available online: https://zoox.com/autonomy/. Accessed 18 Mar 2022

79. Edge AI+ Vision Alliance: AI Advancements Driving Autonomous Vehicle Ubiquity. Available online: https://www.edge-ai-vision.com/2021/07/ai-advancements-driving-autonomous-vehicle-ubiquity/. Accessed 18 Mar 2022

80. Jhung, J., Kim, S.: Behind-the-scenes (Bts): Wiper-occlusion canceling for advanced driver assistance systems in adverse rain environments. Sensors. **21**(23), 8081 (2021)

81. Mishra, A., Kim, J., Cha, J., Kim, D., Kim, S.: Authorized traffic controller hand gesture recognition for situation-aware autonomous driving. Sensors. **21**(23), 7914 (2021)

82. Mishra, A., Lee, S., Kim, D., Kim, S.: In-cabin monitoring system for autonomous vehicles. Sensors. **22**(12), 4360 (2022)

83. Nvidia: Solutions for Self-Driving Cars and Autonomous. Available online: https://www.nvidia.com/en-us/self-driving-cars/. Accessed 18 Mar 2022

84. Nvidia: NVIDIA DRIVE Hyperion Autonomous Vehicle Development Platform. Available online: https://developer.nvidia.com/drive/drive-hyperion/. Accessed 18 Mar 2022

85. Intel: Electronic Vehicles. Available online: https://www.intel.co.kr/content/www/kr/ko/automotive/products/programmable/electric-vehicles.html. Accessed 18 Mar 2022

86. Xilinx: ZYNQ. Available online: https://www.xilinx.com/products/silicondevices/soc/zynq-7000.html. Accessed 18 Mar 2022

87. Intel: Automotive Applications. Available online: https://www.intel.co.kr/content/www/kr/ko/automotive/products/programmable/applications.html. Accessed 18 Mar 2022

88. Talpes, E., Sarma, D.D., Venkataramanan, G., Bannon, P., McGee, B., Floering, B., Jalote, A., Hsiong, C., Arora, S., Gorti, A., Sachdev, G.S.: Compute solution for Tesla's full self-driving computer. Published in IEEE Micro **40**(2) (2020, March–April 1)

89. Tesla: Tesla AI Day. Available online: https://youtu.be/j0z4FweCy4M. Accessed 4 May 2022

90. Tesla: Tesla Dojo System. Available online: https://www.tesla.com/AI. Accessed 4 May 2022

91. Leela, S.K.: NVIDIA Jetson AGX Orin Series Technical Brief: a giant leap forward for robotics and edge AI applications. Online : https://www.nvidia.com/content/dam/en-zz/Solutions/gtcf21/jetson-orin/nvidia-jetson-agx-orin-technical-brief.pdf. Accessed 4 Mar 2022

92. Nvidia: NVIDIA Deep Learning Accelerator (NVDLA) Primer. Online: https://nvdla.org/primer.html. Accessed 4 Mar 2022
93. EETimes: Mobileye's New EyeQ5: How Open is Open? (2018). Available online: https://www.eetimes.com/mobileyes-new-eyeq5-how-open-is-open/. Accessed 20 May 2022
94. Khaveen Investments: Intel: Mobileye And Moovit Valued At \$35.2 Bln (2020), Available online: https://seekingalpha.com/article/4350012-intel-mobileye-and-moovit-valued-35_2-bln. Accessed 20 May 2022
95. Intel Mobileye: CES 2022 Under the Hood an Hour with Amnon (2022). Available online: https://youtu.be/1mXy0oi8d60. Accessed 20 May 2022
96. Intel Mobileye: EyeQ, The System-on-Chip for Automotive Applications. Available online: https://www.mobileye.com/eyeq-chip/. Accessed 20 May 2022
97. Intel: New Mobileye EyeQ Ultra will Enable Consumer AVs (2022). Available online: https://www.intel.com/content/www/us/en/newsroom/news/mobileye-ces-2022-tech-news.html#gs.1kpgn5. Accessed 20 May 2022
98. Qualcomm: Snapdragon Ride SDK. Online: https://www.qualcomm.com/news/onq/2022/01/05/snapdragon-ride-sdk-premium-solution-developing-customizable-adas-and-autonomous. Accessed 4 Mar 2022
99. Samsung: Samsung Exynos V7. Online: https://news.samsung.com/global/samsung-introduces-three-new-logic-solutions-to-power-the-next-generation-of-automobiles. Accessed 4 Mar 2022
100. NXP: S32S Processor Block Diagram. Online: https://www.nxp.com/products/processors-and-microcontrollers/arm-processors/s32-automotive-platform/s32s-microcontrollers-for-safe-vehicle-dynamics:S32S24. Accessed 4 Mar 2022

CNN Hardware Accelerator Architecture Design for Energy-Efficient AI

Jaekwang Cha and Shiho Kim

1 Introduction

Deep learning technology shows a revolutionary advance in various applications (image and video, speech and language, medical, gameplay, robotics, etc.). The performance of deep learning-based models already exceeds that of humans in laboratory environments and practical use. However, as the performance increases, the system requires massive energy, which hinders the deep learning technique from becoming simple and convenient. To be a commercialized technique, the energy consumption issue in the deep learning area becomes one of the essential considerations in designing a deep learning model. The energy efficiency issue in the edge device embedded in the wearable device, hand-held mobile devices, or even the autonomous vehicle is more critical. A deep learning hardware (HW) accelerator is one of the solutions to deal with the problem [1–4]. Accelerating HWs, such as field programmable gate arrays (FPGA) and application-specific integrated circuits (ASIC), usually are more energy-efficient than a central processing unit (CPU) or graphics processing unit (GPU). A convolutional neural network (CNN) is the most preferred network structure in most image processing applications as well as natural language processing. Nowadays, transformer architecture shows better results in some cases. However, it requires enormous data and trainable parameters to achieve good performance through a training process, which results in a lack of energy efficiency. Therefore, developing energy-efficient CNN is still essential to deep neural network (DNN) research. This chapter covers the CNN

J. Cha
Yonsei University, Incheon, South Korea
e-mail: chajae42@yonsei.ac.kr

S. Kim (✉)
School of Integrated Technology, Yonsei University, Incheon, South Korea
e-mail: shiho@yonsei.ac.kr

© The Author(s), under exclusive license to Springer Nature Switzerland AG 2023
A. Mishra et al. (eds.), *Artificial Intelligence and Hardware Accelerators*,
https://doi.org/10.1007/978-3-031-22170-5_10

HW accelerator design consideration for developing energy-efficient architecture which may support personalized edge devices and corresponding implementation examples which target FPGA and ASIC. Non-conventional CNN implementations beyond von Neumann architecture, such as neurocomputing or PIM architecture, are not within the boundary of this chapter.

2 CNN Architecture Analysis: An Energy-Efficient Point of View

2.1 Overview of CNN Structure

In 1989, Yann LeCun et al. [5] introduced the first CNN architecture to implement an automatic handwritten zip code recognizer, and AlexNet [6] surprised many researchers by showing overwhelming performance in the 2012 ImageNet Large Scale Visual Recognition Challenge (ILSVRC) [7]. A CNN is an artificial neural network inspired by animal visual cortex architectures. In the visual cortex, each cortex neuron only reacts to the limited part of sight known as the receptive field, and the whole sight is constructed by overlapping each neuron's receptive field. Shared weights structure and translational invariance characteristics are the distinctive features of CNNs that grant an advantageous structure for extracting informative features from image data in general. As shown in Fig. 1, typical traditional CNN architecture consists of four essential layers: convolutional, activation, pooling, and fully connected layers. Some non-traditional layers include deconvolution (upsampling) and dilated convolutional layers. Detailed descriptions of each layer are as follows.

Convolutional Layer Convolution is a kind of mathematical operation widely used in signal processing, image processing, and other fields of science/engineering. In deep neural networks, convolution is used to measure the similarity between

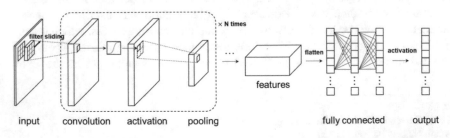

Fig. 1 General implementation of a CNN architecture. A CNN usually comprises several small networks (layers): convolution layer, pooling layer, activation layer, and fully connected layer. Each layer captures distinctive features from its input activation and sends the output activations to the following layers

the input feature maps and the kernel filters (neuron weights). The convolutional operation is a crucial element of CNN biomimicking the synaptic structure of intelligent creature's optic neural networks. It is a convention that we call the cross-correlation operation a convolution in the deep learning research field. The only difference between the two numerical operations is whether an operation reverses an input (convolution) or not (cross-correlation). The reverse operation is not required in measuring similarity, so the convolution layers do a cross-correlation operation to calculate the similarity. Unless otherwise specified, this book uses the term "convolution" as the meaning used in the deep learning field. Even though there are various types of convolution operations according to input and weight dimensions, the simplest case is a 2D convolution operation between an input feature map X and a weight matrix w. It is represented as

$$w^l \times X^l = Y^l (i, j) = \sum_{m=0}^{k-1} \sum_{n=0}^{k-1} w^l (m, n) \cdot X^l (i+m, j+n), \tag{1}$$

where zero bias is supposed. A tuple (i, j) indicates the pixel coordination of the output feature map Y, and l indicates the lth layer component. The size of the weight matrix is $k \times k$, and $w(m, n)$ stands for the corresponding element value of the weight matrix w. The filter (weight) slides along the input image while performing multiply and accumulate (MAC) operations one by one to provide the output feature map, as depicted in Fig. 2. Since the same filter is applied to the entire input while sliding, this is called the shared weight structure. Two-dimensional filter patches can extract the inherent key features in inputs. Specific filter patterns are designed based on the principle of inductive bias, which states that the main input-specific feature is the locality of pixel dependence.

Activation Layer The activation layer at the output of the convolutional layer applies an activation function to a nonlinear function operation to the network output. The concept of activation function stems from the action potential firing mechanism of biological neurons. The nonlinearity allows any multi-layered neural

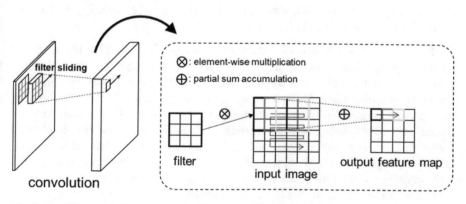

Fig. 2 The 2D convolution operation in the convolutional layer

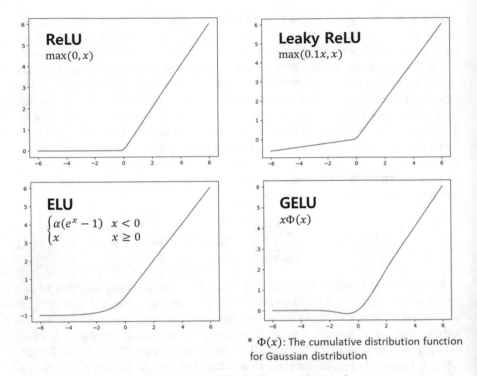

* $\Phi(x)$: The cumulative distribution function
for Gaussian distribution

Fig. 3 Widely used activation functions used in the deep neural networks

network to be a universal function approximator. Rectified Linear Unit (ReLU) is one of the most frequently adapted activation functions, which replaces the negative input value with 0. This simplicity successfully reduces computational resources. Moreover, ReLU makes the output feature maps more sparse than other nonlinearities such as sigmoid or hyperbolic tangent functions. This sparsity gives the network to be more compressible and energy and area saving potential. In recent days, the use of variants of ReLU such as Leaky ReLU [8], ELU [9], and GeLU [10] are increasing to overcome the weakness of ReLU in specific cases. Figure 3 shows some of the famous activation functions.

Pooling Layer The primary role of the pooling layer is to perform the nonlinear down-sampling of the input feature map. This operation squeezes the redundant information, and on the other hand, it enlarges the region of interest (ROI) of the sublayer's filters like the concept of the image pyramid. The down-sampling operation simplifies local feature information, and as a result, this enhances the translation invariance of CNN. As a result of down-sampling, several input pixel values in a window are summarized into a representative value. Max pooling and average pooling are dominant pooling methods. For the former case, the representative value of the pixel values inside the window becomes the maximum value among the pixels. For the latter, the representative becomes the average value

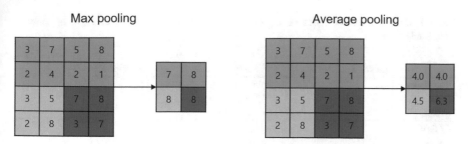

Fig. 4 Simple drawing illustrates how the max pooling and average pooling work

of the group. Global pooling method such as global average pooling makes the input translational invariant so that the network can deal with various input dimensions. The global pooling yields a single value representing all pixels in the input feature map. Figure 4 depicts how the max pooling and average pooling layers work.

Fully Connected Layer The fully connected (FC) layer consists of a network of artificial neurons called the perceptron. All neurons in an FC layer are connected to every neuron in the next FC layer. Typically, the perceptron consists of a MAC operation and activation function. In most CNN designs, the FC layer is located at the last few layers of CNN. Theoretically, more than two successive FC layers could approximate every kind of function to draw the target results, which is why this layer is located at the last of the network. At the position, FC layers take the feature map calculated by convolutional layers as inputs and approximate the target function that extracts the output. According to purposes, the last layer of FC can have a sigmoid function as the activation function (for classification task) or have no activation function (for regression task). The FC layer usually takes a massive amount of memory resources compared to other types of convolutional layers due to the absence of a shared weight structure.

The convolution layer takes a significant portion of the total CNN computational resources. Table 1 shows the number of weights and MAC operations used to construct some popular CNN models. The total amount of MAC operation in a network is usually utilized to measure the computational complexity of a network, and the number of weights indicates the required memory resources to implement the network design. The details regarding MAC operation will be covered in Sect. 3.3. Table 1 shows that the convolution layer takes a significant portion of the total network resource usage. Here, we want to focus on the table's CONV/FC ratio of weights and MACs. In the early day's CNN architecture, the convolution layer takes most of the MACs, and the FC layer takes most of the memory resources. However, recently, the portion of the convolutional layer has been significant in both weights and MACs as the network structure is becoming deeper and deeper to get detailed feature maps. The number of hidden layers increases proportionally as the depth of the network increases, but the number of FC layers does not change relatively much.

Table 1 Computational resource distribution of popular CNN architectures between convolutional and FC layers

	LeNet-5	AlexNet	VGG-16	GoogLeNet (v1)	ResNet-50	EfficientNet-B4
Top-5 error (ImageNet)	n/a	16.4	7.4	6.7	5.3	3.7
CONV layers						
# of weights	2.6K	2.3M	14.7M	6.0M	23.5M	14M
# of MACs	283K	666M	15.3G	1.43G	3.86G	4.4G
F.C. layers						
# of weights	58K	58.6M	124M	1M	2M	4.9M
# of MACs	58K	58.6M	124M	1M	2M	4.9M
Weights: CONV/FC ratio	4.3/95.7	3.8/96.2	10.6/89.4	85.7/14.3	92.2/7.8	74.0/26.0
MACs: CONV/FC ratio	83.0/17.0	91.9/8.1	99.2/0.8	99.9/0.1	99.9/0.1	99.9/0.1
Reference	Lecun, PIEEE 1998 [11]	Krizhevsky, NeurIPS 2012 [6]	Simonyan, ICLR 2015 [12]	Szegedy, CVPR 2015 [13]	He, CVPR 2016 [14]	Tan, ICML 2019 [15]

This fact stands for that we need to focus on making the convolutional layer more energy-efficient to improve the total CNN energy efficiency.

2.2 Convolution Layer Implementation

A basic convolutional layer "2DCONV" has an input data dimension of $B \times H \times W \times C$, where B indicates the batch size, H indicates the input feature map height, W indicates the width, and C indicates the number of channels. And for the convolution operation, $h \times w$ sized weight matrix is adopted, where h is the matrix height and w is the width. The naive implementation of 2DCONV utilizes a six-layered "for" loop architecture. Such a nested loop structure enforces inefficient calculation when a single processing element (PE) deals with it. Therefore, it is essential to exploit parallelization techniques to make the computation more efficient with multiple PEs. The naive way to implement the parallelization is by partitioning the nested loop and mapping the partitions to each PE. Each partition can be assigned to each PE, or each PE can be in charge of specific predefined parts of each partition. The architecture designer should decide the partitioning specifications considering the system purpose and the resource limitations. Another way to implement efficient convolutional layer calculation is to exploit dimension reduction algorithms such as im2col. The details of im2col are covered in Sect. 4.2.3. This kind of method sacrifices memory resources to improve computational efficiency.

2.3 Difference Between Training and Inference

Training and inference are the two main target applications of deep neural network HW accelerators. A large portion of research regarding HW accelerators focuses on implementing an inference task. This separate learning and inference method is popular in the cloud age of ICT nowadays; we perform resource-intensive training for the network weights in a training server and then utilize the server-trained weights in the individual edge devices for the inference tasks. However, there is a demand for training in edge devices in the case of issues for privacy leaks, personalization of devices, and online or lifelong learning requirement. The most significant structural difference between the two tasks is the implementation of a dataflow pipeline for the backpropagation of gradient. The gradient δ^l of the network loss L with respect to the l-th convolutional layer output Y^l can be derived based on (1) and the chain rule.

$$
\begin{aligned}
\delta^l (i, j) = \frac{\partial L}{\partial Y^l(i,j)} &= \sum_{m=0}^{k-1} \sum_{n=0}^{k-1} \frac{\partial L}{\partial Y^{l+1}(i-m,j-n)} \cdot \frac{\partial Y^{l+1}(i-m,j-n)}{\partial Y^l(i,j)} \\
&= \sum_{m=0}^{k-1} \sum_{n=0}^{k-1} w^{l+1}(m,n) \cdot \delta^{l+1}(i-m, j-n),
\end{aligned}
\tag{2}
$$

Similarly, we can calculate the weight gradient with the gradient δ^l and the l-th layer input feature map X^l as

$$
\begin{aligned}
\frac{\partial L}{\partial w^l(m,n)} &= \sum_{i=0}^{H-k} \sum_{j=0}^{W-k} \frac{\partial L}{\partial Y^l(i,j)} \cdot \frac{\partial Y^l(i,j)}{\partial w^l(m,n)} \\
&= \sum_{i=0}^{H-k} \sum_{j=0}^{W-k} \delta^l(i, j) \cdot X^l(i+m, j+n).
\end{aligned}
\tag{3}
$$

Figure 5 depicts forward, backward, and procedures. As shown in Fig. 5 and (3), weight gradient computation requires "large window convolution," which takes more memory and computational resources than the other two propagation procedures.

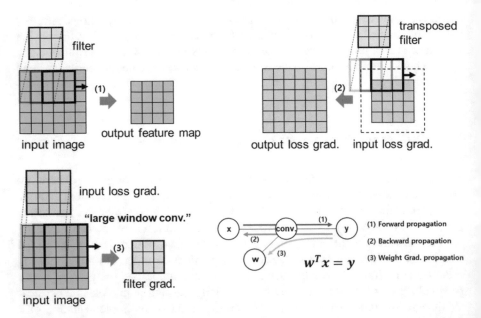

Fig. 5 Forward propagation, backward propagation, and weight gradient computation in the convolutional layer

2.4 Categorization of CNN HW Accelerator Architectures

The typical deep neural network accelerator consists of multiple PEs and memories such as input/output buffer memories, weight buffers, and registers in PEs. PE is a core MAC operation calculator comprising ALU, control units, and registers containing the most frequently accessed information. There can be various accelerator designs with regard to the target purpose. For example, to build an energy-efficient HW accelerator, considering dataflow between on-chip and off-chip memories is one of the most critical issues.

Moolchandani et al. [16] categorized the CNN HW accelerator architecture into four main groups based on the PE arrangement and the dataflow among the accelerator components: 1D-systolic, 2D-systolic, 1D-array, and 2D-array. The first basis, PE arrangements, is how PEs are distributed on the circuit board. Most accelerator designs have 1D (linear) or 2D (planar) shaped PE arrangements. And the second basis stands for how the input and output dataflow is structured to build the accelerator. The systolic PE structure is analogous to the blood pumping process of living bodies. The dataflow is sequential among PEs, so the previous PE's output becomes the following PE's input. On the contrary, the matrix PE structure indicates that each PE is parallelly connected to the global buffer to receive the buffer broadcasted data. There are no direct connections among the neighboring PEs, and these basic architectures are illustrated in Fig. 6.

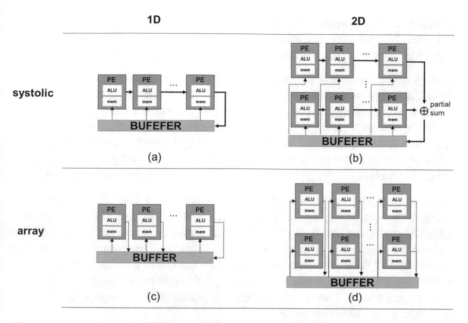

Fig. 6 Various CNN HW accelerator architectures: (**a**) and (**b**) are systolic designs, and (**c**) and (**d**) are array PE arrangements

3   Design Consideration for Energy-Efficient CNN Hardware Accelerator Implementation

3.1   Metrics for Deep Learning HW Accelerator Assessment

Most deep learning accelerator architecture has its own goal and target use case, and we can validate whether the architecture design meets the purpose by exploiting appropriate metrics. Therefore, it is critical to understand the fundamental metrics used for the validation works and apply the right ones to answer the purpose. The following are several important metrics for a successful HW accelerator design.

Accuracy   Accuracy measures the quality of results yielded from the HW accelerator. The inference accuracy metric can be derived from the comparison between ground truth and the output of network results or the comparison between the network outputs using a specific accelerator. It is common to sacrifice the model accuracy to get a high calculation speed in the accelerator; hence, if an accelerator boosts the calculation speed without the accuracy loss, we can say that the proposed architecture is suitable.

Throughput and Latency   Throughput indicates the production rate at a given time; meanwhile, latency implies the time needed to wait until the system processes the input. At first glance, the two concepts seem similar but are different. For example, we can increase the throughput by adding more processors to the system to increase the parallel computing power. More processors output more results at a given time; however, this work cannot improve the latency of the whole system.

Energy and Power It is the primary metric we want to focus on in this chapter. As written, this metric is regarding the amount of energy consumption in the accelerator. The energy or power consumption rate shows a more crucially treated tendency with respect to the implementation target domain, such as the edge devices with limited power capacity. MLPerf [17], a deep learning hardware benchmark, generally evaluate the accelerator's latency, throughput, and power consumption under the condition of achieving the same task, network model, and accuracy.

Aside from these, the cost for HW implementation, the flexibility of application range, or the scalability of the model performance regarding the computational resources are the frequently mentioned metrics for the deep learning HW accelerator architecture assessment. There are always trade-off relations among the mentioned metrics, and all these measures are not mutually exclusive. The energy-efficient HW accelerator design minimizes power consumption while maintaining or maximizing other metric scores. The key design considerations for the purpose are dataflow optimization and MAC operation.

3.2 Dataflow

CNN architecture inherently entails frequent memory read/write operations for the weight update and the activation calculation, and we call these kinds of data movements dataflow in networks. For example, when a perceptron processes the MAC operation, the PE in the hardware accelerator copies the network weights, input activations, and partial sums from the memory to its register. After the MAC computation, the output activations are stored in the memory as an updated partial sum for the subsequent computation. Therefore, if a PE needs to compute N MAC operations, 4N memory read/write would be required in the worst case. The worst case implies a naive implementation without any optimization techniques. These dataflows are well depicted in Fig. 7.

Like other computing systems, memory architecture in most hardware accelerators consists of on-chip (SRAM) and off-chip (DRAM) memory to maximize the efficiency of memory use. The on-chip memory is faster and more energy-efficient than off-chip memory; however, it is too small to store the entire network parameters, so using relatively slow but large off-chip memory is necessary.

Sze et al. [18] reported that AlexNet [6] has 724 million MAC operations and 60 million weights parameters to classify 227 × 227 sized input images into predefined categories. The size of network parameters makes the frequent use of off-chip memory necessary. While the input is being processed, AlexNet needs to access DRAM 3 billion times to get the data, which is approximately 50 times larger

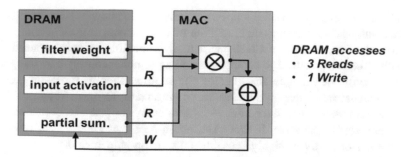

Fig. 7 Dataflow in a perceptron implementation in the hardware accelerator

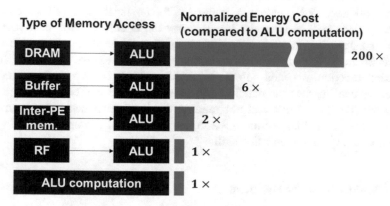

Fig. 8 Normalized energy cost of each computation element in the deep learning HW accelerator

than the total numbers of the network's parameters – an example of the enormous inefficiency of traditional MAC operations.

Moreover, memory read/write accounts for the most significant energy consumption in the hardware accelerator. Sze et al. [18] reported that DRAM spends energy more than 200 times and buffer spends energy more than six times better than ALU spends, as illustrated in Fig. 8. As we can see above, on-chip memory access is more energy-efficient than off-chip memory access. Accordingly, the basic dataflow design maximizes the use of on-chip memory and minimizes the off-chip memory access by optimizing memory usage. Dataflow optimization can be achieved by reducing memory usage or improving the on-chip memory usage rate through two primary considerations: reducing memory access time and memory footprint.

3.2.1   Reduce Memory Access Time

The memory access reducing strategy stems from the fact that the convolution operation in the CNN has inherent computational redundancy when implemented on hardware accelerators. This strategy minimizes redundancy that takes the dominant portion of total energy consumption. Fortunately, the dataflow of CNN architecture has two unique characteristics that facilitate memory access time reduction.

First, the fetched data in CNN has a high probability of being reused, both spatially (reused in different computation units simultaneously) and temporally (reused soon). For example, the filter of convolution layers is repeatedly applied on the input activation map by swiping the whole area of the input. Suppose ALU in the accelerator needs to read the filter weight for every computation even if the filter values are exactly the same. In that case, it can be a massive waste of resources. Therefore, storing these highly reusable data in on-chip memory is reasonable for implementing efficient data flow. The data reuse methods are categorized into four types based on which information to be stationary in the on-chip memory: weight stationary (WS), output stationery (OS), no local reuse (NLR), and row stationary (RS) [19]. Section 4.2.1 will cover the details of data reuse methods and their categories. Second, a significant number of MAC operations in CNN yield ineffectual computations whose output is zero or negative. According to the activation function, the final outputs of these inputs become zero (e.g., ReLU). Hence, we can improve the dataflow by figuring out the ineffectual computation before the MAC operation and not reading the ineffectual data from memories based on the prediction. Section 4.2.2 contains technical examples regarding the ineffectual computation reduction skills.

3.2.2 Reduce Memory Footprint

An inseparable relationship exists between the number of parameters in a network and its memory footprint. It is well known that the number of network parameters increases as the network performance improves. For instance, in 2012, AlexNet had around 60 thousand parameters, but in 2019, EfficientNet-B4 had 20 million, which shows more than a 300-fold increment during 7 years. The large networks trigger frequent off-chip memory access, which drops the computational and energy-wise efficiency. Therefore, reducing the memory footprint is an essential design consideration for the researchers who want to build an energy-efficient DL accelerator. The following techniques are commonly used to reduce memory footprint: pruning, using reduced precision skills or quantization, and utilizing information encoding or data compression. The goal of the mentioned methods is to maximize the memory footprint reduction while minimizing network performance degradation. Section 4.1 describes this memory footprint–reducing method.

3.3 MAC

Multiplication and accumulation (MAC), sometimes called multiply-and-accumulate, is a kind of unit operation in the deep neural network. Every perceptron in the artificial neuron employs MAC operation to mimic the operation of biological neurons such as synaptic transmission or soma. MAC consists of multiplications between the input activations and corresponding weight values and the accumulation that implies the aggregated total of the multiplication results. We can enhance the energy efficiency of the deep learning accelerator by focusing on two considerations. The first consideration is reducing energy per MAC operation. The MAC operation consists of the multiplication and following addition, and the multiplication is more energy-consuming than the addition. Here the point is how to reduce the number of multiplications used for implementing the convolution operation. The second consideration is avoiding unnecessary MAC operations. For instance, the zero-valued input activation or the zero-valued weight does not need to be multiplied because it always outputs zero. Therefore, it is essential to find where zero value exists and summarize its location in this methodology.

4 Energy-Efficient CNN Implementation

This chapter describes the implementation cases of designing the energy-efficient accelerator architecture. The approximation implies the method that improves the energy efficiency by sacrificing a little bit of the network performance. Here, optimization means maximizing efficiency without losing any performance.

4.1 Approximation

4.1.1 Pruning

As the word indicates, the pruning method refers to finding and eliminating ineffective network parts like pruning a tree. This method is a kind of model compression. LeCun et al. [20] verified that parameters or the network connections in the deep neural network are not equally important. Eliminating some parts of the network parameters can increase the efficiency of the network computationally as well as memory resource usage. They measured each network element's degree of importance, called saliency, by computing its impact on yielding the output loss. The basics of pruning are first finding the ineffective parts of the network and second eliminating the detected useless parts. A model can be pruned during or after the training session, which is up to the pruning method. After pruning, we can expect the following: (1) reducing the memory footprint; (2) reducing the computational resources, which accompanies energy efficiency improvement; and (3) improving the generalization performance of the trained network.

Most of the pruning methods are algorithm-based network optimization. However, the network designers should consider the hardware compatibility while adapting pruning methods to their system. The categorization of unstructured pruning and structured pruning is the most popular classification of the pruning methods. The basis of this categorization is whether the algorithm considers the structural location or specific geometry of the pruning target connections. Again, Manikandan et al. [21] define four types of pruning: fine-grain pruning, intra-kernel pruning, kernel pruning, and filter pruning. The former is unstructured, and the others are structured pruning. Generally, the unstructured pruning method shows less hardware compatibility than the structured method due to inducing irregularities in the network. The structural irregularity of networks causes this hardware compatibility issue and prevents efficient dataflow architectures from the hardware accelerator implementation. Similarly, Choudhary et al. [22] classified four pruning methods based on where the pruning happens: weight pruning, neuron pruning, filter pruning, and layer pruning.

Unstructured Pruning Fine-grain pruning is another name for unstructured pruning. This pruning method targets all weight values that contribute little to yielding the output feature map. Han et al. [23] presented a magnitude-based pruning that iteratively performs pruning and fine-tuning in a sequence. They set a specific threshold value to cut off the ineffectual weights, assuming that smaller valued weights possess a more negligible influence on the output instead of using the saliency, which is hard to calculate. Their work achieved three times speedup in the inferencing and a significant compression ratio (nine-fold) while maintaining the network performance. Meanwhile, the pruned network became irregularly sparse, resulting in less hardware compatibility (requires dedicated hardware/libraries to run the pruned network). Guo et al. [24] introduced the dynamic network surgery splicing procedure to the pruning process. This procedure enables the recovery of the already pruned weights but is found to be important later. Their pruning method iterates pruning and splicing to achieve more efficient compression results. Through experimental results, Zhu et al. [25] verified that the pruned large-sparse models outperform small-dense models under the same memory footprint condition.

To optimize the implementation of the unstructured pruning method, several researchers proposed dedicated architectures that support the sparse weights processing in hardware. Most unstructured pruning methods result in the sparse weighted network as a significant portion of the network were eliminated. Han et al. [26] reported the use of sparse weight conserves $20 \sim 30\%$ of the total memory access bandwidth of the network. However, the informational redundancy inside the sparse weight causes inefficient computation when the MAC is processed through matrix-vector multiplication. Dorrance et al. [27] introduced several efficient matrix-vector multiplication implementations, such as compressed sparse row (CSR) and compressed sparse column (CSC). And they suggested using CSC when the output size is smaller than that of the input or the number of filters is not significantly more than the number of weights inside each filter. Figure 9 shows the difference between CSR and CSC. Han et al. [28] support sparse matrix-vector multiplication for the computation of FC layer using CSC, and Parashar et al. [29] presented SCNN, which can compute convolution operation with compressed sparse matrix format.

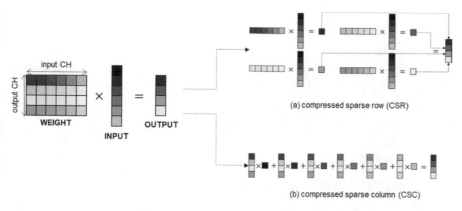

(a) compressed sparse row (CSR)

(b) compressed sparse column (CSC)

Fig. 9 Comparison of CSR and CSC

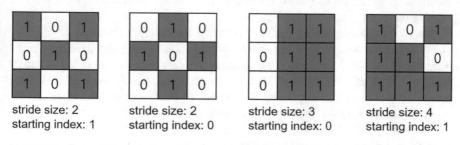

stride size: 2
starting index: 1

stride size: 2
starting index: 0

stride size: 3
starting index: 0

stride size: 4
starting index: 1

Fig. 10 Examples of the intra-kernel pruning granularities introduced in [30]

Structured Pruning Structured pruning methods prune the target network, taking into account the network structure such as channels and even layers. Anwar et al. [30] categorized structured pruning of the convolutional layer according to different scales: intra-kernel, kernel, and channel, which are called pruning granularities. Similar to fine-grain pruning, intra-kernel pruning prunes each weight value in the kernel; however, this method utilizes the sparsity at well-defined locations, called the intra-kernel stridden sparsity. The pruning granularity is predefined by the stride size with the stride starting index of the convolutional layers, as illustrated in Fig. 10. The granularity, discerned as the most affectless, is eliminated among the predefined ones. As it is implemented by exploiting the convolutional layer setting, this method can be utilized in most CNN environments regardless of the hardware/library compatibilities.

Kernel pruning [31] is a kind of structured pruning where any kernel from the output feature maps can be pruned. We can represent kernel pruning as a particular case of inter-kernel pruning where the sparsity is 100%. Each kernel indicates one whole convolution operation between the input feature map and a convolution filter. Similarly, filter pruning [32, 33] targets to delete affectless convolutional filters from the network. The elimination of target channels triggers the elimination of input-

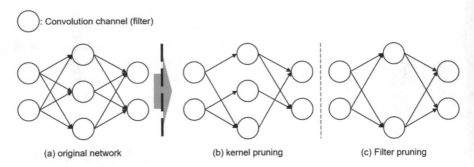

Fig. 11 Comparison of kernel pruning and filter pruning methods

output connections related to the deleted channel. The difference between the two pruning methods is described in Fig. 11.

Above structured pruning methods preserve the original structure of the convolutional layer so that it is applicable without considering the dedicated hardware/libraries. The changes only happened in the networks' architecture. In recent studies, Liu et al. [34] reported that the most significant element of structured pruning, which enhances its performance, is not the fine-tuned weight values but the network structure found by the pruning procedure. Their experiments have shown that randomly initialized pruned networks achieve the same performance level through re-training. Commonly, the size of pruned group shows inverse proportion against the model accuracy [35].

Energy-Aware Pruning In contrast to other pruning methods, which select the useless network parts by their standards such as weight magnitude, Yang et al. [36] pruned the network based on each network part's estimated energy consumption to achieve the energy-efficient network implementation. They proposed an energy estimation technique that estimates the energy considering the memory hierarchy, number of MACs, and data sparsity. The target network was pruned by exploiting the estimation mentioned above skill. Their experimental data accounted for the proposed energy-aware pruning being superior to the magnitude-based pruning in the point of energy efficiency (accuracy was a little bit dropped compared to the magnitude-based pruning).

4.1.2   Reduced Precision and Quantization

The Digital system stores real numbers through a quantization process to approximate the number while suppressing errors. The error between the actual and the quantized value is dominated by the number of bits called the precision level. The higher precision level produces a lower error rate, though adopting high precision could waste not only memory resources but also computational resources of the accelerator. Nowadays, standard computing systems such as PC usually use 32bit floating-point precision to store real numbers with high accuracy; however, the 32bit floating-point precision tends to be a sledgehammer to crack a nut for the case of implementing DNNs. The reason is that, in the practical use of DNNs, even a limited level of precision can deduce proper parameter updating direction. As we can see in most deep learning processes, the direction of the gradient is much more critical than that of the exact value for optimizing the network parameters. The proper quantization can guarantee the network achieves a sufficient level of accuracy with limited bit width (level of precision). Consequently, reduced precision could enhance the system's throughput and save the neural network computation's memory footprint and energy consumption with an affordable amount of additional quantization loss in training as well as inference.

The real number representation method is another consideration for establishing an efficient network. Horowitz et al. [37] presented the overall comparison between the impact of the bit width and the real number representation method on the network energy and area consumption. Their report showed that 8bit fixed point add operation reduced 3.3-fold energy (3.8-fold area) compared to 32bit fixed point and 30-fold energy (116-fold area) compared to 32bit floating point representation, respectively. Additionally, 8bit fixed point multiplication operation saved 15.5-fold energy (12.4-fold area) and 18.5-fold energy (27.5-fold area) compared to the 32bit fixed and floating point representation. The results showed that the fixed point representation could efficiently calculate deep learning parameters in limited computing environments (FPGAs).

We can categorize the quantization method based on the types of mapping functions used for the quantization. A mapping function converts the real numbers to the quantized value determined by predefined quantization intervals, and the precision level decides the resolution of intervals. For example, if we use the linear function as a mapping function, the quantized output will show a proportional relationship between the original and quantized value. We also call this type of quantization technique uniform quantization. In the case of adapting nonlinear mapping functions such as log function or group of clusters, we can highlight or ignore the region of interest from the original value according to the characteristics of the mapping function. The following are implementation cases of each quantization technique.

Linear Quantization This quantization usually focuses on the quantization target (layers) and the level of precision to achieve performance improvement. The quantization methods can be applied to various parts of the network incorporating weights, activations, and even gradients. Since the dynamic range of each layer in DNNs can be diverse, we don't need to apply the same level of precision to the whole network parameters. Dynamic quantization refers to these layer-wise quantization techniques. Judd et al. [38] introduced per-layer precision called fine-grain variations in bit precision. In their research, the weight and activation precision of AlexNet could vary between 4 and 9 with 99–100% relative accuracy loss. Thus, the network throughput was increased (2.24 times speed up), and the network energy efficiency was almost doubled compared to the baseline.

Qiu et al. [39] also presented the dynamic quantization approach. They utilized singular value decomposition (SVD) to enhance the FC layer performance by reducing the number of weights. Their method adapted the layer-wise mapping functions where 8 and 4 bits quantization were used for the convolutional and FC layers, respectively. The reported experimental result showed almost doubled FPS compared to that of 16 bits quantized case when adapted to the VGG16 network. Li et al. [40] presented a ternary weight network that exploits the weight of neural networks constrained to -1, 0, and $+1$. To preserve the representation power of the network parameter, they adopted scaling factors that minimize the Euclidean distance between the original and quantized weight values. As a result, each kernel value became constrained to $-|w|$, 0, and $+|w|$, where w indicates the scaling factor. Zhu et al. [41] proposed another type of ternary quantization technique with more accuracy improvement. First, they normalized the full precision to the range -1 and 1, then applied ternary quantization of the intermediate weights to the range -1, 0, and 1 by thresholding. The threshold value is the same across the network to simplify the process. Further, they introduced two individual quantization scaling factors, which are trainable, for positive and negative weights in each layer. Some linear quantization methods drastically reduced the precision level to binary-level representation. The MAC operations can be replaced with other alternative operations such as XNOR in these cases. This type of technique will be further discussed in Sect. 4.1.4, alternative operations.

Nonlinear Quantization Unlike uniform quantization, nonlinear quantization methods have interest in the fact that the distribution of weights or activations is not uniform in most networks. The log-scale mapping function is one of the candidates for nonlinear quantization. Lee et al. [42] applied to log base-2 quantization to AlexNet and VGG16 and reported that 4-bit log quantization is more accurate than 4-bit uniform quantization (5% and 28% loss, respectively, for the VGG16 ImageNet classification task). The log base-2 quantization technique also supports exploiting barrel shifters, which may enhance the energy efficiency instead of multipliers [43–46]. This alternative operation will be discussed in depth in Sect. 4.1.4. Weight sharing is another prominent solution for nonlinear quantization. In this concept, weights in a network are clustered into several groups with specific representative values, thus lowering the memory footprint. Chen et al. [47] presented HashedNets, which utilizes a low-cost hash function to group weights into hash buckets having predefined sizes (compression ratio). Weights in the same hash bucket share a single parameter, and the parameter value is optimized through a backpropagation algorithm during training as the standard neural network does. Han et al. [23] incorporated the quantization technique into the network compression procedure and the pruning mentioned above. Compared to Chen et al. [47], their work utilized the k-means clustering method to find the shared weights and applied the quantization to the fully trained network.

Usually, batch normalization (BN) could prevent the reduced precision from dropping the network accuracy [48]. Since BN forces the weight and activation values to be centered near zero, it helps reduce the dynamic range of the quantization

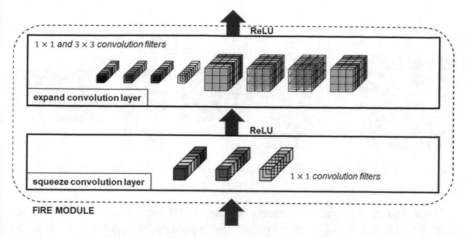

Fig. 12 Fire module of SqueezeNet. (Figure from Ref. [50])

target distribution as well as dependency across layers [49] so that the data can be well represented with reduced precision.

4.1.3 Convolution Filter Decomposition

Filter weights of the convolution layers take a significant portion of the memory footprint. By minimizing the use of memory resources for storing the filter weights, we can handle memory capacity-limited situations such as implementing the CNN-based neural network inference model on compact edge devices. The starting point is to reduce the number of parameters employed to construct filters without losing the efficacy of the original one. For this, the following two strategies are well known.

Using Memory-Efficient Convolution Filter Architecture The authors of VGGNet [12] insisted that using a series of smaller filters achieves memory footprint reduction compared with the original sized filter while maintaining the model accuracy. For instance, double-layered 3x3 filters have the same receptive field as one-layered 5×5 filters with seven less-used parameters. Like this idea, several successive researchers published their own memory-friendly CNN architectures. Iandola et al. [46] presented SqueezeNet architecture, which reduced the number of network parameters 50-fold compared to AlexNet without an accuracy drop. The achievement is thanks mainly to using 1×1 filters as many as possible instead of 3×3 filters. SqueezeNet has its unique structure, the fire module consisting of the squeeze convolution layer, which only contains 1×1 filters, followed by the expand convolution layer, which consists of the combination of 1×1 and 3×3 filters, as shown in Fig. 12. Using the fire module enhances the memory footprint of the network by reducing not only the number of filter parameters but also the number of the filter channel.

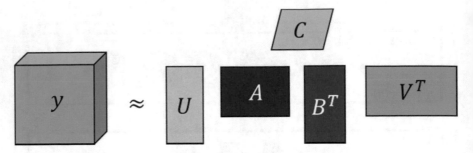

Fig. 13 A simple illustration of CPD with low-rank constraints. (Figure from Ref. [58])

Inception [51] is another memory-efficient CNN implementation. They composed $1 \times N$ and $N \times 1$ filters to effectively implement $N \times N$ filters without the performance drop in the model accuracy. MobileNet [52] introduced the concept of depthwise separable convolution, which consists of depthwise convolution and pointwise convolution. The depthwise convolution layer restricts the number of output filter channels to be one to reduce the memory footprint of a neural network model. Therefore, the number of output channels becomes the same as that of the input channel. The number of output channels is then modified through the pointwise convolution layer, consisting of 1×1 and 3×3 filters. As a result, the pointwise convolution optimizes the weight parameters to learn the cross-channel correlation among output channels. Xception [53] architecture is very similar to MobileNet except for the reversed order of the depthwise and pointwise convolution layers.

Convolution Filter Decomposition Several researchers have focused on concisely representing the huge-sized CNN weight parameters without the performance loss. At this point, if we denote the weight as a 4D tensor, we can apply a tensor decomposition method such as singular value decomposition (SVD) [54] to decompose the high-dimensional tensors with low-rank factor matrices. Canonical polyadic decomposition (CPD) is one of the most frequently employed methods for this purpose (to compress the CNN with little performance degrades) [55–59]. Fig. 13 illustrates how the decomposition works. However, Lebedev et al. discussed the limitation of CPD: There are no stable analytic solutions when the tensor dimension exceeds two. Kim et al. introduced Tucker decomposition instead [55] to handle this problem. We can apply the above-presented methods to the already trained network to minimize the memory footprint, enabling more flexible implementation than modifying the filter architecture.

Both ways (modifying filter architecture and the filter decomposition) do not always guarantee energy efficiency because the reduced memory footprint often requires more computational resources.

4.1.4   Alternative Operations

The convolution operation achieves outstanding feature extracting performance for image inputs. However, as discussed, MAC (multiplication) consumes more computational resources than simpler operations such as add or bit-shift. Therefore, this subchapter introduces the network architecture composed of not convolution operation but alternative operations, which are expected to be energy-efficient with limited performance degradation.

XNOR   Quantization of network parameters or inputs can significantly reduce the memory footage, as shown in Sect. 4.1.2. Moreover, with an extreme level of quantization, in other words, binary quantization enables the use of the XNOR operation instead of the element-wise dot product to implement the convolution operation. The XNOR operation is far more energy-efficient and hardware friendly than the dot product. Rastegari et al. [60] presented the method for implementing the XNOR operation-based energy-efficient CNNs. Their work suggested two versions of a binarized deep neural network: (1) binary-weight-network, in which only weights are binarized, and (2) XNOR-net, where both weights and input are binarized. The proposed method has a concept very similar to BNN [61] (illustrated in Fig. 14) except for determining the scaling factor alpha and ordering the respective operation modules in the convolution block. The binarization module

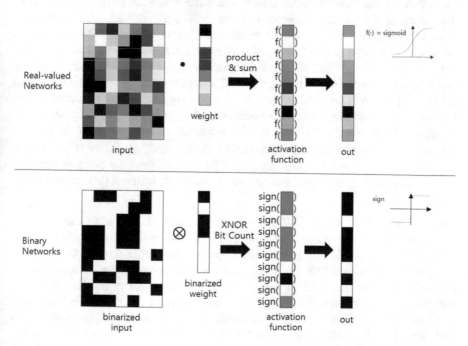

Fig. 14 Comparison between the real-valued networks and the binarized neural network [61] architecture

uses a sign() function so that if each value is positive, the binarized value becomes 1, and if not, the value becomes −1. The scaling factor alpha follows the binarized values. By multiplying the binarized value with the alpha, the binarization error minimizes. The forward and backward pass for the training procedure is the same as usual DNN except for the binarization with the scale factor. They used the real-valued weights for the weight update to ensure the GD algorithm works well without losing the gradient information due to the binarization. The scaling factor equals the average of the weights' absolute value. They compared two scaling factor candidates: (1) the proposed method and (2) the learnable scaling factor, and the former showed better results. They also compared the order of the block structure: (1) C-B-A-P and (2) B-A-C-P (proposed), and the latter showed better results (C: convolution, B: batch norm, A: activation, and P: pooling layers, respectively). The authors insisted that this layer order minimizes the quantization loss.

Bitwise Shift Elhoushi et al. [43] introduced convolutional and fully connected shift (linear shift) operators to reduce the memory footprint for storing CNN filters and energy consumed in multiplication operations. These operations replace resource-consuming multiplication operations with bitwise shift and sign flipping. This kind of bitwise shift operation used network is called the DeepShift model. The paper's authors proposed two methods to train the DeepShift model. The first is the DeepShift-Q, which trains regular weights constrained to powers of 2. And the second is the DeepShift-PS, which trains the values of shifts and sign flips. The proposed model achieved equal or better results by converting the pre-trained model and applying the fine-tuning procedure. Compared to XNORs, the authors insist that DeepShift performs better on larger datasets such as the ImageNet, not Mnist or Cifar10. Applying bitwise shift operation on an element is mathematically equivalent to multiplying it by a power of 2. Incremental Network Quantization (INQ) [62] is a similar concept but can only be applied for the inference of the network, and Accurate Binary Convolution (ABC) [63] uses sign flip only. The DeepShift-Q rounds the weight value to reach the nearest power of 2. For the operation, let P is a shifter and calculated as $P = \text{Round}(\log\_2 \text{abs}(W))$. For the positive P, the left shift happens, and the right shift happens for the negative. The sign is determined by the sign function $s = \text{sign}(w)$ so that the weights are converted, as depicted in Fig. 15.

Adder Only Chen et al. [64] investigated using add operation only instead of exploiting the combination of multiplication and addition to implement the measurement of similarity between the input and the network filter to boost the computation speed and save the computing resources. They focus on the meaning of MAC operation in the CNN: the measurement of similarity between the input and filters. Their motivation is that L1-distance could be an efficient measure to summarize absolute differences (dissimilarity) between the input feature map and the filters. Its measurement could be easily implemented using only adders. The L1-distance in the AdderNet is always negative because it is the measure of dissimilarity. They resort to batch normalization to deal with the negative L1-

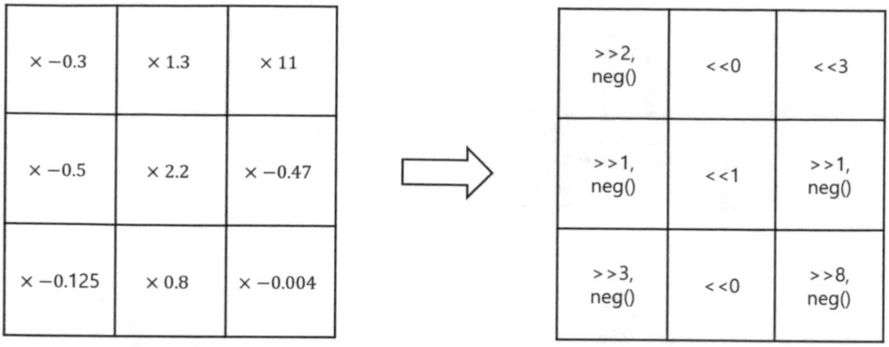

× −0.3	× 1.3	× 11
× −0.5	× 2.2	× −0.47
× −0.125	× 0.8	× −0.004

⟹

>>2, neg()	<<0	<<3
>>1, neg()	<<1	>>1, neg()
>>3, neg()	<<0	>>8, neg()

Fig. 15 Weights of the original and shift convolution operator proposed in DeepShift-Q [43]. (Figure from Ref. [43])

distance output, and the normalization results are forced to stay in an appropriate range. After that, the activation layer can apply the usual activation functions of CNN to get the output. For the gradient of filter weights, they utilized the derivative of L2-norm (full-precision gradient), not l1-norm (sign gradient), to ensure sufficient weights update even for the high-dimensionality computation. To prevent the gradient from exploding (because the gradient of each layer input is accumulated through the network depth), they clip the magnitude of the full-precision gradient to $[-1,1]$. Due to the characteristics of L1-distance, the layer output variance of the AdderNet is smaller than that of usual CNN, which hinders fast convergence of the weight update. The authors proposed a custom local learning rate (layer-wise) to compensate for the small output variance by guaranteeing normalized update steps across filters in each layer. The proposed model uses a double-numbered add operation without multiplication compared to the usual CNN. It performs better than BNN (almost the same as CNN performance). The ablation study showed that using adaptive learning rate and full-precision gradient outputs the best results. Figure 16 describes CNN and its alternative implementations not using multiplication.

4.2 Optimization

There have been various trials to achieve the energy-efficient CNN architecture while maintaining the network performance. The basic is to make the hardware operation as efficient as possible using parallelization techniques. The parallelization technique can be categorized into temporal and spatial architectures.

Temporal architecture is a typical parallelization method used in most CPUs and GPUs. SIMD (Single Input Multiple Data) or SIMT (Single Instruction Multiple Thread) is one of the examples. In Sect. 2.4, 1D and 2D array PEs are the temporal architectures. ALUs in PEs process MAC operation along with the control command

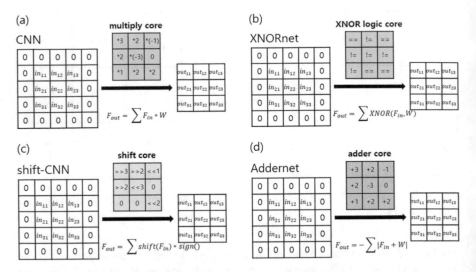

Fig. 16 CNN and its alternative implementations. (Figure from Ref. [65])

from the central primary operation devices and fetch the data from hierarchical memory architecture. In general, there is no communication among PEs. The computational transform is one of the options to optimize the temporal architecture by reducing the number of required multiplications and maximizing the network throughput.

Meanwhile, spatial architecture focuses on optimizing overall system dataflow by exploiting the data processing among ALUs. Processing Engine (PE) is the architecture consisting of ALU, control logic, and local memory, and it supports the data processing between ALUs. It is generally adopted in ASIC and FPGA and exploits its internal components (local memory) to minimize the off-chip memory access for energy-efficient computing.

This chapter will explain the optimization technique for spatial (data reuse, Sect. 4.2.1; computational reduction, Sect. 4.2.2) and temporal architectures (Im2col & GEMM, Sect. 4.2.3; FFT, Strassen, and Winograd, Sect. 4.2.4).

4.2.1 Data Reuse

It is off-chip memory access that consumes the most energy throughout the operation of CNN. It also triggers the bottleneck regarding the processing time point of view. Therefore, we need to minimize the off-chip memory access to make the network more energy-efficient. MAC comprises three memory reads (filter weight, feature map activation, and partial sum) and one memory write (partial sum update), as shown in Fig. 7 of Sect. 3.2. If every memory read happens at DRAM, it will be the worst case (see Fig. 8 of Sect. 3.2.). The hierarchical memory structure of spatial architecture helps to reduce DRAM access by storing the frequently used data in the

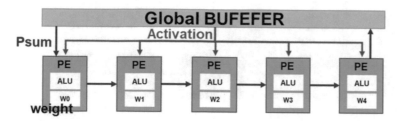

Fig. 17 Weight stationary data reuse architecture

local memory. Most spatial architecture adopts a global buffer, inter-PE network, and RF (Register File) in each PE as local memories. The global buffer is the most far from ALUs and energy-consuming. Still, it has the largest capacity, and RF is the most energy-efficient memory level because it locates in the PE with ALUs. The optimized dataflow design aims to maximize access to energy-efficient memory levels. For this, it is crucial to maximize memory reuse by putting frequently used data into the memory closest to ALUs. Sze et al. [18] categorized the hardware architecture for maximizing data reuse into four major types.

Weight Stationary (WS) WS is a DNN hardware accelerator design concept that minimizes the energy consumption required to read and write the weight value by storing the filter value at RF of PE Feature map activations (input), and partial sums are read from the global buffer. Figure 17 shows one of the basic implementations of WS architecture that weight values are fixed in PE, feature map activations from the global buffer are broadcasted throughout PEs, and partial sums from the global buffer are accumulated through each PE. Finally, PE writes the completed partial sum value on the global buffer at the last position of the partial sum accumulation. WS maximizes the data reuse of filter and has the advantage for the case that frequent read/write of weight values are expected. Google's TPU exploits the WS architecture to build the neural network computational system.

Output Stationary (OS) Like WS stores the filter value in PE, OS stores the partial sum in each PE to minimize the memory access to high-level memory hierarchy such as global buffer or DRAM. Each PE holds the partial sum, the output of the convolution operation, as shown in Fig. 18a. The operation controller streams the input activation from the global buffer to the sequence of PEs in the order of the input activation window sliding. Similarly, the weight values are read from the global buffer and broadcasted to all PEs in the order of filter sliding on the input feature map, as depicted in Fig. 18b. As the unit processing time passes, the input activation moves to the next PE, and as a result, each PE is in charge of each local output space (e.g., output feature map pixel). According to the targeted format of output channels and activations, there are various implementations of OS, for example, OS_A for the convolution layers or OS_C for the fully connected layers. These variations are illustrated in Fig. 19.

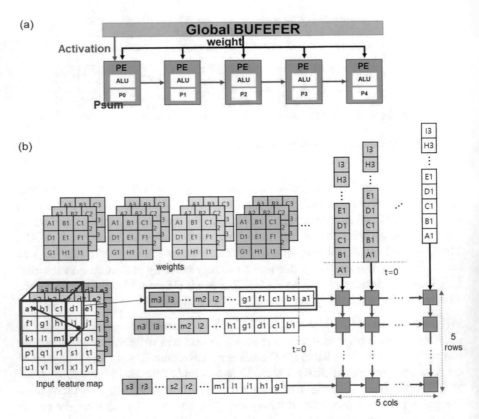

Fig. 18 (**a**) Output stationary data reuse architecture and (**b**) the dataflow example of OS architecture for 2D convolution operation

No Local Reuse (NLR) The two methods above adopt local storage (RF in PE) for data reuse. However, RF is not efficient in the point of chip area. Even though RF shows low energy consumption for the data accesses, it requires a larger physical area than the global buffer to store the same amount of information. NLR is another type of architecture that focuses on maximizing the storage capacity and minimizing the off-chip memory bandwidth by not allocating the local storage in PE and using the global buffer only. Figure 20 shows how NLR is implemented. Each PE reads the weight value, the feature map activation, and the partial sum from the global buffer, and the partial sum is accumulated through the PEs until the accumulation ends. Then, the completed partial sum is again stored in the global buffer. As a result of these dataflows, NLR could suffer from traffic congestion between the global buffer and PEs.

Row Stationary (RS) Instead of former architectures, RS is designed to reuse all read data: weight value, input feature map, and partial sum. As shown in Fig. 21, each PE has RF assigned to the filter, input feature map, and partial sum. Weight

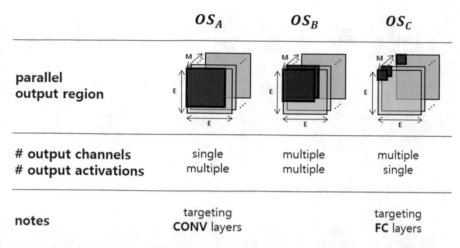

Fig. 19 Variants of output stationary architecture

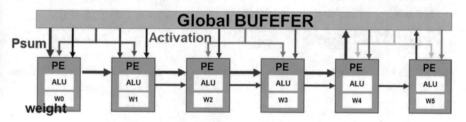

Fig. 20 No local reuse (NLR) architecture for the hardware accelerator implementation

values are stored row by row to the RF; for the example of a 3 × 3 filter, the three 1 × 3 rows are stored in each PE. Then, the controller streams the input row by row to each PE's RF by the weight row-sized sliding window length, as depicted in Fig. 21. For instance, if it is a 5 × 5 input, it is streamed to the PE containing the corresponding weight values by 1 × 5. Then if the weight row length is equal to three, the input is streamed to RF like how a queue works. The weight row length three becomes the length of the queue. There is one memory space for storing the partial sum in each PE, and it is updated every processing time. We can get the final output by accumulating N PE partial sum. Again, for the example of a 3 × 3 filter case, the partial sums must be merged as the calculation proceeds by three parallel PEs. The number of PEs required for the above operation becomes the filter column size multiplied by the output activation column size. The PEs for the operation can be arranged two-dimensionally, as illustrated in Fig. 22a.

Figure 22b shows how RS architecture deals with multiple feature maps, filters, and channels to output appropriate partial sums. The input can be split as scheduled by the system controller. This control can be managed by the optimization compiler (Fig. 23a) that considers the CNN configurations and hardware resources. Figure 23b is the Eyeriss architecture proposed by Chen et al. [19] in 2016. It adopted

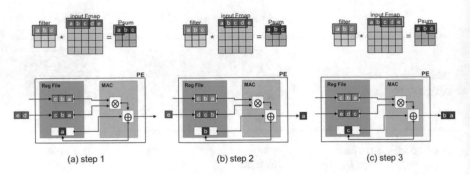

Fig. 21 1D row stationary data reuse architecture

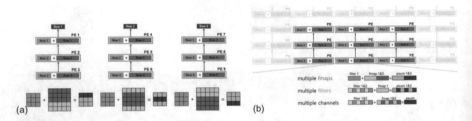

Fig. 22 2D RS architecture and variants implementation according to its feature map, filter, and channel design

the RS architecture and has a 12 × 14 two-dimensional PE array. Replication and folding are utility techniques that maximize PE computing resources when the shape of the PE array to computing the convolution operation and the hardware PE array are mismatched. For example, RS architecture needs a 3 × 13 sized PE array to compute an AlexNet layer 3–5. However, if we use only one 3 × 13 PE array at a processing unit time, there are many unused PEs in the 12 × 14 PE array of the Eyeriss accelerator. The replication technique fills out the unused PEs by parallelly arranging the other three 3 × 13 PE arrays to use 12 × 13 PEs, and the 12 × 1 PEs left are clock gated to prevent the not intended use of these PEs. For another example, RS architecture requires a 5 × 17 sized PE array to process an AlexNet layer 2. And this exceeds the width of the Eyeriss accelerator PE array. The folding technique cuts these exceeding parts of the required PE array and pastes the left into other vacant rows of the hardware resource. These examples are illustrated in Fig. 24.

The RS method saves energy the most among the above four methods in terms of energy efficiency. Sze et al. [18] have illustrated the energy breakdown across memory hierarchy and the data type for AlexNet, as shown in Fig. 25. However, it is necessary to consider a hardware accelerator structure according to the desired target application for the optimization because the optimum method may vary depending on the target network structure and input data.

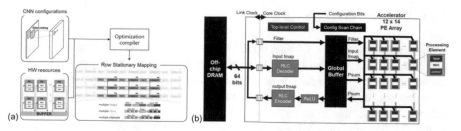

Fig. 23 The Eyeriss architecture for implementing the energy-efficient HW accelerator

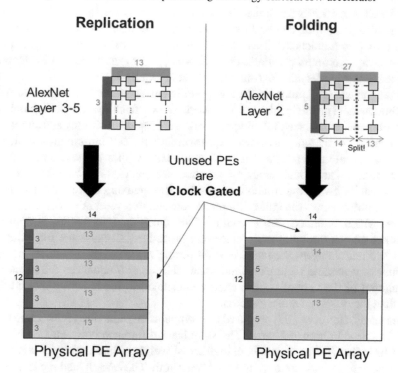

Fig. 24 Replication and folding in the RS architecture

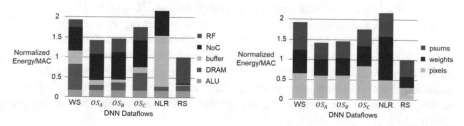

Fig. 25 Energy breakdown across memory hierarchy and data type

4.2.2 Computation Reduction

While computing the convolution operation, near zero-valued filter or input activation have almost no contribution to producing the output activation. These small values are called ineffectual data. We can save the computational resources by replacing the ineffectual data with 0 and not computing MAC regarding these values. However, this replacement could cause sparsity in the input feature map. In the parallelized convolutional operation using multiple PEs, this data sparsity may degrade the parallelization efficiency. Because the sparsity is not uniformly distributed, not all PEs can finish the calculation simultaneously, and there will be unwished waiting time among PEs. Therefore, if the distribution of sparsity in the feature map is predictable, it can help to improve performance (uniform sparsity achieves the maximum parallelization efficiency). For this, a method for recording the location of ineffectual data and utilizing it is required.

Zhang et al. [66] proposed an energy-efficient accelerator architecture named Cambricon-X. The main idea of this work is not fetching the ineffectual input activation and, conversely, fetching only effectual input activations and suppressing the unnecessary memory accesses. As shown in Fig. 26, Cambricon-X utilizes the indexing module to fetch the effectual inputs only with a pre-calculated effective weight index. The module finds the practical weight value by reading the weight values. The indexing information is stored in the indexing buffer at each PE. The indexing buffer stores the index difference between two successive non-zero weight values, which is called step encoding. This indexing information is applied to input activations through the step indexing procedure. Because the filter weight is fixed and not changed, the index is valid during the network inference. Although the indexing module requires an additional chip area, it enhances the latency and throughput of the network. Also, energy consumed in the indexing module is far less than the off-chip memory access.

Similarly, Han et al. [67] proposed the computation reduction architecture named EIE, Efficient Inference Engine. The significant difference between Cambricon-X and EIE is that EIE considers both ineffectual weight values and input values. EIE stores the index of non-zero weight and inputs both. The zero-valued input activation is detected when it is read from memory, and this information is broadcasted throughout all PEs to prevent unnecessary computations. Zhou et al. [68] presented Cambricon-S. Like Cambricon-X, it also creates the indexing vector from weights and input activations. These are called synapse indexes and activation indexes, respectively. Neuron selector module (NSM) is an indexing module in Cambricon-X. This module compresses the memory footprint required for storing the indexes using Huffman encoding and decoding. Additionally, Cambricon-S exploits the activation sparsity information caused by the ReLU activation function at the network inference runtime.

Alberico et al. [69] proposed Cnvlutin, which encodes the input activation with the practical value and the locational information of the effective value to reduce the memory footprint and computation resource consumption. The proposed encoding method is called ZFNAf (zero-free neuron array format). As depicted in Fig. 27, this

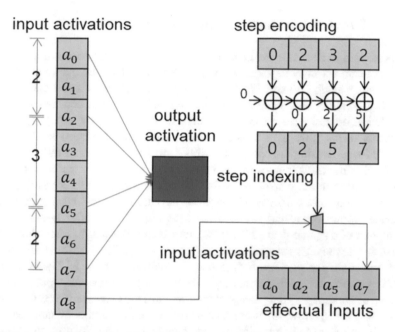

Fig. 26 Indexing module of Cambricon-X architecture

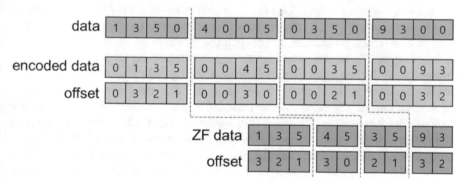

Fig. 27 ZFNAf encoding format

encoding method divides the input into the predetermined data chunk size N. The offset indicates the data locations inside the chunk in descending order. Zero-valued data receive zero offsets and are not stored in the encoding except for the case that no non-zero values in the data chunk.

4.2.3 GEneral Matrix Multiplication-Based Convolution

As mentioned in Sect. 2.2, the general way to implement 2D convolution operation requires using six-layered *for* loops, which degrades the computational efficiency. On the other hand, GEneral Matrix Multiplication (GEMM)-based convolution is a kind of convolution operation optimization technique that converts a high-cost convolution operation to a low-cost matrix multiplication by applying computational transforms to the data. This transform reduces the number of multiplications required for processing the operation without any loss of bitwise results.

Im2col is the most famous method for implementing the transform. It converts the convolution to the matrix multiplication of the input and filter matrices. Here, let's see how im2col works step by step for the case of im2col between $C_{in} \times H \times W$ shaped input tensor and $C_{in} \times K \times K \times C_{out}$ shaped filter tensor. With the assumption of a square-shaped 2D filter, filter height is equal to filter width so that we can denote the filter tensor shape as $C_{in} \times K^2 \times C_{out}$. First, im2col transforms the input tensor into a relaxed form of the Toeplitz matrix, as shown in Fig. 28. While the K^2 filters sweep the input feature maps, each sliding window takes pixels from the K^2 input feature maps. These sampled feature maps are converted to column vectors and are stacked in a row direction. As a result, when the number of sampled feature maps (the number of sliding windows) is equal to X, the im2col algorithm converts the input tensor to $(K^2 * C_{in}) \times X$ shaped matrix. Second, im2col also transforms the filter tensor into a matrix. Due to this transformation, each 2D filter becomes a row vector, $C_{in} \times K^2 \times C_{out}$ shaped filter tensor becomes $C_{out} \times (K^2 * C_{in})$ shaped matrix. The matrix multiplication between the transformed filter matrix and input feature map matrix results in the same output as that of the convolution operation between the filter and input tensors. Moreover, each pixel of the output feature map implies the inner product between a row vector of a filter matrix and a column vector of an input matrix. Therefore, we can exploit the result as a similarity measurement between the two vectors. However, this method requires additional storage space to transform input feature map tensors into input matrices instead of getting the computational efficiency, as shown in Fig. 28. Specifically, a larger input feature map requires more memory footprint. Additionally, this method can cause more irregular data access patterns.

There are various GEMM-based convolution algorithms, and different algorithms have different advantages depending on the input feature map size. Anderson et al. [70] compared the computation speed of various GEMM-based convolution algorithms according to their input dimensions, as depicted in Fig. 29. Each GEMM library chooses appropriate im2col methods according to the input feature map size to show the best computing results.

GEMM (GEneral Matrix Multiplication) is a group of software libraries designed to calculate the matrix multiplication between two matrices. As it is optimized to the target hardware, GEMM shows high-performance computing. Open-BLAS [71] and Intel's MKL [72] are typical GEMMs used in CPUs, and cuBLAS [73] and cuDNN [74] are mainly used in GPUs. In particular, cuDNN is well known because

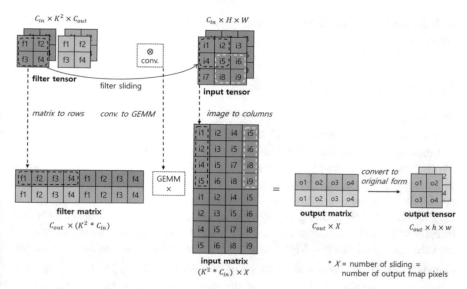

Fig. 28 Basic im2col algorithm

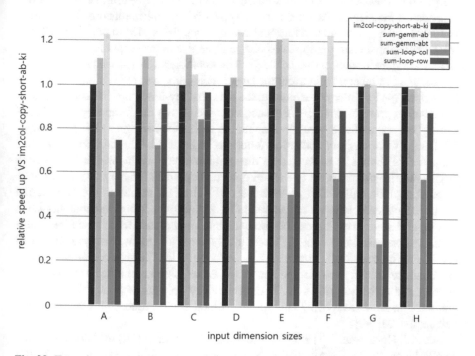

Fig. 29 Execution time of GEMM-based convolutional layer according to its implementation algorithm. (Adapted from Ref. [70])

of TensorFlow or PyTorch, recently widely used python deep learning libraries. The combination of im2col and GEMM enables practical convolution computation.

4.2.4 FFT, Strassen, and Winograd

This chapter deals with other convolution optimization techniques like FFT (Fast Fourier Transform), Strassen, and Winograd. First, FFT-based convolution optimization exploits the fact that convolution in the spatial domain is as same as multiplication in the frequency domain. Mathieu et al. [75] presented the convolution optimization technique utilizing the FFT-based convolution. It transforms the spatial filter and input feature map into frequency domain matrices, as depicted in Fig. 30. Then, the algorithm gets the convolution results by only multiplying the two transformed matrices. Inverse FFT (IFFT) restores the frequency domain output to the original spatial domain. This domain transform reduces the number of multiplications required for implementing the convolution as $O\left(N_o^2 N_f^2\right)$ to $O\left(N_o^2 \log_2 N_o\right)$, where the output size is $N_o \times N_o$ and the filter size is $N_f \times N_f$. However, there are several drawbacks to this FFT-based method. (1) The optimization efficiency depends on the size of the filter kernel. Typically, it is advantageous for the filter size larger than 5×5 due to additional computation demanded to perform FFT and IFFT, called the transformation overhead. (2) The size of the output feature map also affects the transformation overhead. (3) The coefficients in the frequency domain are complex numbers, which require larger memory space to be stored and more bandwidth. (4) FFT hinders exploiting the data sparsity-based optimization methods discussed in previous chapters. Because filter weights in the frequency domain are usually not sparse (zero value in a spatial domain often triggers high frequency). (5) To apply pooling and activation functions in the spatial domain, the frequency domain data should be recovered to the spatial domain again, which causes high inefficiency in computations. Meanwhile, several optimization methods for the FFT-based convolution alleviate the drawbacks. (1) The FFT results of the filter kernel can be calculated in advance and stored in the memory for further computations. (2) The input feature map also can be pre-calculated and reused to generate the multi-channeled convolution outputs. (3) Because the input images are always real-valued pixels, Fourier transform outputs are always symmetric. The accelerator can exploit this characteristic to save storage space. (4) Some methods are devised that apply pooling and activation functions in the frequency domain without using IFFT, as presented in Ko et al. [76]. This type of method could reduce the transformation overhead drastically.

Second, Cong et al. [77] adapted the Strassen algorithm [78] to reduce the computational workload of the convolution operation. This method helps the accelerator to reduce the number of multiplication while processing matrix multiplications. For $N \times N$ sized filter kernel, $O(N^3)$ computational complexity is reduced to $O(N^{2.807})$ by applying the customized Strassen algorithm. However, it requires additional storage and reduces numerical stability instead of the computational advantage.

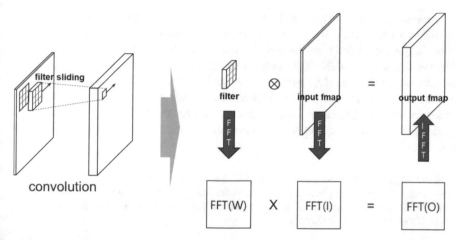

Fig. 30 FFT-based convolution optimization

Finally, Lavin et al. [79] introduced the Winograd algorithm-based convolution optimization method. The Winograd algorithm [80] is the minimal filtering algorithm, which is able to reduce the arithmetic complexity of a convolution operation in the neural network up to a factor of four compared to the original convolution. While the FFT-based convolution is advantageous for the larger filter kernels, the Winograd-based optimization benefits smaller filter sizes such as the 3 × 3 filter, the most frequently used filter size nowadays. Specifically, this algorithm boosts the calculation speed to 2.25 times for the 3 × 3-sized filter. However, it requires specialized processing according to the size of the filter kernel.

There is no almighty convolution optimization solution. Many popular DNN accelerating libraries, such as MKL [72] and cuDNN [74], adaptively select the most proper optimization algorithm, considering the input feature map and filter dimensions to maximize the hardware performance, similar to the im2col cases mentioned in the previous chapter.

5 Conclusion

The DNN technology has drastically grown over the last decade. Currently, it almost entered a mature stage, including the field of image processing, natural language processing, and robotics technologies. However, nowadays, DNNs require high computational complexity and colossal memory storage for such outstanding results. And these points become obstacles to the application of DNN to more diverse and practical applications. Consequently, it has become essential to implement DNN applications to improve energy efficiency and throughput, not sacrificing accuracy with cost-effective hardware.

Moreover, it is an inevitable issue that improves CNN's energy efficiency, one of the most frequently used DNN structures. In order to deal with these issues, this chapter has briefly introduced the design method and existing technologies for constructing a DNN HW accelerator, specifically, an energy-efficient CNN accelerator. Unfortunately, there is still no perfect DNN hardware accelerator architecture that achieves the best performance in all situations. Therefore, the accelerator designer should consider the input data's characteristics and the target application's performance goal to construct the energy-efficient and high-performance HW accelerator.

Acknowledgments This work was supported by the Institute of Information & Communications Technology Planning & Evaluation (IITP) grants funded by the Korean government (MSIT) (No. 2022-0-00966, Development of AI processor with a Deep Reinforcement Learning Accelerator adaptable to Dynamic Environment).

References

1. Kim, S., Deka, G.C.: Hardware Accelerator Systems for Artificial Intelligence and Machine Learning. Academic, Cambridge, MA (2021)
2. Park, H., Kim, S.: Hardware accelerator systems for artificial intelligence and machine learning. Adv. Comput. **122**, 51–95 (2021)
3. Park, H., Kim, S.: Hardware accelerator for training with integer backpropagation and probabilistic weight update. Adv. Comput. **122**, 343–365 (2021)
4. Park, H., Kim, D., Kim, S.: TMA: Tera-MACs/W neural hardware inference accelerator with a multiplier-less massive parallel processor. Int. J. Circuit Theory Appl. **49**, 1399–1409 (2021)
5. LeCun, Y., Boser, B., Denker, J.S., Henderson, D., Howard, R.E., Hubbard, W., Jackel, L.D.: Backpropagation applied to handwritten zip code recognition. Neural Comput. **1**, 541–551 (1989)
6. Krizhevsky, A., Sutskever, I., Hinton, G.E.: Imagenet classification with deep convolutional neural networks. In: NeurIPS 2012 (2012)
7. Deng, J., Dong, W., Socher, R., Li, L.-J., Li, K., Fei-Fei, L.: Imagenet: A large-scale hierarchical image database. In: CVPR 2009 (2009)
8. Maas, A.L., Hannun, A.Y., Ng, A.Y.: Rectifier nonlinearities improve neural network acoustic models. In: ICML 2013 (2013)
9. Clevert, D.-A., Unterthiner, T., Hochreiter, S.: Fast and accurate deep network learning by exponential linear units (elus). arXiv preprint arXiv:1511.07289 (2015)
10. Hendrycks, D., Gimpel, K.: Gaussian error linear units (gelus). arXiv preprint arXiv:1606.08415 (2016)
11. LeCun, Y., Bottou, L., Bengio, Y., Haffner, P.: Gradient-based learning applied to document recognition. Proc. IEEE 86. **11**, 2278–2324 (1998)
12. Simonyan, K., Zisserman, A.: Very deep convolutional networks for large-scale image recognition. In: ICLR 2015 (2015)
13. Sezegedy, C., Liu, W., Jia, Y., Sermanet, P., Reed, S., Anguelov, D., Erhan, D., Vanhoucke, V., Rabinovich, A.: Going deeper with convolutions. In: CVPR 2015 (2015)
14. He, K., Zhang, X., Ren, S., Sun, J.: Deep residual learning for image recognition. In: CVPR 2016 (2016)
15. Tan, M., Le, Q.: Efficientnet: Rethinking model scaling for convolutional neural networks. In: ICML 2019 (2019)

16. Moolchandani, D., Kumar, A.: Sarangi, SR: accelerating CNN inference on ASICs: a survey. J. Syst. Archit. **113**, 101887 (2021)
17. MLPerf – AI Benchmark: https://mlcommons.org/
18. Sze, V., Chen, Y.-H., Yang, T.-J., Emer, J.S.: Efficient processing of deep neural networks: a tutorial and survey. Proc. IEEE. **105**, 2295–2329 (2017)
19. Chen, Y.-H., Krishna, T., Emer, J.S., Sze, V.: Eyeriss: An energy-efficient reconfigurable accelerator for deep convolutional neural networks. IEEE J. Solid State Circuits. **52**, 127–138 (2016)
20. LeCun, Y., Denker, J.S., Solla, S.A.: Optimal brain damage. In: NeurIPS 1990 (1990)
21. Manikandan, N., Priyanka, M., Sasikumar, R.: Approximation computing techniques to accelerate CNN based image processing applications–a survey in hardware/software perspective. Int. J. **9**, 3828–3846 (2020)
22. Choudhary, T., Mishra, V., Goswami, A., Sarangapani, J.: A comprehensive survey on model compression and acceleration. Artif. Intell. Rev. **53**, 5113–5155 (2020)
23. Han, S., Pool, J., Tran, J., Dally, W.J.: Learning both weights and connections for efficient neural network. In: NeurIPS 2015 (2015)
24. Guo, Y., Yao, A., Chen, Y.: Dynamic network surgery for efficient DNNs. In: NeurIPS 2016 (2016)
25. Zhu, M., Gupta, S.: To prune, or not to prune: Exploring the efficacy of pruning for model compression. arXiv preprint arXiv:1710.01878 (2017)
26. Han, S., Mao, H., Dally, W.J.: Deep compression: Compressing deep neural networks with pruning, trained quantization and huffman coding. In: ICLR 2016 (2016)
27. Dorrance, R., Ren, F., Marković, D.: A scalable sparse matrix-vector multiplication kernel for energy-efficient sparse-blas on FPGAs. In: FPGA 2014 (2014)
28. Han, S., Liu, X., Mao, H., Pu, J., Pedram, A., Horowitz, M.A., Dally, W.J.: EIE: Efficient inference engine on compressed deep neural network. In: ISCA 2016 (2016), 44, 243
29. Parashar, A., Rhu, M., Mukkara, A., Puglielli, A., Venkatesan, R., Khailany, B., Emer, J., Keckler, S.W., Dally, W.J.: SCNN: An accelerator for compressed-sparse convolutional neural networks. In: ISCA 2017 (2017)
30. Anwar, S., Hwang, K., Sung, W.: Structured pruning of deep convolutional neural networks. ACM J. Emerg. Technol. Comput. Syst. **13**, 1–18 (2017)
31. Anwar, S., Sung, W.: Compact deep convolutional neural networks with coarse pruning. arXiv preprint arXiv:1610.09639 (2016)
32. He, Y., Zhang, X., Sun, J.: Channel pruning for accelerating very deep neural networks. In: ICCV 2017 (2017)
33. Liu, B., Wang, M., Foroosh, H., Tappen, M., Pensky, M.: Sparse convolutional neural networks. In: CVPR 2015 (2015)
34. Liu, Z., Sun, M., Zhou, T., Huang, G., Darrell, T.: Rethinking the value of network pruning. In: ICLR 2018 (2018)
35. Mao, H., Han, S., Pool, J., Li, W., Liu, X., Wang, Y., Dally, W.J.: Exploring the regularity of sparse structure in convolutional neural networks. arXiv preprint arXiv:1705.08922 (2017)
36. Yang, T.-J., Chen, Y.-H., Sze, V.: Designing energy-efficient convolutional neural networks using energy-aware pruning. In: CVPR 2017 (2017)
37. Horowitz, M.: 1.1 computing's energy problem (and what we can do about it). In: ISSCC 2014 (2014)
38. Judd, P., Albericio, J., Hetherington, T., Aamodt, T.M., Moshovos, A.: Stripes: Bit-serial deep neural network computing. In: MICRO 2016 (2016)
39. Qiu, J., Wang, J., Yao, S., Guo, K., Li, B., Zhou, E., Yu, J., Tang, T., Xu, N., Song, S.: Going deeper with embedded fpga platform for convolutional neural network. In: FPGA 2016 (2016)
40. Li, F., Zhang, B., Liu, B.: Ternary weight networks. arXiv preprint arXiv:1605.04711 (2016)
41. Zhu, C., Han, S., Mao, H., Dally, W.J.: Trained ternary quantization. In: ICLR 2017 (2017)
42. Lee, E.H., Miyashita, D., Chai, E., Murmann, B., Wong, S.S.: Lognet: Energy-efficient neural networks using logarithmic computation. In: ICASSP 2017 (2017)

43. Elhoushi, M., Chen, Z., Shafiq, F., Tian, Y.H., Li, J.Y.: Deepshift: Towards multiplication-less neural networks. In: CVPR 2021 (2021)
44. Miyashita, D., Lee, E.H., Murmann, B.: Convolutional neural networks using logarithmic data representation. arXiv preprint arXiv:1603.01025 (2016)
45. You, H., Chen, X., Zhang, Y., Li, C., Li, S., Liu, Z., Wang, Z., Lin, Y.: Shiftaddnet: A hardware-inspired deep network. In: NeurIPS 2020 (2020)
46. Zhou, A., Yao, A., Guo, Y., Xu, L., Chen, Y.: Incremental network quantization: Towards lossless cnns with low-precision weights. In: ICLR 2017 (2017)
47. Chen, W., Wilson, J., Tyree, S., Weinberger, K., Chen, Y.: Compressing neural networks with the hashing trick. In: ICML 2015 (2015)
48. Ioffe, S., Szegedy, C.: Batch normalization: Accelerating deep network training by reducing internal covariate shift. In: ICML 2015 (2015)
49. Lin, D., Talathi, S., Annapureddy, S.: Fixed point quantization of deep convolutional networks. In: ICML 2016 (2016)
50. Iandola, F.N., Han, S., Moskewicz, M.W., Ashraf, K., Dally, W.J., Keutzer, K.: SqueezeNet: AlexNet-level accuracy with 50× fewer parameters and <0.5 MB model size. In: ICLR 2017 (2017)
51. Szegedy, C., Vanhoucke, V., Ioffe, S., Shlens, J., Wojna, Z.: Rethinking the inception architecture for computer vision. In: CVPR 2016 (2016)
52. Howard, A.G., Zhu, M., Chen, B., Kalenichenko, D., Wang, W., Weyand, T., Andreetto, M., Adam, H.: Mobilenets: Efficient convolutional neural networks for mobile vision applications. arXiv preprint arXiv:1704.04861 (2017)
53. Chollet, F.: Xception: Deep learning with depthwise separable convolutions. In: CVPR 2017 (2017)
54. Klema, V., Laub, A.: The singular value decomposition: its computation and some applications. IEEE Trans. Autom. Control. **25**, 164–176 (1980)
55. Kim, Y.-D., Park, E., Yoo, S., Choi, T., Yang, L., Shin, D.: Compression of deep convolutional neural networks for fast and low power mobile applications. In: ICLR 2016 (2016)
56. Lebedev, V., Ganin, Y., Rakhuba, M., Oseledets, I., Lempitsky, V.: Speeding-up convolutional neural networks using fine-tuned Cp-decomposition. In: ICLR 2015 (2015)
57. Phan, A.-H., Sobolev, K., Sozykin, K., Ermilov, D., Gusak, J., Tichavský, P., Glukhov, V., Oseledets, I., Cichocki, A.: Stable low-rank tensor decomposition for compression of convolutional neural network. In: ECCV 2020 (2020)
58. Phan, A.-H., Tichavský, P., Sobolev, K., Sozykin, K., Ermilov, D., Cichocki, A.: Canonical polyadic tensor decomposition with low-rank factor matrices. In: ICASSP 2021 (2021)
59. Zhang, X., Zou, J., He, K., Sun, J.: Accelerating very deep convolutional networks for classification and detection. IEEE Trans. Pattern Anal. Mach. Intell. **38**, 1943–1955 (2015)
60. Rastegari, M., Ordonez, V., Redmon, J., Farhadi, A.: Xnor-Net: Imagenet classification using binary convolutional neural networks. In: ECCV 2016 (2016)
61. Courbariaux, M., Hubara, I., Soudry, D., El-Yaniv, R., Bengio, Y.: Binarized neural networks: Training deep neural networks with weights and activations constrained to +1 or −1. arXiv preprint arXiv:1602.02830 (2016)
62. Zhou, A., Yao, A., Guo, Y., Xu, L., Chen, Y.: Incremental network quantization: Towards lossless cnns with low-precision weights. arXiv preprint arXiv:1702.03044 (2017)
63. Lin, X., Zhao, C., Pan, W.: Towards accurate binary convolutional neural network. In: NeurIPS 2017 (2017)
64. Chen, H., Wang, Y., Xu, C., Shi, B., Xu, C., Tian, Q., Xu, C.: AdderNet: Do we really need multiplications in deep learning? In: CVPR 2020 (2020)
65. Wang, Y., Huang, M., Han, K., Chen, H., Zhang, W., Xu, C., Tao, D.: AdderNet and its minimalist hardware design for energy-efficient artificial intelligence. arXiv preprint arXiv:2101.10015 (2021)
66. Zhang, S., Du, Z., Zhang, L., Lan, H., Liu, S., Li, L., Guo, Q., Chen, T., Chen, Y.: Cambricon-X: An accelerator for sparse neural networks. In: MICRO 2016 (2016)

67. Han, S., Liu, X., Mao, H., Pu, J., Pedram, A., Horowitz, M.A., Dally, W.J.: EIE: Efficient inference engine on compressed deep neural network. ACM SIGARCH Comput. Archit. News. **44**, 243–254 (2016)
68. Zhou, X., Du, Z., Guo, Q., Liu, S., Liu, C., Wang, C., Zhou, X., Li, L., Chen, T., Chen, Y.: Cambricon-S: Addressing irregularity in sparse neural networks through a cooperative software/hardware approach. In: MICRO 2018 (2018)
69. Albericio, J., Judd, P., Hetherington, T., Aamodt, T., Jerger, N.E., Moshovos, A.: Cnvlutin: ineffectual-neuron-free deep neural network computing. ACM SIGARCH Comput. Archit. News. **44**, 1–13 (2016)
70. Anderson, A., Vasudevan, A., Keane, C., Gregg, D.: High-performance low-memory lowering: GEMM-based algorithms for DNN convolution. In: SBAC-PAD 2020 (2020)
71. Open-BLAS Library: https://www.openblas.net/
72. Intel Math Kernel Library: https://software.intel.com/en-us/mkl
73. cuBLAS Library: https://docs.nvidia.com/cuda/cublas/
74. Chetlur, S., Woolley, C., Vandermersch, P., Cohen, J., Tran, J., Catanzaro, B., Shelhamer, E.: cudnn: Efficient primitives for deep learning. arXiv preprint arXiv:1410.0759 (2014)
75. Mathieu, M., Henaff, M., LeCun, Y.: Fast training of convolutional networks through ffts. In: ICLR 2014 (2014)
76. Ko, J.H., Mudassar, B., Na, T., Mukhopadhyay, S.: Design of an energy-efficient accelerator for training of convolutional neural networks using frequency-domain computation. In: DAC 2017 (2017)
77. Cong, J., Xiao, B.: Minimizing computation in convolutional neural networks. In: ICANN 2014 (2014)
78. Strassen, V.: Gaussian elimination is not optimal. Numer. Math. **13**, 354–356 (1969)
79. Lavin, A., Gray, S.: Fast algorithms for convolutional neural networks. In: CVPR 2016 (2016)
80. Winograd, S.: Arithmetic Complexity of Computations. Siam, Philadelphia (1980)

Printed in the United States
by Baker & Taylor Publisher Services